A Laboratory Textbook of

Anatomy and Physiology

CAT VERSION

EIGHTH EDITION

Anne B. Donnersberger
Anne Lesak Scott

Moraine Valley Community College

122-123

JONES AND BARTLETT PUBLISHERS
Sudbury, Massachusetts
BOSTON   TORONTO   LONDON   SINGAPORE

World Headquarters
Jones and Bartlett Publishers
40 Tall Pine Drive
Sudbury, MA 01776
978-443-5000
info@jbpub.com
www.jbpub.com

Jones and Bartlett Publishers Canada
2406 Nikanna Road
Mississauga, ON L5C 2W6
CANADA

Jones and Bartlett Publishers International
Barb House, Barb Mews
London W6 7PA
UK

PHOTO CREDITS

Kathleen Ahearn: Figures 1.5 (left), 1.7 (left), 1.8 (left), 5.4, 5.7–15, 5.17, 6.2, 6.7, 6.9, 6.10, 6.11 (left), 6.13 (left), 7.12, 7.13, 7.16, 8.5, 8.9, 10.12, 10.16, 10.19, 10.23, 10.24, 10.26, 11.7–10, 12.2, 12.4–7, 13.5, 13.8, 13.9, 15.2, 15.5–8, 15.21–25
Richard A. Scott, M.D.: Figures 3.2, 4.4, 4.6, 4.8, 4.12, 4.14, 4.15, 4.17, 4.20, 4.21, 7.2, 7.4, 7.6–8, 10.6, 10.9, 10.10, 10.13, 10.14, 11.1, 11.3, 11.4, 12.14, 12.15, 12.17–19, 12.21–24, 12.28, 12.29, 12.33, 12.34, 15.8, 19.5, 19.7, 19.8
Ames, Inc., Division of Miles Laboratories, Inc.: Figures 13.12, 13.13
Applied Science Laboratories, Division of Applied Science Group, Inc.: Figure 10.35
Adapted with permission of the Biological Sciences Curriculum Study from *Basic Genetics: A Human Approach* (1983). Dubuque, IA: Kendall/Hunt Publishing Co. Adaptation with permission from NABT, Reston, VA: Figure 4.29
Carolina Biological Supply Company/Phototake: Figures 3.4, 12.16, 12.23, 12.25, 12.26, 15.12
Denoyer-Geppert, Chicago, IL: models in Figures 15.22–25
Leica Instruments: Figure 10.30
Oakton Instruments, Vernon Hills, IL: Figure 14.2
Reichert Scientific Instruments: Figure 2.1
Roche Diagnostics: Figure 9.9
Smiths Medical PM, Inc.: Figure 10.31
With permission from Field, H. E. and M. E. Taylor (1950, 1969). *An Atlas of Cat Anatomy*. Chicago: University of Chicago Press, 136pp.: Figures: 6.25, 6.31, 6.34, 12.8, 12.9
Turtox, Inc.: Figures 4.23–25, 15.19
Visuals Unlimited: Color Plates 1–36, 38–48

ILLUSTRATION CREDITS

Cecile Duray-Bito: Figures 1.5 (right), 1.6, 1.7 (right), 1.8 (right), 5.2, 10.7, 10.8, 11.9, 15.1, 15.2, 15.5, 15.6
Elizabeth Morales: Unnumbered figure p. 37, 6.25, 6.31, 6.34, 7.3, 10.21, 10.22, unnumbered figures on pp. 335, 416, 418, and 422, 13.14
Penelope J. Nicholls: Figures 5.21, 6.3–6, 6.8–9, 6.11, 6.12, 6.13 (right), 10.8, 10.25, 10.29, 12.3, 15.3, 15.20
Pat Oakes: Figures 5.3, 5.5, 5.6, 5.16, 5.18, 5.22–29, 5.31–39, 5.41–44, 5.47, 5.48, 7.11, 7.19, 8.10, 10.15, 10.17, 10.32, 11.3, 13.1, 13.7, 13.11
Mickey Senkarik: Figures 3.1, unnumbered figure p. 39, drawn portions of 4.1–3, 4.5, 4.7, 4.9–11, 4.13, 4.16, 4.18, 4.19, 4.22–28, 5.19, 5.30, 5.45, 5.46, 6.1, 6.14, 6.16–24, 6.26–30, 6.32, 6.33, 6.35, 6.36, 10.A–C, 15.16
TecDocPub, Inc: Unnumbered figures pp. 136, 137, 269, 270, 350

PRODUCTION CREDITS

Chief Executive Officer: Clayton Jones
Chief Operating Officer: Don W. Jones, Jr.
President, Higher Education and Professional Publishing: Robert W. Holland, Jr.
V.P., Design and Production: Anne Spencer
V.P., Sales and Marketing: William Kane
V.P., Manufacturing and Inventory Control: Therese Bräuer
Executive Editor, Science: Stephen L. Weaver
Managing Editor, Science: Dean W. DeChambeau
Associate Editor, Science: Rebecca Seastrong
Senior Production Editor: Louis C. Bruno, Jr.
Marketing Manager: Matthew Payne
Marketing Associate: Laura M. Kavigian
Text and Cover Design: Anne Spencer
Printing and Binding: Courier Westford
Cover Printing: Courier Westford
Cover: *Sprinter*, sculpture by Dominic Sutton, installed at the Sydney Olympic Park after the 2000 Olympics. ©Chris Ivin/Alamy Images.

Library of Congress Cataloging-in-Publication Data

Donnersberger, Anne B.
    A laboratory textbook of anatomy and physiology / Anne B. Donnersberger, Anne Lesak Scott.—8th ed.
        p. cm.
    Includes index.
    ISBN 0-7637-2656-7 (alk. paper)
1. Physiology—Laboratory manuals.   2. Anatomy—Laboratory manuals.   I. Scott, Anne Lesak.   II. Title.
QP44.D677 2005
612'.0078—dc22   2004056790

Printed in the United States of America

08 07 06 05 04   10 9 8 7 6 5 4 3 2 1

Contents

Preface

A Laboratory Textbook of Anatomy and Physiology, Eighth Edition, is the product of two authors who have been teaching human anatomy and physiology for over 40 years. Today, as always, our students have provided a foundation and direction for this book. The study of anatomy and physiology is not only interesting but also challenging and time-consuming. Recognizing the many demands on students, we have organized and presented the content in a comprehensive, easy-to-read style with many supportive illustrations. The *Laboratory Textbook* is intended for a one- or two-term course for students planning a healthcare-related career. This edition, like the previous seven, provides a strong foundation for students in nursing, allied health, and preprofessional curricula.

A Laboratory Textbook of Anatomy and Physiology is designed to be used with any primary textbook. It is also comprehensive enough to be used alone. Instructors can easily adapt the *Laboratory Textbook* to accommodate courses of differing lengths or levels. Units may be presented in almost any order and are suitable to self-paced, individualized learning with little supervision needed. Each of the 15 units progresses from Purpose to Objectives, Materials, Procedure, Self-Test, and Case-Studies. Longer units are divided into Anatomy, Physiology, Human, and Cat Exercises. The text's many new interactive learning tools and exercises encourage active learning and student collaboration.

NEW EXERCISES AND EXPANDED, UPDATED CLASSICS!

- Exercises on muscle tone, graded contraction, and muscle fatigue, with students using themselves as subjects in Unit 6C
- Exercise on afterimages in Unit 8A
- Revised endocrine physiology exercises in Unit 9B
- Determination of pulse rate and oxygen saturation in Unit 10B
- Expanded exercises on digestive chemistry in Unit 12B
- Exercise on nutrition in Unit 12B
- Revised urine physiology exercises in Unit 13B
- Expanded acid-base balance exercise in Unit 14
- Revised exercise on testing for hCG in urine 15B

NEW ILLUSTRATIONS!

- Isotonic, hypotonic, and hypertonic solutions in Unit 3B
- More color photomicrographs of tissue types
- More cat muscles in Unit 6B
- Motor neuron in Unit 7A
- Cat arterial and venous systems in color in Unit 10A

- More cat digestive organs in Unit 12A
- Protein, fat, and carbohydrate digestion in Unit 12B
- Urine crystals in Unit 13B
- Color photos of major organs

KEY FEATURES!

Question

- Clear instructions to students promote learning, with little direct supervision needed
- Thought questions within each exercise, as indicated by an icon
- Self-test at the end of each unit for review and assessment
- Case studies at the end of each unit test application of knowledge
- Tables and illustration of human bones, bone markings, and articulations
- Tables and photographs of human cadaver and cat muscles
- Extensive use of photographs and line drawings to aid dissection
- Separate histology unit, with colored photomicrographs and illustrations; histology continues as a thread through the remaining units
- Detailed descriptions of anatomy, with major structures boldfaced to aid organization and dissection
- Selected exercises enable students to reference and use their own bodies as subjects
- Tables in the reproductive unit reference stages of human development
- Tables in the endocrine unit categorize glands and their secretions
- Labeling of photographs and photomicrographs; magnifications of photomicrographs are stated as photographed
- Glossary/index of terms
- An appendix of solutions and measurements

MORE INTERACTIVE FEATURES TO REINFORCE KNOWLEDGE!

- Eighteen identification illustrations for you to label throughout the *Laboratory Textbook*, as indicated by an icon
- Chart for human muscles origins, insertions, and actions in Unit 6
- Chart for cranial nerves and their functions in Unit 7
- Learning tree in Unit 10
- Enzyme chart and nutrition log in Unit 12

Students!

Studying anatomy and physiology will help you understand how systems serve the whole body and how the body serves its systems in an integrated manner. You will also develop an understanding of the relationship between structure and function.

You will benefit from reading through the assigned unit or exercise *before* each lab period, so you will know what is expected and will be able to complete each exercise within the class time.

Some features that will help you learn from this *Laboratory Textbook* include:

- The Safety Rules. Review the safety rules on the inside front cover of the *Laboratory Textbook* and note the safety icons that appear throughout the book. Failure to do so may result in injury to yourself or others. It is also important that you come to class with a "prepared mind" and follow written and verbal instructions exactly.
- Consistent Organization. Each of the fifteen integrated units consists of a Purpose, Objectives, Materials, Procedures, Self-Test, and one or more Case Studies.
- Review Questions. Short answer questions alongside each exercise will enhance your understanding
- Labeling Exercises. Many units have illustrations for you to label or a learning tree to be completed, to reinforce knowledge, and allow you to quiz yourself as you progress through the unit
- Illustrations and Photographs. More than 290 illustrations and photographs of human and animal structures reinforce text material, dissections, and experimental procedures
- Units Emphasizing Terminology. Every discipline has its own "language." Medical terminology and microscope units provide the tools you will need to understand anatomy and physiology
- Phonetic Guides. Selected units contain phonetic guides for pronunciation of anatomical terms
- Cat Dissection Aides. Cat orientation icons, photographs, and illustrations of cat muscles, blood vessels, and internal organs will help you with dissections
- Boldfaced Structures. Structures that you need to identify are keyed with boldface type to help you correlate them with tables and illustrations.

We wish you much success in using this book, since our goals are for it to provide an excellent foundation for further study in your career, and for you to learn about your own body, which you will be living with for the rest of your life. Enjoy!

Both faculty and students may be interested in the following additional resources:

Ancillaries

FOR THE INSTRUCTOR

An online *Instructor's Guide to Accompany A Laboratory Textbook of Anatomy and Physiology,* by Anne B. Donnersberger and Anne Lesak Scott, provides an answer key for questions found in exercises, unlabeled illustrations, self-tests, and case studies; information for the set-up of all labs, including quantities of supplies and equipment needed; an updated list of major vendors; unlabeled figures with lead lines; and answers to unlabeled figures.

FOR THE STUDENT

Human Anatomy Flash Cards, by Robert K. Clark, includes over 90 text and illustrated cards on the human skeletal and muscle systems. Designed to test and reinforce your understanding of these structures, this set of cards is an effective study tool for your course.

The *Human Anatomy and Physiology Coloring Workbook and Study Guide*, by Paul D. Anderson and Victor M. Spitzer, is designed to help you learn introductory anatomy and physiology, and virtually every structure of the human body typically studied in an introductory course is examined. Chapters are short, concise and complete, enabling you to master smaller sections of information in a cohesive manner.

The *Cross-Sectional Anatomy Tutor: An Interactive Course for Anatomy Education and Evaluation*, CD-ROM, by Marianne Bouvier, Ann L. Bushyhead, and Anthony N. Benson, contains images from both the Visible Human Male and Female Dataset. These images, along with clinical case studies, allow you to broaden your understanding of cross-sectional anatomy. The updated atlas section offers you an essential tool for mastering anatomical concepts. Built-in practice tests are also included to allow you to test your knowledge.

Acknowledgements

We wish to thank our students, who expressed interest in a revised, more interactive *Laboratory Textbook*, to our colleagues for their support, and to Kathleen M. Ahearn for many of the gross anatomy photographs. We would like to remember the late Richard A. Scott, M.D. for his encouragement and for many of the black and white photomicrographs. We are also grateful to David R. Donnersberger J.D., M.D. for his suggestions and for contributing the case studies. A special thank you to our survey respondents for their assistance and suggestions.

Finally, we thank the editorial and production staff at Jones and Bartlett Publishers, especially Louis C. Bruno, the Senior Production Editor and copyeditor, Stephen L. Weaver, the Executive Editor, Dean W. DeChambeau, the Science Managing Editor, Rebecca Seastrong, the Associate Editor, and Laura Kavigian, the Marketing Associate.

Anne B. Donnersberger

Anne Lesak Scott

REVIEWERS OF PREVIOUS EDITIONS

Jan Ackerman
    Mott Community College
William Dunscombe
    Union County College
Marsha Jones
    Southwestern Community College

Karen A. McMahon
    University of Tulsa
SuEarl McReynolds
    San Jacinto College Central
Roberta M. O'Dell-Smith
    University of New Orleans

Patricia M. O'Mahoney-Damen
  University of Southern Maine
Tim Roye
  San Jacinto College South
Judy Sullivan
  Antelope Valley College
Donald L. Terpening
  Ulster County Community College

Richard G. Thomas
  Mohawk Valley Community
  College
Kathy Webb
  Bucks County Community College
Callie A. Vanderbilt White
  San Juan College

SURVEY RESPONDENTS

Don Alsum
  St. Mary's University of Minnesota
  Winnona, Minnesota
P. Bagavandoss
  Kent State University
  Canton, Ohio
Bobby Baldridge
  Ashbury College
  Wilmore, Kentucky
Susan T. Baxley
  Troy State University
  Montgomery, Alabama
Jenene Blodgett
  Black Hawk College—East
  Campus
  Kewanee, Illinois
Teresa Brandon
  Dona Ana Community College
  Las Cruces, New Mexico
Alese Bruce
  University of Massachusetts
  Lowell, Massachusetts
Nishi Bryska
  University of North Carolina
  Charlotte, North Carolina
Jackie Butler
  Grayson County College
  Denison, Texas
Barabara Cohen
  Delaware County Community
  College
  Medic, Pennsylvania
Suzanne Cormier
  La Cite Collegiate
  Ottowa, Ontario, Canada

Roisheen Doherty
  University College of the Fraser
  Valley
  Abbotsford, British Columbia,
  Canada
Lesley Flemming
  University of New Brunswick
  Fredericton, New Brunswick,
  Canada
Chaya Gopalan
  St. Louis Community College
  St. Loius, Missouri
Michael Glasgow
  Anne Arundel Community College
  Arnold, Maryland
Keith Graham
  Lutheran College of Heath
  Professions
  Fort Wayne, Indiana
Ann Henninger
  Wartburg College
  Waverly, Iowa
Harry Holden
  Northern Essex Community
  College
  Haverhill, Massachusetts
Herbert House
  Elon College
  Elon College, North Carolina
James Janik
  Miamai University
  Oxford, Ohio
Stephen Kenny
  Yakima Valley Community College
  Grandview, Washington

Donald Kisiel
  Suffolk Community College
  Selden, New York
Mary Losi
  Erie Community College South
  Orchard Park, New York
John Martin
  Clark College
  Vancouver, Washington
Richard McCloskey
  Boise State University
  Boise, Idaho
Charles McKinley
  Albany College of Pharmacy
  Albany, New York
Margaret Merkley
  Delaware Technical and
  Community College
  Georgetown, Delaware
Bernard Murphy
  Hagerstown Junior College
  Hagerstown, Maryland
William Nicholson
  University of Arkansas, Monticello
  Monticello, Arkansas
John Noble
  Murray State College
  Tishomingo, Oklahoma
Claire Oakley
  Rocky Mountain College
  Billings, Montana
John Pasto
  Middle Georgia College
  Cochran, Georgia
Mary Lou Percy
  Navarro College
  Corsicana, Texas
Ralph Porter
  Fresno City College
  Fresno, California

Kenneth Saladin
  Georgia College and State
  University
  Milledgeville, Georgia
Martha DePecol Sanner
  Middlesex Community Technical
  College
  Middletown, Connecticut
Janice Yoder Smith
  Tarrant County Junior College—
  NW Campus
  Fort Worth, Texas
Ron Stephens
  Ferrum College
  Ferrum, Virginia
Sandra Stewart
  Vincennes University
  Vincennes, Indiana
Thea Trimble
  College of the Sequoias
  Visalia, California
Charles Wert
  Linn-Benton Community College
  Albany, Oregon
Deborah Wiepz
  Madison Area Technical College
  Madison, Wisconsin
Peter Wilken
  Purdue University—North Central
  Westville, Indiana
Diana Wymann
  New Hampshire Community
  Technical College
  Claremont, New Hampshire
Paul Yancey
  Whitman College
  Walla Walla, Washington
John Yrios
  Edgewood College
  Madison, Wisconsin

Medical Terminology

PURPOSE

Unit 1 will give you a foundation in medical terminology, the language of anatomy and physiology.

OBJECTIVES

After completing Unit 1, you will be able to

- demonstrate proficiency in using terms describing body directions, planes, cavities, and abdominal regions.
- utilize prefixes that denote location, direction, tendency, and number.
- use prefixes that identify organs.
- use suffixes that are commonly used in various clinical procedures.
- construct medical terms from common descriptions.
- build a medical-scientific vocabulary.

MATERIALS

human torso
anatomical charts

PROCEDURE

Correct use of medical terminology is required in any field of study that refers to body structures, clinical procedures, or medical applications.

EXERCISE 1 **Terms Describing Human Body Directions**

Anatomical position refers to the human body in a standing position with the palms facing forward. The following terms are used to describe human body directions in anatomical position:

Example

Superior (or **cranial**): toward the head end of the body.

The shoulder is superior to the hip.

Inferior (or **caudal**): away from the head.

The knee is inferior to the elbow.

Anterior (or **ventral**): front.

The nose is anterior to the back of the head.

Posterior (or **dorsal**): back.

The shoulder blades are on the posterior surface of the body.

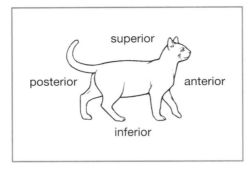

Medial: toward the midline of the body.

The great toe is medial to the little toe.

Lateral: away from the midline of the body.

The little toe is lateral to the great toe.

Proximal: toward or nearest the trunk or point of origin of a part.

The elbow is proximal to the wrist.

Distal: away from or farthest from the trunk or point of origin of a part.

The foot is distal to the knee.

In a quadruped organism, such as the cat, the cranial portion of the body is referred to as *anterior* and the caudal portion as posterior. Likewise, in a standing position, the quadruped dorsal, or upper, surface is superior, and the ventral, or belly, surface is *inferior.* It is important for you to remember these meanings when you are making comparisons with the human body.

EXERCISE 2   **Terms Designating Planes of the Body**

The following terms designate planes of the body (use **Figures 1.1–1.4** for reference):

Sagittal (SA-ji-tal): a plane that runs from front to back and divides the body or any of its parts into right and left sides (not necessarily equal).
Midsagittal: a plane that divides the body or its parts into right and left halves.
Coronal (or **frontal**): a lengthwise plane running from side to side that divides the body or any of its parts into anterior and posterior portions; or, in the quadruped, into superior and inferior portions.
Transverse (or **horizontal**): a crosswise plane that divides the body or any of its parts into superior and inferior parts; or, in the quadruped, into anterior and posterior portions.

Question 1.1

If you had a cut across the tip of your nose, which body plane would describe your cut?

EXERCISE 3   **Terms Designating Major Body Cavities and Regions**

Use a torso or **Figures 1.5** and **1.6** to help you identify body cavities and regions.

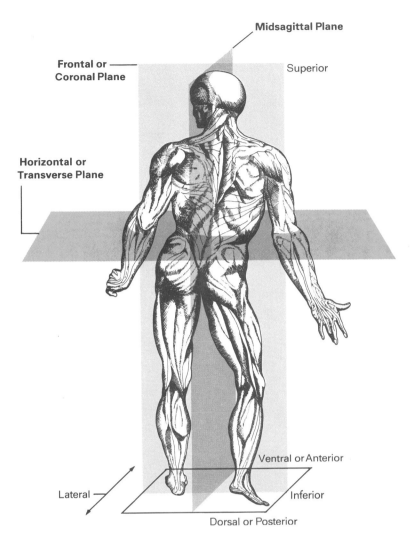

Midsagittal Plane

Frontal or
Coronal Plane

Superior

Horizontal or
Transverse Plane

Ventral or Anterior

Lateral

Inferior

Dorsal or Posterior

Figure 1.1

Planes of the body

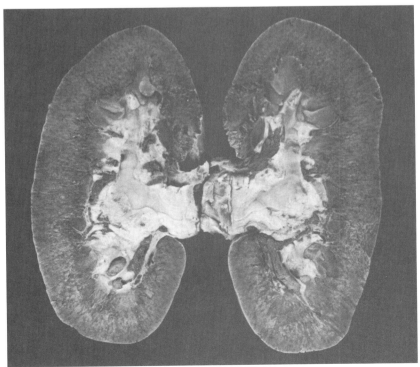

Figure 1.2

Frontal or coronal section of right kidney (Color Plate 48)

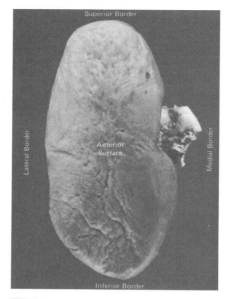

Figure 1.3

Anterior view of right kidney

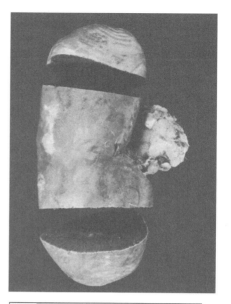

Figure 1.4

Transverse sections of right kidney

Figure 1.5

Major ventral body cavities

1. pleural cavities
2. mediastinum
3. diaphragm
4. abdominal cavity
5. pelvic cavity
6. thoracic cavity
7. abdominopelvic cavity

Question 1.2

Name one organ found in the:

mediastinum:

dorsal superior cavity:

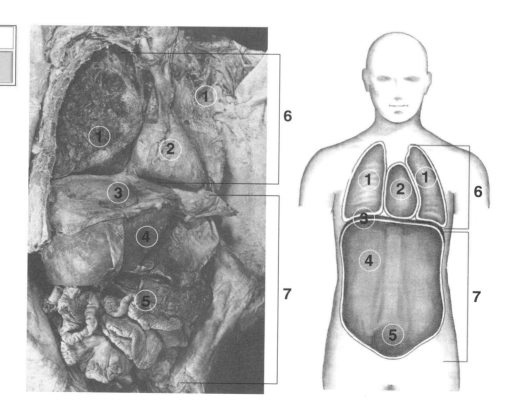

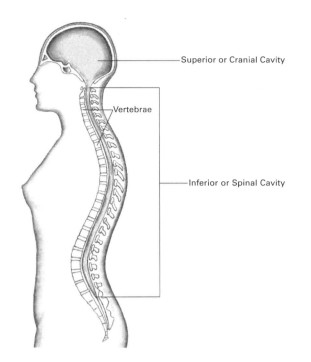

Superior or Cranial Cavity

Vertebrae

Inferior or Spinal Cavity

Figure 1.6

Major dorsal body cavities

Question 1.3

Using the anatomical method, identify one structure found in each region:

left hypochondriac:

right iliac:

The diagrams and terms in **Figures 1.7** and **1.8** describe abdominal regions. Identify at least one organ found in each region.

Figure 1.7

Viscera in relation to anatomical abdominal regions

Anatomical Method

1. right **hypochondriac** region
2. **epigastric** region
3. left **hypochondriac** region
4. right **lumbar** region
5. **umbilical** region
6. left **lumbar** region
7. right **iliac** region
8. **hypogastric** region
9. left **iliac** region

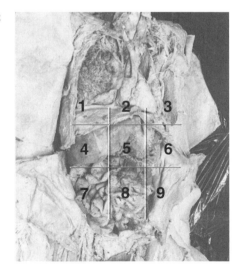

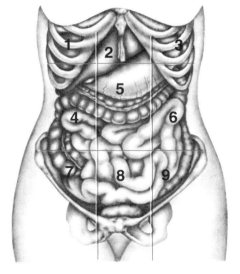

Figure 1.8

Viscera in relation to clinical abdominal regions

Clinical Method

1. right upper quadrant
2. left upper quadrant
3. left lower quadrant
4. right lower quadrant
5. thoracic cavity

6. diaphragm
7. abdominopelvic cavity
8. left pleural cavity
9. mediastinum

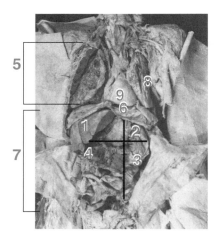
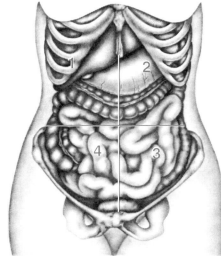

Question 1.4

Using the clinical method, identify one structure found in each region:

left upper quadrant:

right lower quadrant:

Identification 1.1

 1.1*

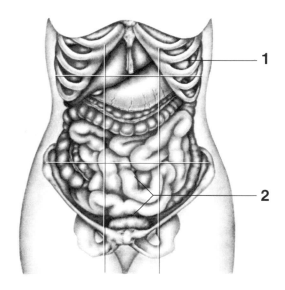

*Figure IDs appear throughout the *Laboratory Textbook* for you to label.

EXERCISE 4 Prefixes Indicating Location, Direction, and Tendency

These prefixes are found at the beginning of many common everyday words. Can you think of another example in addition to the one given?

Prefix	Meaning	Example
ab-	from, away	abnormal: away from normal
ad-	to, near, toward	adrenal: adjoining the kidney
ante-	before	antepartum: before delivery
anti-	against	antiseptic: against infection
brady-	slow	bradycardia: slow heart rate
circum-	around	circumocular: around the eye
co-	with, together	coordination: working together
con-	with, together	congenital: with birth
contra-	against	contraindicated: not indicated
counter-	against	counterirritant: nonirritating
dis-	apart from	disarticulation: taking a joint apart
ect-	outside	ectonuclear: outside the nucleus of a cell
end-	within	endocardium: membrane lining the inside of the heart
epi-	upon	epidermis: upon the dermal layer of skin
ex-	out from	exhale: to breathe out
hyper-	above	hypertension: higher than normal blood pressure
hypo-	under	hypodermic: under the skin
im-	not	immature: not mature
in-	not	incurable: not curable
infra-	under	infrapatellar: under the kneecap
media-	middle	mediastinum: region of the chest cavity between the lungs
peri-	around	pericardium: membrane around the heart
post-	after	postmortem: after death
pre-	before	prenatal: before birth
pro-	before	prognosis: a prediction
super-	above	superciliary: above the eyebrow
supra-	above	suprapubic: above the pubic bone
sym-	with, together	symphysis: a fusion
syn-	with, together	synanthrosis: a bony union
tach-	rapid	tachycardia: rapid heart rate
trans-	through	transurethral: through the urethra

Question 1.5

Give an example (not the one given in the table) for each of the following prefixes:

ad:

con:

ex:

hyper:

super:

EXERCISE **5**   Prefixes Denoting Organs and Structures

It is essential to know the prefixes in the following list. Check off those with which you are already familiar and learn those that are new to you.

Question 1.6

Think of a word that uses each of the following medical terms:

gastro:

hepato:

derm:

Prefix	Meaning	Example
abdomin/o-	abdomen	abdominal: pertaining to the abdomen
acr/o-	extremity	acromegaly: having unusually large extremities
aden/o-	gland	adenitis: inflammation of a gland
angi/o-	vessel	angiogram: visualization of blood vessels
arthr/o-	joint	arthritis: inflammation of a joint
cardi/o-	heart	cardiology: study of the heart
chol/e-	gallbladder	cholecystitis: inflammation of the gallbladder
chondr/o-	cartilage	chondroma: a cartilaginous tumor
cyst/o-	bladder	cystoscopy: examination of the inside of the bladder
cyt/o-	cell	cytokinesis: cytoplasmic division
dent/o-	tooth	dental: referring to the teeth
dermat/o- or derm/o-	skin	dermatologist: physician specializing in skin diseases
duoden/o-	duodenum	duodenal: having to do with the duodenum (first portion of small intestine)
enter/o-	intestine	enteric: relating to the intestine
gastr/o-	stomach	gastroscopy: viewing the interior surface of the stomach
hepat/o-	liver	hepatitis: inflammation of the liver
laryn/go-	larynx	laryngoscope: instrument for viewing inside the larynx
my/o-	muscle	myocardium: heart muscle
nephr/o-	kidney	nephrology: study of the kidneys
neur/o-	nerve	neurologist: physician specializing in diseases of the nervous system
ocul/o- or opt/o-	eye	optical: pertaining to vision
oste/o-	bone	osteocyte: bone cell
ot/o-	ear	otology: study of the ear
path/o-	disease	pathological: relating to disease
pneumon/o-	lung	pneumonia: inflammation of the lung
rhin/o-	nose	rhinitis: inflammation of the nasal passages
stomat/o-	mouth	stomatitis: inflammation of the mouth
thorac/o-	thorax (chest)	thoracentesis: puncture of the thorax for the removal of fluid

EXERCISE 6 — Prefixes Denoting Number and Measurement

The following quantitative terms are common; review their meanings.

Prefix	Meaning	Example
uni-	one	unicellular: consisting of one cell
mon-	one	mononuclear: having one nucleus
bi-	two	bilateral: affecting two sides
bin-	two	binocular: two-eyed
di-	two	dicephalic: two-headed
ter-	three	tertiary: of the third stage
tri-	three	trilobar: having three lobes
quadr-	four	quadriceps femoris: group of four muscles in the thigh
tetra-	four	tetralogy of Fallot: heart anomaly having four features
poly-	many	polydactyly: having (abnormally) many digits
macr-	large	macrocephalic: having an exceptionally large head
mega-	great	megadontia: having exceptionally large teeth
micro-	small	microscope: an instrument for viewing small objects
oligo-	few	oliguria: excretion of small amounts of urine

EXERCISE 7 — Suffixes Denoting Relations, Conditions, and Agents

Suffixes add specificity to words. Think of additional examples for the following suffixes.

Suffix	Meaning	Example
-ac	related to	cardiac: relating to the heart
-ious	related to	contagious: communicable by touch
-ic	related to	pyloric: relating to the pyloric valve
-ism	condition	mutism: inability to speak
-osis	condition	tuberculosis: infection by tuberculosis bacteria
-tion	condition	constipation: condition of passing infrequent and hard stools
-ist	agent (one who practices)	ophthalmologist: medical specialist who treats eye disorders
-or	agent	operator
-er	agent	examiner
-ician	agent	physician

EXERCISE 8 Suffixes Used in Operative Terminology

These suffixes are essential to clinical practice. Some may be more familiar to you than others.

Question 1.7

How does an -ectomy differ from an -ostomy and -otomy?

Suffix	Meaning	Example
-centesis	to puncture	amniocentesis: puncture of the amniotic sac
-ectomy	to cut out or remove	appendectomy: excision of the appendix
-ostomy	to cut to form an opening	colostomy: opening formed in the large intestine and draining to the outside
-otomy	to cut into	tracheotomy: cut into the trachea
-pexy	to fix or repair	gastropexy: repair of the stomach
-plasty	to repair or reform	rhinoplasty: repair of the nose
-(r)rhaphy	to suture	arteriorrhaphy: suture of an artery
-scopy	to view	otoscopy: to view the ear canal

EXERCISE 9 Miscellaneous Suffixes

These suffixes are relevant to your study. Check off those you know and learn those that you do not know.

Question 1.8

Write another example using each of the following suffixes:

_____ algia

_____ itis

_____ ology

Suffix	Meaning	Example
-algia	pain	neuralgia: nerve pain
-emia	of the blood	bacteremia: bacteria in the blood
-gram	writing	electrocardiogram: tracing of the electrical activity of the heart
-itis	inflammation of	appendicitis: inflammation of the appendix
-ology	study of	ophthalmology: study of the eye
-orrhea	flow	amenorrhea: cessation of menstrual flow
-phobia	fear of	claustrophobia: fear of confined spaces

SENTENCE COMPLETION

Choose the correct response for Questions 1–4 from the answers listed below:

Name _____

Section _____

Date _____

a. anterior	e. distal
b. posterior	f. proximal
c. medial	g. coronal
d. lateral	h. sagittal

_____   1. The plane of the body dividing it into right and left sections is _____

_____   2. The hip is _____ to the ankle.

_____   3. The navel is _____ to the spinal cord.

_____   4. The great toe is _____ to the little toe.

MATCHING

Column A

Column B

_____   5. Study of the heart   a. cardiology

_____   6. Inflammation of the skin   b. cholecystectomy

_____   7. Removal of the gallbladder c. dermatitis

_____   8. Pain in a muscle   d. myalgia

MULTIPLE CHOICE

_____   9. Which of the following is located in the midsagittal plane of the body?
   a. foot   c. leg
   b. toe   d. spinal column

_____   10. The kidneys are located primarily in the _____ regions of the abdomen.
   a. hypochondriac   c. epigastric
   b. hypogastric   d. lumbar

_____   11. When a surgeon amputates a leg, he or she makes a _____ cut through the bone.
   a. sagittal   c. transverse
   b. coronal   d. longitudinal

_____ 12. A term describing a structure that surrounds an organ is
   a. transurethral.   c. antepartum.
   b. pericardium.   d. superciliary.

_____ 13. In a quadruped organism, such as the cat, the cranial portion of the body is referred to as
   a. anterior.   c. dorsal.
   b. posterior.   d. superior.

_____ 14. A cytological examination involves observing
   a. the inside of the bladder. c. tissues.
   b. cells.   d. behavior.

_____ 15. A person with pneumonitis would most likely have which of these symptoms?
   a. painful urination   c. difficulty breathing
   b. inability to swallow   d. nausea and vomiting

_____ 16. What is the medical term for inflammation of a kidney?
   a. neuritis   c. cystorrhaphy
   b. nephrotomy   d. nephritis

_____ 17. Viewing the inside of the duodenum is known as
   a. duodenology.   c. duodenectomy.
   b. duodenoscopy.   d. duodenitis.

_____ 18. A person with otitis may have difficulty
   a. hearing.   c. smelling.
   b. seeing.   d. tasting.

_____ 19. When puncturing the chest cavity with a needle, a surgeon performs a
   a. pneumonectomy.   c. thoracentesis.
   b. pneumonostomy.   d. thoracotomy.

_____ 20. Which abdominal region is inferior to the umbilical region?
   a. left hypochondriac   c. hypogastric
   b. epigastric   d. right iliac

WORD RELATIONSHIPS

Which is an _incorrect_ word relationship in each of the following questions?

_____ 21. a. _anti_ = before   c. _brady_ = slow
          b. _bi_ = two   d. _cyst_ = bladder

_____ 22. a. _cyto_ = cell   c. _epi_ = top
          b. _entero_ = intestine   d. _hyper_ = below

_____ 23. a. _algia_ = pain   c. _gram_ = tracing, mark
          b. _ectomy_ = cut out   d. _itis_ = draining

_____ 24. a. _centesis_ = puncture c. _gastro_ = kidney
          b. _ology_ = study   d. _pexy_ = repair

_____ 25. a. _derm_ = skin   c. _aden_ = gland
          b. _arthr_ = joint   d. _opt_ = ear

26. List three advantages of using medical terminology.

a. _____

b. _____

c. _____

27. List two disadvantages of using medical terminology.

a. _____

b. _____

CASE STUDY

BODY REGIONS

Not having suited up, Johnny Bell ran out to play a game of street football. His team was ahead until Johnny got hit in the back in the waist region. He was in excruciating pain and when brought to the ER had blood in his urine.

1. What abdominopelvic region was most likely involved in the injury?

2. What organ(s) were involved?

3. Describe what probably occurred?

4. Could these symptoms be explained wholly by skeletomuscular trauma?

CASE STUDY

MEDICAL TERMINOLOGY

Jenny was experiencing an aching pain in her right hip. She had slipped on the ice running to her anatomy class the day before. When the pain became unbearable she sought medical help from her family doctor. Dr. D. had just hired a new receptionist who took notes and transcribed them into the record. He ordered x-rays, determined that Jenny had a fracture in the flaring, lateral bone of her hip. When asked to pay the bill, the insurance company refused. They said that no x-ray views of the small intestine had been taken and that no digestive symptoms were reported.

What error did the new receptionist make that resulted in the insurance company refusing to pay for the hip x-rays? Explain your answer.

The Microscope

PURPOSE

Unit 2 will familiarize you with the structure and function of the microscope so you can better understand cytology and histology.

OBJECTIVES

After completing Unit 2, you will be able to
- name and identify the major parts of the microscope.
- state and demonstrate the functions of microscope parts.
- list and follow the directions for proper care of the microscope.
- demonstrate use of the microscope.

MATERIALS

compound microscopes
lens paper
roll of Scotch Brand Magic Tape

prepared letter *e* slides and prepared
slides of other specimens

PROCEDURE

The microscope is an essential tool; to use it effectively, perform the following exercises.

EXERCISE 1 **The Structure of the Compound Microscope**

The microscope is an invaluable tool in your study of anatomy and physiology. Using it properly will enable you to see very small structures such as cells, which contribute to the total structure and function of living organisms. To maximize the benefits of using the microscope, it is essential that you know each of its parts and how to use them.

The **compound microscope** is commonly used in laboratory studies and consists of two lenses or lens systems, the **ocular** and the **objective**. The ocular, or eyepiece, magnifies an object that is again magnified by the objective. Total magnification is the product of ocular magnification times objective magnification.

Your instructor may assign a microscope to you. If so, use this same microscope for your studies throughout this course. Before obtaining the microscope, read and follow these rules:

- Place one hand beneath the scope for support.
- Put the other hand around the curved arm.
- Carry the microscope in an upright position.

Figure 2.1

Optical and mechanical features of the microscope

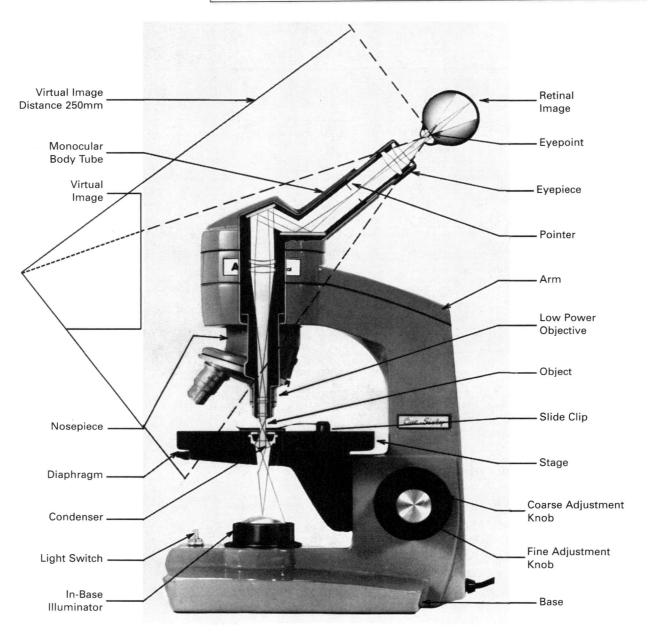

Place the microscope in front of you at your work area and plug the cord into the electrical outlet. Using **Figure 2.1** and the table entitled Microscope Parts and Their Functions, identify the following parts:

1. ocular, or eyepiece
2. revolving nosepiece
3. objectives
4. arm
5. stage
6. coarse adjustment knobs
7. fine adjustment knobs
8. condenser
9. diaphragm
10. base with illuminator

EXERCISE 2 Functions of the Microscope Parts

Reexamine the microscope parts that you identified in Exercise 1. Note the function of each part. Keep in mind that your primary purpose is to adjust the microscope to achieve the desired magnification and clarity (**resolution**) of the object being viewed. Learn the functions of each part, as described in the following table.

Question 2.1

Which microscope part(s) contain(s) magnifying lenses?

Microscope Parts and Their Functions   (Figure 2.1)

Parts	Function
1. **Ocular**, or **eyepiece**	A lens of a given magnification, which is probably engraved on the rim—for example, 10×. (You may also see a pointer embedded in the ocular. This is used to aid you in indicating specific locations within the microscopic field.)
2. **Revolving nosepiece**	This plate, which is capable of rotation, allows you to utilize objectives of different magnifications.
3. **Objectives**	Lenses of varying magnifications. The values (such as 4×, 10×, 43×, and 100×) are usually engraved on the objectives.
Scanning	Lens with the least magnification (often 4×).
Low-power	Lens with greater magnification than scanning objective (often 10×).
High-power	Lens with greater magnification than high power objective (often 43×).
Oil-immersion	Lens with greatest magnification (often 100×). You will infrequently use this lens in this course. Proper techniques must be demonstrated by your instructor prior to its use.
4. **Arm**	Handle for holding and positioning the microscope.
5. **Stage**	Platform on which the slide is positioned for focusing. Note that your scope may have clips for anchoring the slide or may have a mechanical slide holder. Obtain a blank slide and fit it into the holder. The slide can be moved forward, backward, and from side to side by rotating the small knobs on the underside or on top of the stage. If your microscope is not equipped with a mechanical stage, you will need to move the slide manually.
6. **Coarse adjustment knobs**	Large knobs on both sides of the base of the arm that allow for initial focusing of the object to be viewed.
7. **Fine adjustment knobs**	Small knobs on both sides of the base of the arm that allow for refinement of detail in focusing.
8. **Condenser**	A lens system that concentrates light from the illumination source so that a cone of light fills the aperture of the objective. After checking to see that the light bulb is on, move the condenser up and down and note the varying intensities of light visible through the ocular. Your microscope may not be equipped with a movable condenser.
9. **Diaphragm**	A plate with an aperture allowing for varying amounts of light to pass through the specimen. Open and close the diaphragm by adjusting it with its handle so that varying intensities of light are visible through the ocular.
10. **Base with illuminator**	Platform on which the microscope is structured, usually containing an electric light source.

EXERCISE 3   Proper Care of the Microscope

Care must be exercised when removing the microscope from its storage area and carrying it to your work area (Exercise 1). Care must also be exercised when cleaning the lenses and returning the microscope to its storage area.

Do not remove the lenses to clean them. Gently wipe only the external surface of the ocular and objective lenses with lens paper. Do not use any other material for wiping, because the lenses are easily damaged.

When returning the microscope to its storage area:

1. Turn off the light source.
2. Rotate the low power objective down toward the specimen stage.
3. Remove all slides from the stage.
4. Rewind the electrical cord loosely around the arm.

EXERCISE 4   Microscopic Observation of Prepared Slides

Obtain a slide of the letter *e* from the slide tray and clean it with lens paper if it appears smudged. Place the slide securely in the slide holder or mechanical stage. Rotate the revolving nosepiece so the low power objective is down. You may hear it click into place.

Be sure the light source is on. Rotate the coarse adjustment knob slowly away from you, bringing the low power objective down to its limit. While looking through the ocular, adjust the diaphragm until you see a bright field. The condenser under the stage may also be used to adjust the light intensity. Look into the ocular with both eyes open. At first you may be distracted, but with practice you will become unaware of the surroundings and will concentrate on what is being observed. Slowly rotate the coarse adjustment knob toward you until you see the letter *e*. Move the slide on the stage if necessary, so that what you are viewing is in alignment with the objective lens.

Now focus with the fine adjustment knob. The edges of the letter should be sharp and well defined. It may again be necessary to readjust the light to make it dimmer or more intense.

To obtain greater magnification, it will be necessary to rotate the nosepiece until the high power objective is in place. If your microscope is **parfocal**, the image under high power should be in focus. If it is not, turn the fine adjustment knob until the image is clear. *Do not use the coarse adjustment knob under high power.* Under high power, more light is also often necessary. As the power of the objective increases, less light enters the objective through the diaphragm. Thicker specimens also require more light than thinner ones. Open the diaphragm to allow more light to penetrate the specimen, if needed.

Continue to practice focusing, using another prepared slide. Draw and label the images that you have viewed under both low- and high power magnification in the spaces provided.

Clean the microscope lenses and replace the microscope in the storage area. Return the slides that you have used to their proper places in the slide trays.

Letter *e* Low power magnification	Letter e High power magnification
Prepared slide Low power magnification	Prepared slide High power magnification

Remove the slide from the stage. Hold the slide up to room light. How does the orientation of the letter *e* on the slide compare to what you have drawn?

Which objective requires the most illumination?

EXERCISE 5   Microscopic Observation of a Human Hair

You can also use a strand of your own hair to help you learn to focus the microscope. Remove a single long hair from your scalp. Tape the hair on the stage under low power. Adjust the diaphragm for light, and focus with the coarse adjustment knob. Once the hair is in focus, switch the objective to high power. With the fine adjustment knob, make the image as clear and defined as possible. You may need to allow for more light. Draw your hair in the space below.

Why is it necessary to check that the objectives are locked into position on the nosepiece before viewing a slide?

The Microscope

SELF-TEST

MATCHING

Name _____

Section _____

Date _____

Column A	Column B
_____ 1. Condenser	a. magnifying lens found on the revolving nosepiece
_____ 2. Diaphragm	
_____ 3. Fine adjustment	b. eyepiece
_____ 4. Ocular	c. used for refinement of detail in focusing an image
_____ 5. Objective	d. a plate with an aperture allowing for varying amounts of light to pass through the slide
	e. a lens system concentrating light from an illumination source

MULTIPLE CHOICE

_____ 6. A scientific instrument that is used for visual study of tissues and cells is the
a. centrifuge.
b. pH meter.
c. microtome.
d. microscope.

_____ 7. If the letter *p* is placed under the microscope in the normal reading position, which of the following orientations of the letter would the viewer see?
a. *b*
b. *p*
c. *q*
d. *d*

_____ 8. The part of the compound microscope that controls the amount of light penetrating the specimen is the
a. objective.
b. diaphragm.
c. eyepiece.
d. nosepiece.

_____ 9. If a viewer wished to observe a particular region of a specimen in greater detail, which of the following objectives would provide the greatest magnification?
a. 10×
b. 20×
c. 45×
d. 100×

_____ 10. The objective on a compound microscope that requires the least amount of light would be the one marked
a. 100×.   c. 20×.
b. 10×.   d. 45×.

_____ 11. Which of the following magnifications represents the "total magnification" that would be seen in the field of vision if a 5× eyepiece and a 20× objective were used?
a. 100×   c. 200×
b. 1000×   d. 20×

_____ 12. A _____ microscope contains a lens system made up of an eyepiece and an objective.
a. simple   c. compound
b. complex   d. magnifying

_____ 13. Which of the following operations should be performed first if the specimen image appears blurred to the viewer?
a. Turn the substage light on.   c. Clean the eyepiece and
b. Change to a higher power   objective lenses with lens
   objective.   paper.
                                 d. Adjust the stage.

_____ 14. Which of the following procedures should be done just prior to viewing through the compound microscope?
a. Clean all lenses with lens   c. Adjust the diaphragm.
   paper.   d. All of the above.
b. Turn the light on.

_____ 15. If the power of the ocular is 10× and total magnification is 1000×, what is the magnification of the objective being used?
a. 10×   c. 43×
b. 100×   d. 1010×

_____ 16. The revolving nosepiece holds which microscope part?
a. ocular   c. low power objective
b. condenser   d. diaphragm

_____ 17. Which statement is _incorrect?_
a. More light is necessary   c. The compound microscope
   when using high power   consists of two lens
   than when using low power.   systems.
b. When using the coarse   d. The oil immersion objective
   adjustment, the objective   allows for greater
   should touch the slide.   magnification than the low
                                    power objective.

_____ 18. Two parts of the microscope that regulate the amount of light visible through the ocular are the
a. fine adjustment and coarse   c. illuminator and revolving
   adjustment knobs.   nosepiece.
b. ocular lens and objective   d. condenser and diaphragm.
   lens.

_____ 19. Initial focusing of any slide is done under the
  a. scanning or low power objective.
  b. high power objective.
  c. oil immersion objective.
  d. objective demonstrating the greatest detail.

_____ 20. Which of the following is a structure that focuses light from the light source through the slide?
  a. ocular
  b. objective
  c. diaphragm
  d. condenser

_____ 21. Which of these represents poor technique when using a microscope?
  a. storing the microscope with the low power objective toward the specimen stage
  b. turning the coarse adjustment knob with the objective on high power while looking through the ocular
  c. looking through the ocular with both eyes open
  d. wiping the external surfaces of the ocular and objective lenses with lens paper prior to using the microscope

_____ 22. The term _resolution_ refers to
  a. magnification.
  b. clarity.
  c. light intensity.
  d. all of the above.

_____ 23. Which of these objectives has the longest barrel length?
  a. low power
  b. high power
  c. oil immersion
  d. all barrel lengths are equal

_____ 24. Under which magnification would there be the greatest field of vision?
  a. 100×
  b. 430×
  c. 900×
  d. 1000×

_____ 25. Under which magnification would the distance between the objective and the slide being observed be the least?
  a. 100×
  b. 430×
  c. 900×
  d. 1000×

26. If your field of magnification under low power appears too dark, what should you do to lighten it?

27. What is meant by the term *parfocal?*

28. Does a diaphragm aperture of small diameter clarify or distort an image focused under high power? Explain.

CASE STUDY

MICROSCOPE

Rhonda was having difficulty observing cells in mitosis. On her parfocal microscope she focused from low to high power. She used the fine adjustment knob to clarify her image, yet the entire field lacked brightness. Rhonda was unable to distinctly identify the chromosomes.

How would you suggest that Rhonda correct her focusing problem?

Cells

A. Cell Structure

PURPOSE

Unit 3A will familiarize you with cell structure and cell division.

OBJECTIVES

After completing Unit 3A, you will be able to

- name and identify the organelles found in a cell.
- prepare a smear of oral mucosa cells and identify the nucleus, cytoplasm, and plasma membrane.
- recognize the stages of mitosis.

MATERIALS

clean microscope slides	coverslips
medicine droppers	prepared whitefish mitosis slides
compound microscope	distilled water
flat toothpicks	animal cell model
methylene blue	

PROCEDURE

Cells are the basic structural and functional units of living organisms. You need to know the structural components of cells to understand their functions. Use models and diagrams as references for study, because you will not be able to see most of the cellular structures under your compound microscope.

Question 3A.1

Which cellular structures can you see and identify using a compound microscope?

EXERCISE 1   Animal Cell Structure

Study the illustration of an animal cell as seen under an electron microscope (**Figure 3.1**). Use the illustration as a guide for finding these structures on an animal cell model.

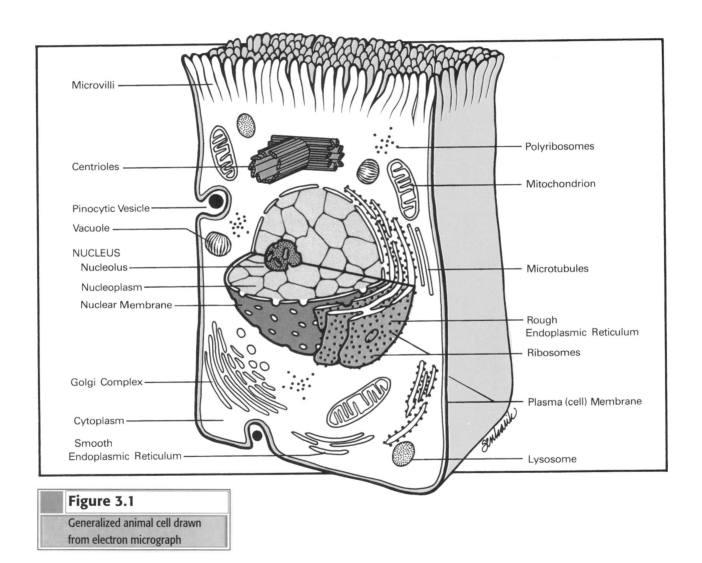

Microvilli

Centrioles

Pinocytic Vesicle

Vacuole

NUCLEUS
  Nucleolus
  Nucleoplasm
  Nuclear Membrane

Golgi Complex

Cytoplasm

Smooth
Endoplasmic Reticulum

Polyribosomes

Mitochondrion

Microtubules

Rough
Endoplasmic Reticulum

Ribosomes

Plasma (cell) Membrane

Lysosome

Figure 3.1

Generalized animal cell drawn from electron micrograph

Question 3A.2

List the functions of the structures you identified.

This table lists the basic cell structures that can be seen with the electron microscope.

Cell Structures and Their Functions   (Figure 3.1)

Structures	Function
PLASMA (PLAZ-ma) **MEMBRANE**	A thin (7.5–10 nanometer) envelope around the entire cell, consisting of a double layer of lipid molecules into which protein and glycoprotein molecules are embedded. The membrane is differentially permeable, discriminating which molecules gain entry to or exit from the cell.
NUCLEUS (NOO-klē-us)	A spherical body of nucleoplasm containing **chromosomes** and DNA.
Nuclear membrane	A double-membraned envelope consisting of lipids with protein molecules embedded in it; it is porous and allows easy movement of molecules between the nucleoplasm and cytoplasm.
Nucleolus (noo-KLĒ-ō-lus)	A spherical structure found in the nucleus, where ribosomal RNA is produced. There may be more than one nucleolus in a nucleus.
Chromatin (KRŌ-ma-tin) **material**	Proteins containing DNA; condense during cell division to become chromosomes.
CYTOPLASM (SI-tō-plazm) or **CYTOSOL**	A colloidal solution between the plasma membrane and the nucleus, containing dissolved organic substances, larger particles, and organelles: a **cytoskeleton** consisting of microtubules provides a framework for cytoplasmic structures.
ORGANELLES (or-gan-ELZ)	Structures within the cytoplasm that contribute to cellular *vitality*.
Mitochondria (mī'tō-KON-drē-un)	Bilayered spherical or elongated self-replicating structures containing DNA and oxidative enzymes that are found on the inner, involuted layer **(cristae);** function in liberation of energy that is used to synthesize high energy molecules of ATP.
Ribosomes (RĪ-bō-sōmz)	Granules composed of RNA and protein; serve as site for protein synthesis.
Endoplasmic reticulum (en'dō-PLAZ-mik re-TIK-yoo-lum) or **ER**	An extensive network of vesicular structures composed of a double layer of protein that traverses the entire cell; functions as a cellular circulatory system; may be of rough or smooth type.
▪ **Rough** or **granular ER**	That part of the ER to which ribosomes are attached on the outer surface; functions in protein synthesis.
▪ **Smooth** or **agranular ER**	That part of the ER lacking ribosome attachment; functions in lipid synthesis and other metabolic processes.
Lysosomes (LĪ-sō-sōmz)	Bilayered lipid-covered structures containing hydrolytic enzymes that, when released, act to digest cellular materials.
Golgi (GŌL-jē) **apparatus**	A stacked, vesicular structure continuous with the ER and located near the nucleus; contains membranes similar to smooth ER and functions in the synthesis, processing, and transportation of secretory products. Release of secretory products is through the plasma membrane in units called secretory granules or vesicles.
Centrosome (SEN-trō-sōm)	An organelle located near the nucleus containing two **centrioles,** which are small structures necessary for the formation of **spindle fibers** during cell division.
INCLUSIONS	Cytoplasmic components that are *nonvital* and vary in size, number, and type, depending on the specific cell type.
Vacuoles (VAK-yoo-ōl)	Voids in the cytoplasm that vary as to content, size, number, and location but can fill with food or waste products.
Microtubules (mī-krō-TOOB-yool) and **microfilaments** (mī-krō-FIL-a-ment)	Aggregates of elongated protein molecules that contribute to cell shape, support, and function.
Microvilli (mī'-krō-VIL-ī)	Minute cytoplasmic protrusions found in certain cells; serve to increase cellular absorptive capacity.

Question 3A.3

Can you observe mitochondria in the slide of your oral mucosa cells?

Explain your answer.

EXERCISE 2 **Observation of Oral Mucosa Cells**   *(Color Plate 1)*

Make a slide of your own cells and using the compound microscope, observe them. Follow these directions.

Place a small drop of water on a clean slide. Scrape the inside of your cheek (oral mucosa) with a flat toothpick and mix with water on the slide. The thinner the smear the better. Allow the slide to air dry. Add one small drop of methylene blue to your smear and cover with a coverslip. Observe under a microscope and draw what you see. Label the nucleus, cytoplasm, and plasma membrane, as shown in **Figure 3.2** and **Color Plate 2**.

Figure 3.2

Human oral mucosa cells (430×)

1. nucleus
2. cytoplasm
3. plasma (cell) membrane
4. cytoplasmic granules
5. nuclear membrane

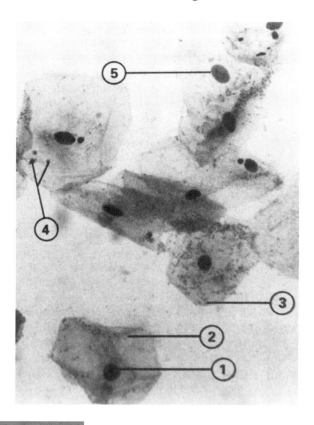

Question 3A.4

How many chromosomes are present in a somatic cell of a human?

EXERCISE 3 **Mitosis**

Mitosis is the process by which somatic, or body, cells divide. In mitosis, the two daughter cells that are produced have the same number of chromosomes as the parent cell. In human beings, the normal number of chromosomes in somatic cells is 23 pairs, or 46.

Obtain a whitefish mitosis slide that shows the blastula embryological stage. Observe first under low power, and then switch to high power. Look for the following stages of mitosis (**Figure 3.3** and **Color Plate 1**).

1. **Prophase:** Chromatin shortens and thickens, spindle fibers appear, and the nucleolus and nuclear membrane disappear.
2. **Metaphase:** Chromosomes are aligned across the equator of the cell.
3. **Anaphase:** Homologous chromosomes move apart to opposite poles.

4. **Telophase**: Spindle fibers disappear, cytokinesis occurs, and the nuclear membrane reappears.

When a cell is not actively dividing, it is in **interphase**. In interphase, **chromatin** is not condensed into chromosomes. DNA replicates during interphase; however, this process cannot be observed microscopically.

A prepared slide of human chromosomes is shown in **Figure 3.4**.

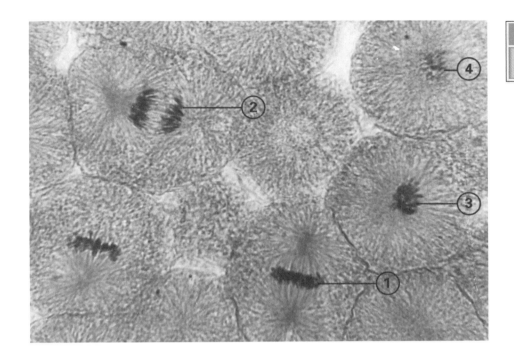

Figure 3.3

Whitefish mitosis (430×)

1. metaphase
2. anaphase
3. prophase
4. interphase (nonmitotic phase)

Question 3A.5

How many chromosomes are present in each of the two daughter cells produced by mitosis?

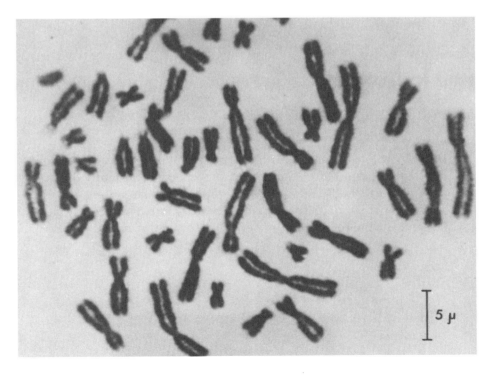

Figure 3.4

Normal human somatic chromosomes (1000×).

5 μ

B. Cell Physiology

PURPOSE

Unit 3B will give you an understanding of some basic concepts of cell physiology.

OBJECTIVES

After completing Unit 3B, you will be able to
- observe various types of diffusion.
- demonstrate the principles of osmosis and dialysis.
- define a hypertonic, an isotonic, and a hypotonic solution.

EXERCISE 1   Diffusion

Diffusion is the movement of molecules from a highly concentrated area to a less concentrated area until equilibrium is achieved.

MATERIALS

perfume	sharp probes
distilled water	medicine dropper
beakers	test tubes containing gelatin
potassium permanganate crystals	methylene blue

PROCEDURE

There are three major types of diffusion:

1. *Diffusion of gases.* Open a bottle of perfume at the front of the room. Why can the scent soon be detected in all parts of the room?

2. *Diffusion within a liquid.* Your body contains fluids such as blood and interstitial (tissue) fluid. Substances such as salts constantly diffuse through these fluids. To observe this process in a lab setting, drop a crystal of potassium permanganate ($KMnO_4$) into a beaker of water. Observe during the laboratory period. What happens?

3. *Diffusion through a colloid.* Cytoplasm is a colloid through which substances such as glucose diffuse. To observe this process in a lab setting, add one or two drops of methylene blue to a test tube containing 8–10 cc of gelatin. Then, using a sharp probe, stab through the dye into the gelatin. Observe during the laboratory period and at the beginning of the next period. What is the result?

EXERCISE 2 Osmosis

In order to understand these exercises, you need to be familiar with the following terms: A **solute** is a substance that is dissolved in a liquid. A **solvent** is a liquid in which the solute is dissolved. **Osmosis** is the movement of water through a selectively permeable membrane, such as the plasma membrane, from an area of higher concentration to an area of lower concentration. The solvent, water, moves toward the area of higher concentration of the solute.

MATERIALS

large white potatoes that have been soaked in water
1/2"- or 1/3"-diameter drill bit
molasses
capillary tubing, 60 cm long
rubber stoppers

large beakers
distilled water
ring stands
single adjustable burette clamps
meter sticks

PROCEDURE

Use a drill bit to bore a hole approximately two thirds of the way into the length of a large white potato. This exercise will give better results if the potato has been immersed in water overnight or for at least 2 hours. Fill the hole to within ¾" to 1" of the top with molasses. Insert a piece of glass tubing (capillary tubing is preferable) into a rubber stopper, and insert the stopper securely into the potato containing the molasses. The tubing should extend nearly to the bottom of the hole.

Place the prepared potato into a large beaker nearly full of water, and place the beaker on a ring stand, anchoring the tubing with a test tube clamp. It is a good idea to wrap a paper towel around the tubing at the point where the clamp will be fastened to hold it more tightly.

Observe the apparatus during the laboratory period. The molasses should rise into the tubing as a result of water passing through the potato. Using a meter stick, measure the height of the molasses solution in the capillary tubing in centimeters after 1 hour: _____ cm.

Question 3B.1

As a group, design an experiment to demonstrate osmosis.

EXERCISE **3**   Dialysis

Dialysis is the separation of crystalloids from colloids through a selectively permeable membrane. This separation occurs naturally in the kidneys. You have probably heard of renal dialysis. To observe this process in the lab setting, complete and observe the following.

MATERIALS

distilled water
5% sodium chloride solution
10% glucose solution
albumin (1 tsp/400 mL water) or
    two raw egg whites
dialysis membrane or tubing
beakers
test tubes

graduated cylinders
dilute nitric acid
1% silver nitrate solution
Benedict's solution
safety glasses
Bunsen burners or hot plates
string or thread

PROCEDURE

This exercise should be set up early in the lab period.

Place water, sodium chloride (NaCl), glucose, and albumin or raw egg white from one egg into a bag of dialysis membrane or cellophane, tightly secure the ends with string or thread, and place in a beaker of water. Allow the bag to be immersed in the water. Let stand for 1 to 2 hours; then run the following tests on the water in the beaker and the solution in the bag.

Use care when working with nitric acid.

1. *Test for albumin.* Pour about 5 mL of fluid from the beaker into a test tube and add a few drops of nitric acid (HNO_3). If albumin is present, it will be coagulated by the HNO_3 and turn white. What is your conclusion?

Use care when pouring silver nitrate.

2. *Test for NaCl.* Pour about 5 mL of fluid from the beaker into another test tube and add a drop of silver nitrate ($AgNO_3$). If the solution turns cloudy or white, silver chloride (AgCl) has been formed, and the test is positive. What is your conclusion?

3. *Test for glucose.* Put 5 mL of Benedict's solution into a test tube and add four or five drops of the beaker fluid. Boil for 2 minutes over a Bunsen burner or in a water bath. Cool slowly. If a green, yellow, or red precipitate forms, the presence of glucose is indicated. What is your conclusion?

Repeat the above tests using 5-mL amounts of solution from within the dialysis bag.

EXERCISE 4   Hypertonic, Isotonic, and Hypotonic Solutions

If cells are placed in a **hypertonic solution,** in which the solute concentration is greater outside the cell than inside, the cells will lose water and become shriveled, appear to be indented, or shrink in size. In red blood cells or erythrocytes, this process is known as **crenation.** If cells are placed in an **isotonic solution,** in which the solute concentration outside the cells is the same as inside, there will be no net change in the amount of water in the cells. Therefore with red blood cells, they will retain their original round, biconcave shape. If cells are placed in a **hypotonic solution,** in which the solute concentration is greater inside the cell than outside, there will be a net gain of water in the cells, causing a rise in intracellular pressure, and the cells will appear swollen or will burst. In the case of red blood cells, remnants or "ghosts" of cell membranes will remain. This process is known as **hemolysis.**

Crenation	Maintains shape	Hemolysis
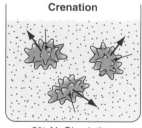	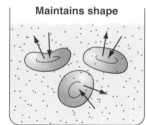	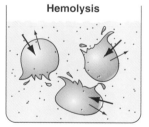
3% NaCl solution (Hypertonic)	0.9% NaCl solution (Isotonic)	0.2% NaCl solution (Hypotonic)

MATERIALS

sterile lancets (hemolets)
dropper bottles of 1.5% saline solution
dropper bottles of 0.9% saline solution
dropper bottles of distilled water
alcohol wipes
precleaned microscope slides
coverslips

compound microscopes
flat toothpicks
paper towels
disposable gloves
disinfectant solution
sheep red blood cells (optional)
sealed, puncture-proof container

CAUTION
BLOODBORNE PATHOGENS

Use a new, sterile lancet or hemolet. Exercise care when working with body fluids.

Because of the possibility of infections being transmitted from one student to another during exercises involving human body fluids, it is important that you take great care in collecting and handling such fluids. *The body fluids used in the exercises should be your own. Follow proper procedures for collecting handling, and disposing of body fluids as directed by your instructor.* Because of particular concern over procedures involving human blood, it may be preferable for the instructor to perform the following procedure as a demonstration.

PROCEDURE

Use either your own blood or sheep red blood cells. If using your own blood, place a clean paper towel over the area on which you will be working. Without touching their surfaces, place three slides and three coverslips on the paper towel. Swab the distal end of the middle or fourth finger of your non-dominant hand with an alcohol wipe. Holding the sterile hemolet near the tip in your gloved hand, quickly puncture the distal end of the swabbed finger. Squeeze the tip of the punctured finger until a distinct drop of blood appears. Working quickly, place one drop of blood on each of the slides, again squeezing the tip of your finger between the first and second drops if necessary, to obtain sufficient blood.

Using your own blood or sheep red blood cells, place one large drop of 1.5% saline solution on the first slide, a drop of 0.9% saline on the second slide, and a drop of distilled water on the third slide. Using the broad end of a flat toothpick, gently mix the solutions on each slide for 3–5 seconds. It is important to complete this process before the blood dries.

After 3 minutes, apply a coverslip over each slide and observe under high power for crenation, normal cells, or hemolysis. Draw the appearance of the red blood cells on each of the slides in the table below:

Question 3B.2

Does this exercise demonstrate diffusion, osmosis, or dialysis?

Slide 1	Slide 2	Slide 3

Remember to dispose of the hemolets in a sealed, puncture-proof container and to clean your work surface with a disinfectant solution after you have finished this exercise. Dispose of the slides you prepared in a pan of disinfectant solution.

Cells

MULTIPLE CHOICE

Name _____

Section _____

Date _____

_____ 1. Which type of diffusion would occur most *slowly?*
a. diffusion of gases
b. diffusion within a liquid
c. diffusion through a colloid
d. all types occur at the same rate

_____ 2. A membrane is said to be _____ if it will allow some but not all substances to diffuse through.
a. impermeable
b. permeable
c. parapermeable
d. selectively permeable

_____ 3. A solution of 0.9% sodium chloride, sometimes termed *normal saline,* is said to be _____ to the cells of humans and other mammals.
a. hypertonic
b. hypotonic
c. isotonic
d. osmotic

_____ 4. Protein synthesis in human cells occurs
a. on ribosomes.
b. in lysosomes.
c. in the nucleus.
d. in the Golgi apparatus.

_____ 5. If this figure represents osmosis, what is the net reaction that would occur?
a. NaCl molecules would move from compartment B to compartment A.
b. Water would move from compartment A to compartment B.
c. Water and NaCl molecules would cross the selectively permeable membrane.
d. Neither water nor NaCl would cross the membrane.

_____ 6. The correct sequence of mitotic stages during cell division is
a. interphase, metaphase, anaphase, and telophase.
b. anaphase, metaphase, telophase, and prophase.
c. prophase, metaphase, anaphase, and telophase.
d. metaphase, anaphase, interphase, and telophase.

_____ 7. During a normal cell cycle, a cell spends most of its time in
   a. prophase.
   b. interphase.
   c. anaphase.
   d. telophase.

_____ 8. During which stage of mitosis do the chromosomes line up across the equator of a cell?
   a. telophase
   b. prophase
   c. metaphase
   d. anaphase

_____ 9. If a bag of dialysis membrane containing a 10% NaCl solution were placed in a beaker of water, what would be the net reaction?
   a. Water would go in and salt would go out of the bag
   b. Water would go into the bag.
   c. Water would go out of the bag.
   d. Nothing would happen.

_____ 10. How would you determine whether NaCl molecules moved across the dialysis membrane in question 9?
   a. Add HNO_3 to 5 mL of solution from the beaker.
   b. Add methylene blue to 5 mL of solution from the beaker.
   c. Add $AgNO_3$ to 5 mL of solution from the beaker.
   d. Add Benedict's solution to 5 mL of solution from the beaker and boil for 2 minutes.

_____ 11. Enzymes are
   a. lipids.
   b. carbohydrates.
   c. proteins.
   d. nucleic acids.

_____ 12. DNA replication takes place during
   a. interphase.
   b. prophase.
   c. metaphase.
   d. anaphase.

_____ 13. —NH_2 and —COOH are characteristic of
   a. organic acids.
   b. amino acids.
   c. carbohydrates.
   d. nucleic acids.

_____ 14. How many chromosomes are in an oral mucosa cell?
   a. 32 pairs
   b. 46
   c. 23
   d. 46 pairs

_____ 15. Which of these is *not* a mechanism that enables substances to enter cells?
   a. active transport
   b. phagocytosis
   c. pinocytosis
   d. cytokinesis

_____ 16. Cytokinesis occurs during
   a. interphase.
   b. metaphase.
   c. anaphase.
   d. telophase.

_____ 17. During which stage of mitosis are the chromosomes migrating toward opposite poles of the cell?
   a. prophase
   b. metaphase
   c. anaphase
   d. telophase

_____ 18. Which of these functions in mitosis?
    a. cell membrane   c. centriole
    b. cell wall   d. vacuole

_____ 19. Which organelles function as the major sites of ATP production within cells?
    a. centrioles   c. mitochondria
    b. lysosomes   d. ribosomes

_____ 20. Which of these would *not* pass through a dialysis membrane because of large molecular size?
    a. salt   c. proteins
    b. simple sugars   d. water

_____ 21. A physiological process that separates high molecular weight proteins from smaller electrolytes is
    a. osmosis.   c. dialysis.
    b. equilibrium.   d. active transport.

_____ 22. In the osmosis exercise of this unit (Exercise B2), what would be the concentration of the molasses in the tubing with respect to the molasses first put into the potato?
    a. It would be more concentrated.   c. It would be of the same concentration.
    b. It would be more dilute.   d. Concentration differences would be impossible to determine.

_____ 23. Which of these cannot be observed under a compound microscope?
    a. cell membrane   c. endoplasmic reticulum
    b. nucleus   d. cytoplasm

_____ 24. Which of these would yield a positive reaction with Benedict's solution?
    a. albumin   c. egg white
    b. NaCl   d. glucose

_____ 25. What would happen to red blood cells if they were immersed in a 10% salt water solution?
    a. They would multiply more rapidly.   c. They would lose water.
    b. They would burst.   d. There would be no effect.

DISCUSSION QUESTIONS

26. State the function of each of these cellular organelles:

   a. chromatin

   b. nucleolus

   c. centrioles

   d. plasma membrane

   e. endoplasmic reticulum

   f. ribosomes

   g. Golgi apparatus

   h. lysosomes

   i. mitochondria

   j. nuclear membrane

27. Using your textbook or other references, name the stage(s) of mitosis which each of the following occurs in the cell:

   a. Cell is not actively dividing.

   b. Cytokinesis occurs.

   c. Chromosomes are lined up at the equator of the cell.

   d. Chromatin condenses into chromosomes.

   e. Nuclear membrane is not present.

   f. DNA is replicating.

28. What physical principle determines whether various types of molecules will pass through a dialysis membrane?

CASE STUDY

CELLS

Paul was asked to identify common structures of cells during Week 1 of anatomy class. He used a compound microscope and high power magnification. He made slides of his red blood cells, oral mucosa, and flakes of his skin. He reported that he was able to identify the cell membrane, nucleus, and cytoplasm in all three stained slides.

Was he correct? Explain your answer.

Tissues

PURPOSE

Unit 4 will familiarize you with the various tissue types.

OBJECTIVES

After completing Unit 4, you will be able to

- name the general categories of tissues found in the body.
- state the location of various tissue types in the body.
- identify general and specialized tissue types microscopically.
- microscopically differentiate between pigmented and nonpigmented skin.
- identify the epidermal, dermal, and hypodermal layers of skin.
- compare the dermal ridges of fingerprints with respect to patterns.

MATERIALS

Slides of the following tissue types:
  simple squamous epithelium
  simple cuboidal epithelium
  simple columnar epithelium
  stratified squamous epithelium
  pigmented and nonpigmented skin
  pseudostratified ciliated columnar
    epithelium
  transitional epithelium
  dense fibrous connective tissue
  areolar tissue

  adipose tissue
  reticular tissue
  hyaline cartilage
  fibrocartilage
  elastic cartilage
  osseous tissue
  skeletal muscle
  smooth muscle
  cardiac muscle
compound microscope
model of human skin

PROCEDURE

Tissues are groups of cells that perform a common function. There are four primary tissue types: **epithelial tissue** that covers a surface and functions in protection, secretion, and absorption; **connective tissue** that binds and supports; **muscle** tissue that contracts; and **nervous tissue** that conducts impulses.

These four basic tissue types can be further subdivided as shown in the Classification of Tissues table. Epithelial tissue (or **epithelium**) is composed of many cells and very little intercellular material. It contains no blood vessels. Epithelium can be simple, meaning one cell layer thick, or **stratified**, which is composed of several cell layers. Because it covers or lines a surface, epithelium always has a free border or a layer of cells next to an open area. If the cells

Primary Tissue	Types	Divisions	Examples

Epithelium (Color Plates 2–10, 34, 35)	Covering external body surface or lining internal surface	Simple
		  Squamous
		  Cuboidal
		  Columnar
			Glomerular capsule (kidney)
			Collecting tubule (kidney)
			Gallbladder (nonciliated)
			Uterine tube (ciliated)
			Intestinal mucosa

		Pseudostratified Columnar	Male urethra (nonciliated) Trachea (ciliated)
		Stratified Squamous	Skin (keratinizing) Vagina (nonkeratinizing) Cornea

		Cuboidal
		  Columnar
		  Transitional
			Sweat glands
			Male urethra
			Urinary bladder

	Multicellular glands	Exocrine
		  Simple
		  Compound
		Endocrine
			Gastric, sweat
			Salivary
			Thyroid, adrenal

Muscle (Color Plates 26–30)	Smooth (involuntary) Striated (voluntary) Cardiac (involuntary)		Intestinal tract, blood vessels Skeletal muscle Heart muscle
Connective Tissue	General (Color Plates 11–14) Loose	 Mesenchyme Mucoid Areolar Adipose Reticular	 Embryonic and fetal tissue Wharton's jelly (umbilical cord) Found in most organs and tissues Subcutaneous tissue Bone marrow, lymph nodes
	Dense	Irregular Regular	Dermis, capsules of organs Tendon, cornea

	Special (Color Plates 15–25)	Cartilage
		  Hyaline
		  Fibrous
		  Elastic
			
			Costal cartilage, trachea
			Intervertebral disc
			External ear, epiglottis

		Bone
		  Cancellous
		  Compact
			
			Epiphyses of long bones
			Shaft of long bone

		Hemopoietic
		  Myeloid
		  Lymphoid
			
			Bone marrow
			Spleen, lymph node

		Blood Lymph	

Nervous Tissue (Color Plates 31, 32)		Central nervous system
		  Gray matter
		  White matter
		Peripheral nervous system
		  Gray matter
		  White matter
		Special receptors
			Brain, spinal cord
			Nuclei
			Tracts
			Ganglia
			
			Nerves (nonmyelinated)
			Nerves (myelinated)
			Eye, ear, nose, cutaneous

comprising the epithelium are flat, they are known as **squamous**; if they are cube shaped, they are **cuboidal**; and if they are taller than they are wide, they are referred to as **columnar**. Occasionally, the nuclei of a single cell layer are at different levels, giving the appearance of stratified epithelium. This type is referred to as **pseudostratified**. Another epithelial type consists of stratified cells, the surface layer of which appears balloon-shaped. This is known as **transitional epithelium**, which is capable of stretching, and is found in parts of the urinary tract.

Connective tissue, in contrast to epithelial tissue, contains relatively few cells and much intercellular material. It may be classified as **dense fibrous**, as found in tendons and ligaments, or **loose areolar**, which forms the "packing" in and around organs. There are several specialized types of connective tissue. **Adipose**, or fat, consists of cells in which most of each cell's volume is occupied by fat rather than cytoplasm. This feature imparts a characteristic "signet ring" appearance to adipose cells. Cartilage and bone are other specialized connective tissue types. Cartilage may be **hyaline**, **elastic**, or **fibrous**. Bone, or **osseous** tissue, may be **compact** or **spongy**. Blood is a specialized connective tissue. You will be studying blood in Unit 10.

There are three types of muscle tissue: **skeletal**, **cardiac**, and **smooth**. The primary functional cells in nervous tissue are known as **neurons**. Muscle and nerve slides will be studied superficially in this unit and in greater depth in later exercises.

Using the labeled black-and-white photographs and the color plates, examine the slides of tissue types your instructor has made available to you.

After studying the slides and the summary table, you should be able to answer the questions in the Self-Test.

EXERCISE 1 Microscopic Identification of Tissue Types

Microscopic examination of tissue types requires proficient use of your microscope. Focus first under low power and after you become oriented to the general field, observe under high power. Observe the details visible under high power.

Identify each tissue type using the illustrations in this unit as a guide (**Figures 4.1–4.21** and **Color Plates 1–48**). Make drawings and note the distinguishing characteristics of each tissue you observe. Tissue slides are often made from sections through organs that are composed of several tissues. Therefore, there may be several different tissue types on the same slide. You will need to search the field in order to identify the particular tissue under investigation.

Figure 4.1

Human squamous epithelial cells; smear from inner surface of cheek; cells are flat with round, centrally located nuclei (430×) (Color Plate 2)

1. cytoplasm
2. nucleus
3. cell membrane

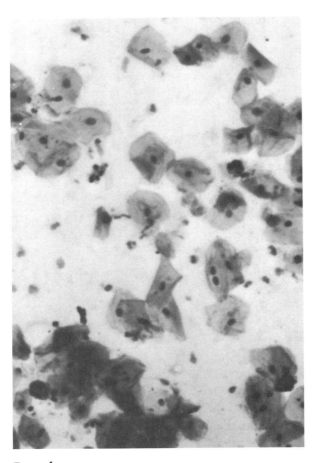

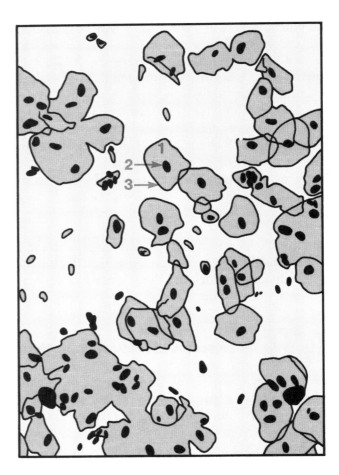

Drawing:

Figure 4.2

Simple columnar epithelium showing goblet cells and brush border as found in intestinal mucosa; cells are tall and slender with oval-shaped nuclei (430×) (Color Plate 6)

Question 4.1

What is the function of goblet cells?

1. goblet cells
2. brush border
3. nuclei of columnar epithelial cells

 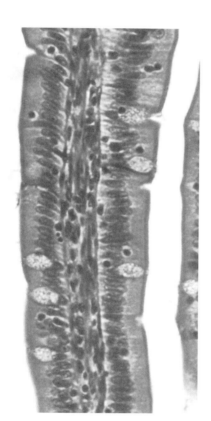 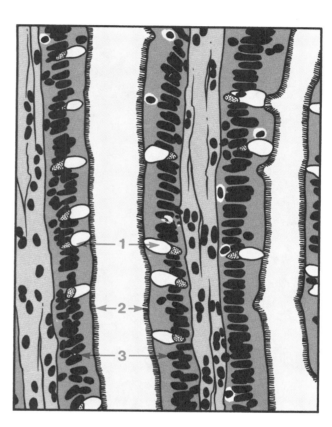

Drawing:

Figure 4.3

Cuboidal epithelium as found in the collecting tubules of the kidney; epithelium is one cell layer thick; nuclei are large and round (430×) (Color Plates 3 and 4)

1. cuboidal cells
2. nucleus of cuboidal cell
3. lumen of tubule

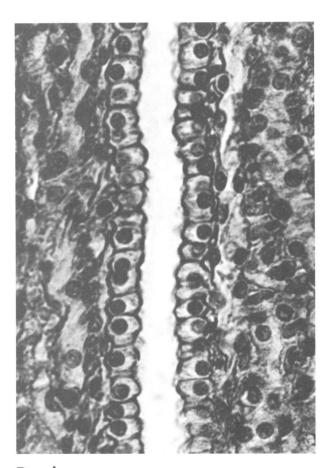
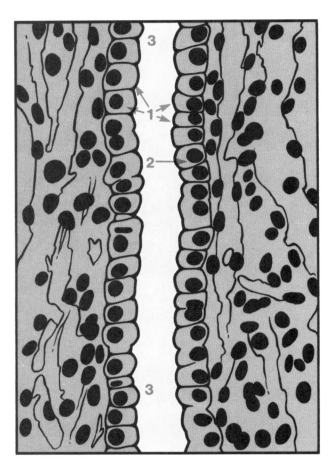

Drawing:

Figure 4.4

Pseudostratified ciliated columnar epithelium; this tissue type is found primarily in the respiratory tract (400×) (Color Plate 7)

1. cilia
2. pseudostratified columnar cells
3. goblet cell
4. basement membrane
5. loose areolar connective tissue

Question 4.2

Which major tissue type underlies a basement membrane?

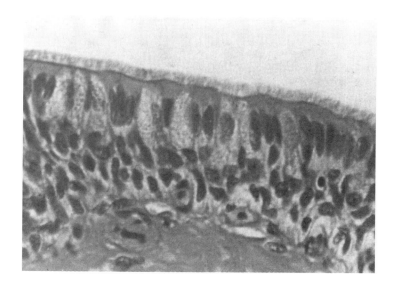

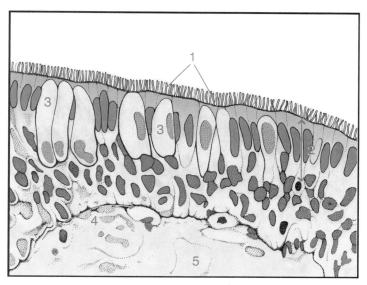

Drawing:

Figure 4.5

Stratified squamous epithelium; layers of flattened cells serve to protect underlying tissues
(430×) (Color Plate 8)

1. stratified squamous epithelium
2. basement membrane
3. loose connective tissue

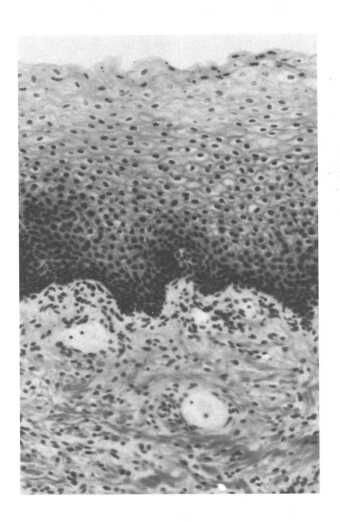
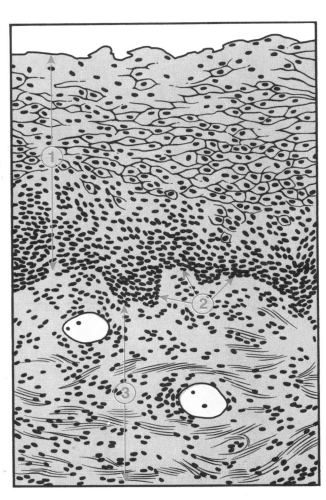

Drawing:

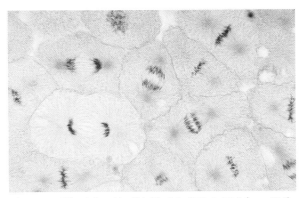

Plate 1 Mitosis in white fish blastula (195×) (© Science VU/ Visuals Unlimited)

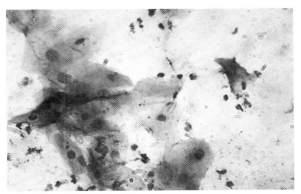

Plate 2 Simple squamous epithelium from human oral mucosa smear (450×) (© Michael Abbey/Visuals Unlimited)

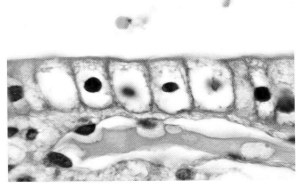

Plate 3 Simple cuboidal epithelium of human kidney tubules (360×) (© Carolina Biological/Visuals Unlimited)

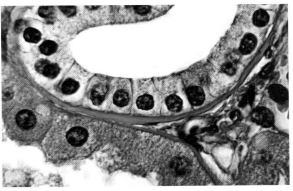

Plate 4 Simple cuboidal epithelium (360×) (© Carolina Biological/Visuals Unlimited)

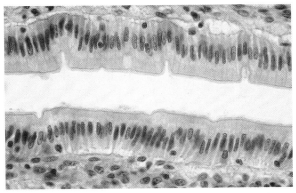

Plate 5 Simple columnar epithelium (160×) (© Dr. Fred Hossler/Visuals Unlimited)

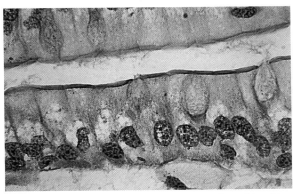

Plate 6 Simple columnar epithelium of intestine, with goblet cells (250×) (© Science VU/Visuals Unlimited)

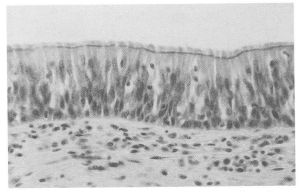

Plate 7 Pseudostratified ciliated columnar epithelium of larynx (200×) (© Dr. Fred Hossler/Visuals Unlimited)

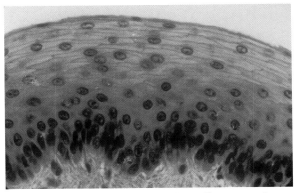

Plate 8 Stratified squamous epithelium of human esophagus (400×) (© Dr. Donald Fawcett/Visuals Unlimited)

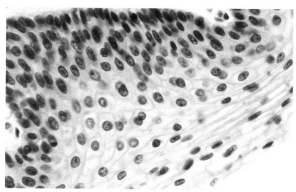

Plate 9 Stratified squamous epithelium of urethra (100×) (© Dr. Richard Kessel/Visuals Unlimited)

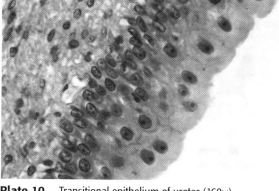

Plate 10 Transitional epithelium of ureter (160×) (© Dr. Richard Kessel/Visuals Unlimited)

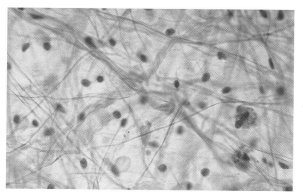

Plate 11 Loose areolar connective tissue (170×) (© Science VU/Visuals Unlimited)

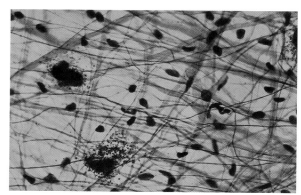

Plate 12 Areolar connective tissue (150×) (© Science VU/Visuals Unlimited)

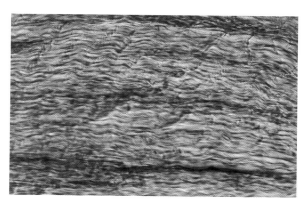

Plate 13 Collagenous fibers of human tendon (125×) (© Carolina Biological/Visuals Unlimited)

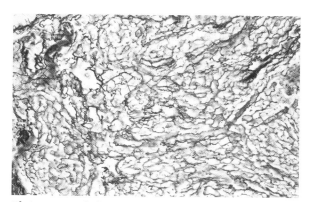

Plate 14 Reticular tissue, silver stain (100×) (© Dr. Richard Kessel/Visuals Unlimited)

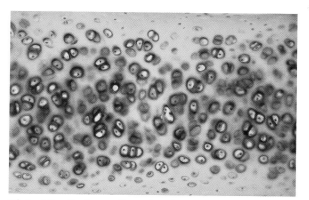

Plate 15 Hyaline cartilage of trachea (230×) (© Dr. David Phillips/Visuals Unlimited)

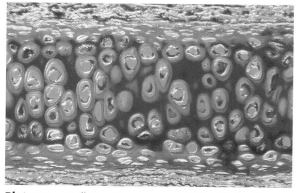

Plate 16 Hyaline cartilage, mammal (360×) (© Carolina Biological/Visuals Unlimited)

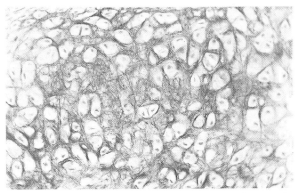

Plate 17 Elastic cartilage of epiglottis (95×) (© Science VU/ Visuals Unlimited)

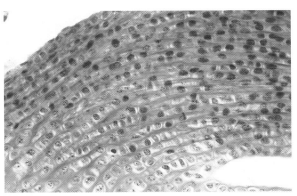

Plate 18 Fibrous cartilage, tendon insertion of knee (450×) (© Robert Calentine/Visuals Unlimited)

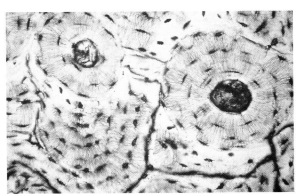

Plate 19 Human compact (ground) bone (64×) (© Dr. Fred Hossler/Visuals Unlimited)

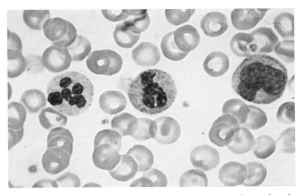

Plate 20 Human blood cells, combination. Showing red and nucleated white cells and platelets (400×) (© Dr. Fred Hossler/Visuals Unlimited)

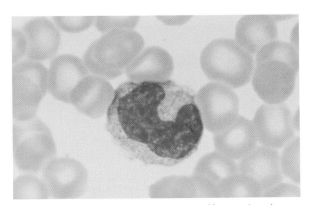

Plate 21 Monocytes and erythrocytes of human (710×) (© Science VU/Visuals Unlimited)

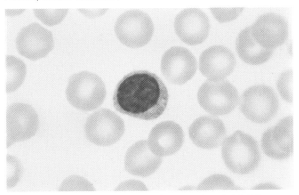

Plate 22 Lymphocyte and erythrocytes (710×) (© Science VU/Visuals Unlimited)

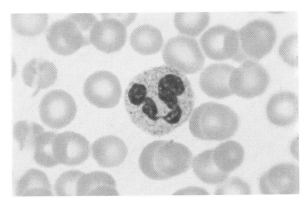

Plate 23 Neutrophil and erythrocytes (710×) (© Science VU/Visuals Unlimited)

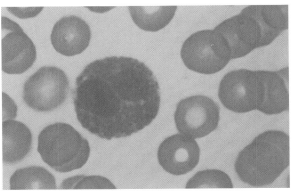

Plate 24 Eosinophil and erythrocytes (710×) (© Science VU/ Visuals Unlimited)

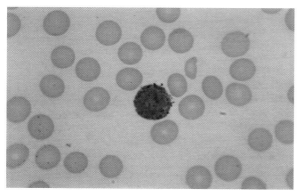

Plate 25   Basophil and erythrocytes (400×) (© Science VU/ Visuals Unlimited)

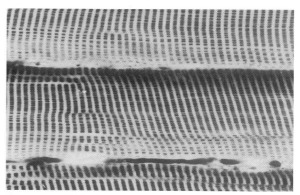

Plate 26   Skeletal muscle (600×) (© Dr. Donald Fawcett/Visuals Unlimited)

Plate 27   Cardiac muscle, l.s. (185×) (© Science VU/Visuals Unlimited)

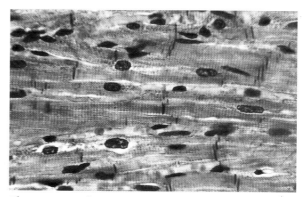

Plate 28   Cardiac muscle showing intercalated discs (185×) (© Science VU/Visuals Unlimited)

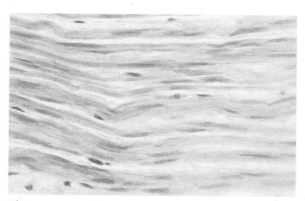

Plate 29   Smooth muscle, l.s. (185×) (© Science VU/Visuals Unlimited)

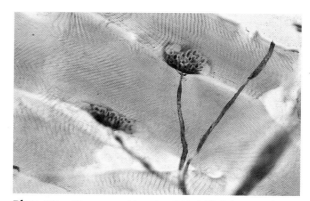

Plate 30   Myoneuronal junction (160×) (© Science VU/ Visuals Unlimited)

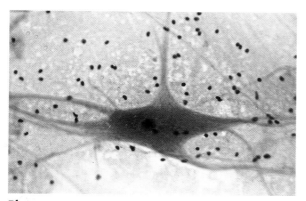

Plate 31   Motor neuron, spinal cord (175×) (© Carolina Biological/Visuals Unlimited)

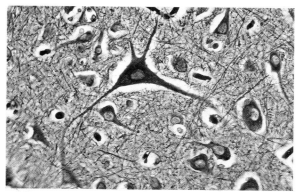

Plate 32   Human cerebral cortex (75×) (© Carolina Biological/Visuals Unlimited)

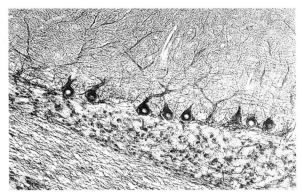

Plate 33 Pukinje cells of cat cerebellum (40×) (© Biodisc/
Visuals Unlimited)

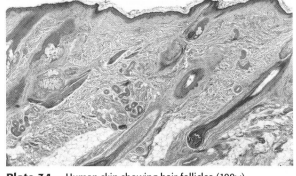

Plate 34 Human skin showing hair follicles (100×)
(© Carolina Biological/Visuals Unlimited)

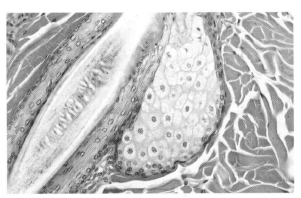

Plate 35 Human skin showing sebaceous glands at base of
hair shaft (75×) (© Carolina Biological/Visuals Unlimited)

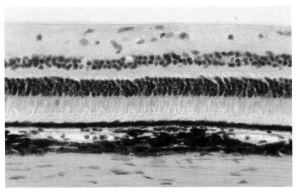

Plate 36 Human eye showing retinal details (64×)
(© Science VU/Visuals Unlimited)

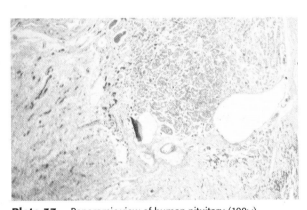

Plate 37 Panoramic view of human pituitary (100×)
(author provided)

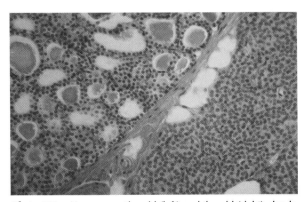

Plate 38 Human parathyroid (left) and thyroid (right) glands
(10×) (© Dr. David Phillips/Visuals Unlimited)

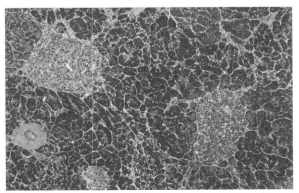

Plate 39 Pancreas showing islets of Langerhans (100×)
(© Gladden Willis, M.D./Visuals Unlimited)

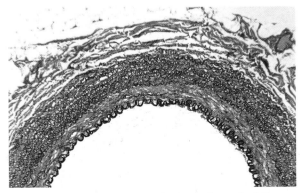

Plate 40 Large human artery demonstrating tunica intima,
media, and adevntitia layers (30×) (© Biodisc/Visuals Unlimited)

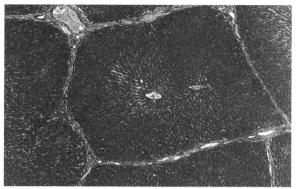

Plate 41   Liver showing lobule (25×) (© Carolina Biological/Visuals Unlimited)

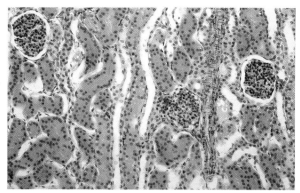

Plate 42   Kidney glomeruli (40×) (© Dr. Fred Hossler/Visuals Unlimited)

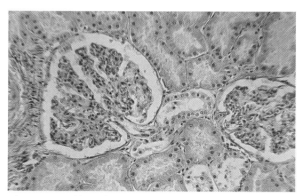

Plate 43   Kidney glomeruli (80×) (© Biodisc/Visuals Unlimited)

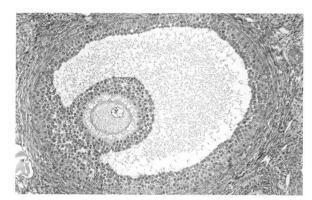

Plate 44   Ovary with Graafian follicle (350×) (© Carolina Biological/Visuals Unlimited)

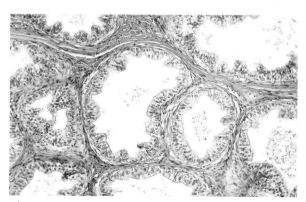

Plate 45   Testis with seminiferous tubules (36×) (© Carolina Biological/Visuals Unlimited)

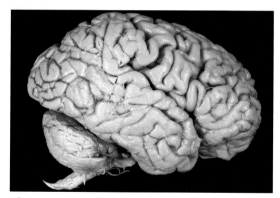

Plate 46   Brain, lateral view (© Dr. Fred Hossler/Visuals Unlimited)

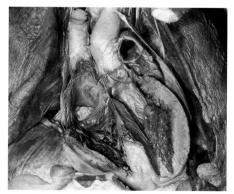

Plate 47   Heart, frontal section (© Dr. L. Bassett/Visuals Unlimited)

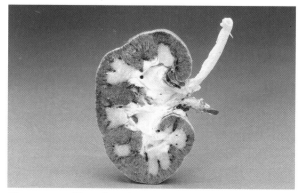

Plate 48   Kidney, frontal section (© Carolina Biological/Visuals Unlimited)

Figure 4.6

Transitional epithelium; this tissue type is found in the urinary tract (100×) (Color Plate 10)

Question 4.3

What is a lumen?

1. transitional epithelial layer
2. balloon-shaped cap cells next to lumen
3. lumen
4. basement membrane
5. loose areolar connective tissue

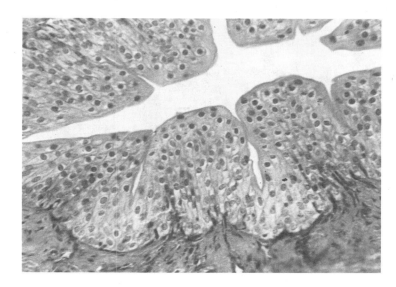

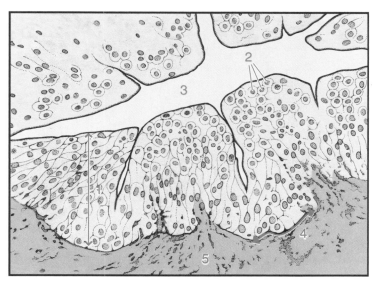

Drawing:

Figure 4.7

Loose areolar connective tissue; contains cells and fibers; functions as binding and packing substance around organs (430×) (Color Plates 11 and 12)

1. elastic fibers
2. collagenous fiber
3. macrophage cell
4. fibroblast cells

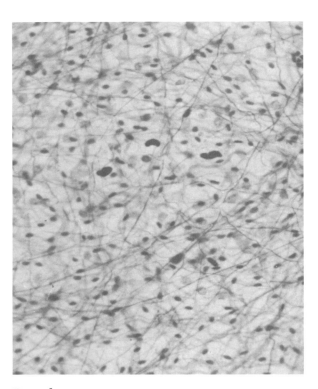

Drawing:

Figure 4.8

Dense, fibrous connective tissue (regular); closely aligned parallel rows of wavy fibers are separated by single rows of darker staining cells; found in tendons and ligaments (100×) (Color Plate 13)

1. cells (fibrocytes)
2. fibers

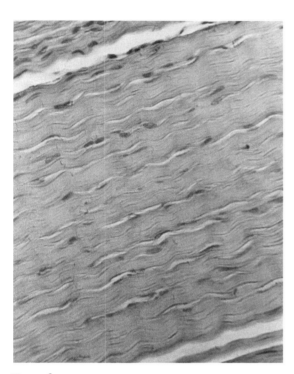
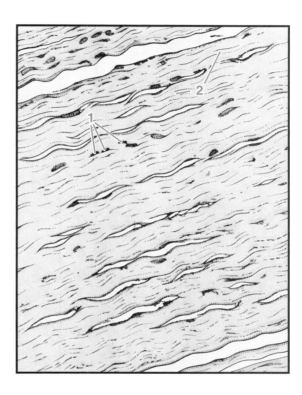

Drawing:

Figure 4.9

Hyaline cartilage of the trachea; when stained, this tissue type has a bluish, glasslike appearance; also found covering ends of bones to reduce friction (100×) (Color Plates 15 and 16)

1. chondrocytes
2. intercellular matrix
3. perichondrium
4. lacunae

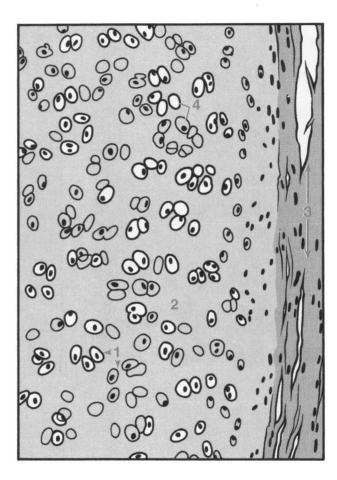

Drawing:

Figure 4.10

Hyaline cartilage detail; cells (chondrocytes) are contained within cavities known as lacunae (430×) (Color Plates 15 and 16)

1. intercellular matrix
2. nucleus of chondrocyte
3. lacuna containing chondrocyte

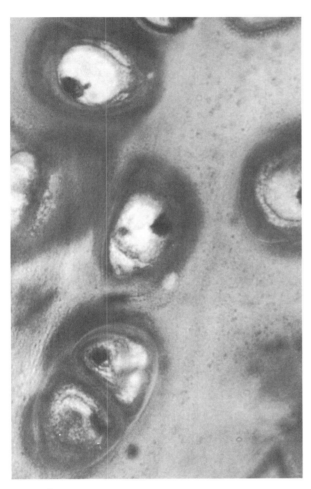
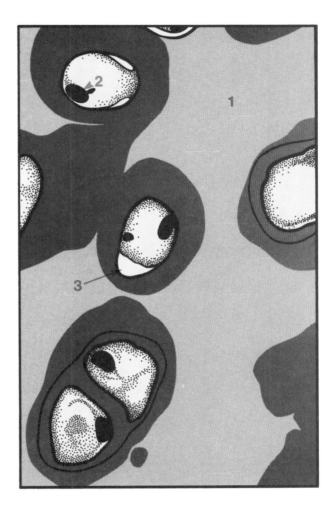

Drawing:

Figure 4.11

Elastic cartilage and perichondrium, as seen in the epiglottis of the larynx; matrix contains elastic fibers, which give this tissue type its flexibility (430×) (Color Plate 17)

1. perichondrium
2. intercellular matrix
3. elastic fibers
4. chondrocyte
5. lacuna

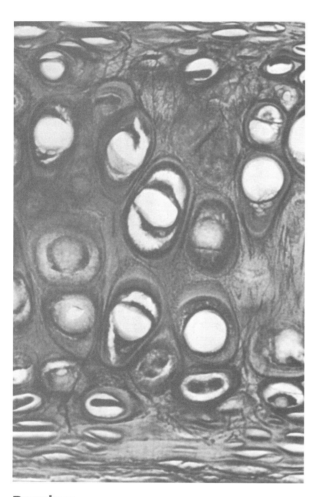

Drawing:

Figure 4.12

Ground bone (compact bone); this type of osseous tissue contains osteons (Haversian systems), which contribute to its rigidity (100×) (Color Plate 19)

Which osteon structure is most central?

1. osteon (Haversian system)
2. osteonic canal (Haversian canal)
3. lacunae containing osteocytes
4. concentric lamella
5. interstitial matrix
6. canaliculi

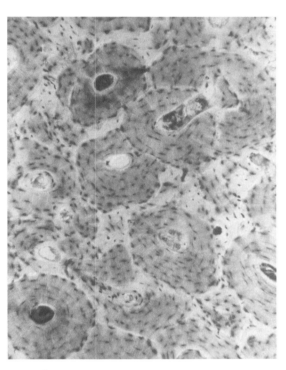

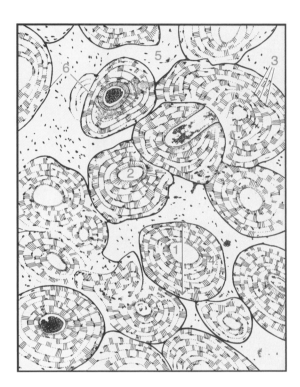

Drawing:

Figure 4.13

Endochondral bone formation as seen in a fetal metacarpal bone; cartilage is replaced by bone as an individual matures (100×)

1. subcutaneous layer
2. perichondrium
3. periosteum

4. proliferating cartilage
5. spicules of bone

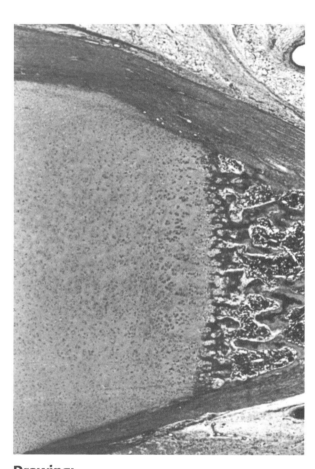

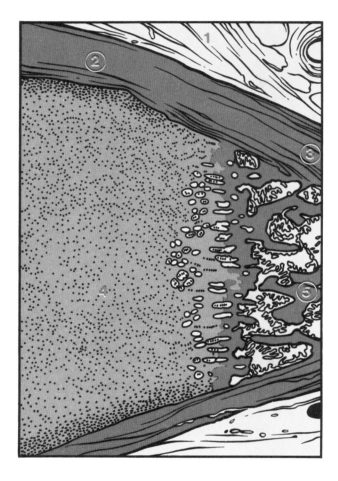

Drawing:

Question 4.5

What does the term "ossification" mean?

Figure 4.14

Adipose tissue is composed of adipocytes containing fat (100×)

1. adipocyte (adipose cell)
2. fat vacuole
3. nucleus

4. cell membrane
5. capillary

Question 4.6

In the illustration below, what comprises the greatest volume of an adipose cell?

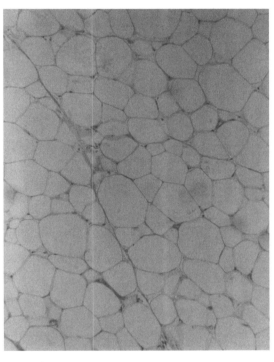

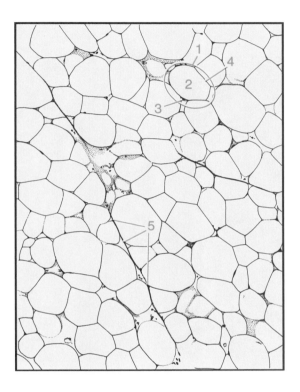

Drawing:

Figure 4.15

Adipose tissue; nucleus and cytoplasm of adipocytes are displaced toward periphery of the cell due to the volume of fat within each cell (400×)

1. cell membrane
2. nucleus
3. cytoplasm
4. vacuole containing fat

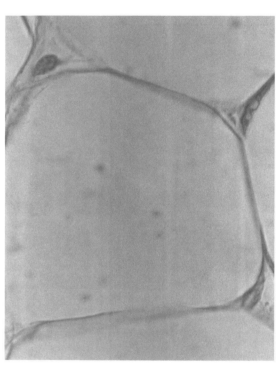
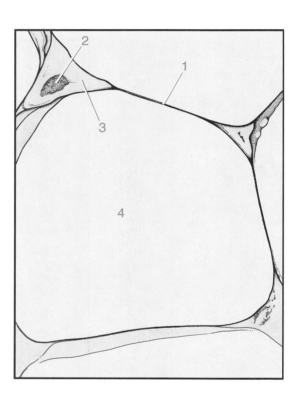

Drawing:

Figure 4.16

Skeletal muscle; fibers (cells) appear striated with peripheral oval nuclei; comprises voluntary muscles which move bones (430×) (Color Plate 26)

1. nucleus
2. skeletal muscle fiber
3. striations
4. sarcolemma

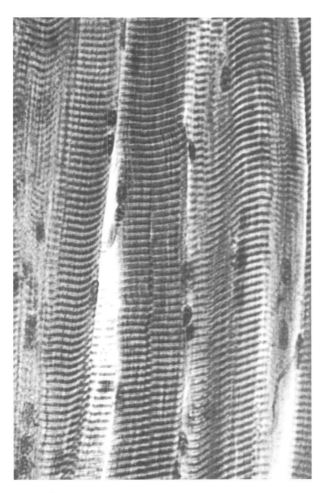 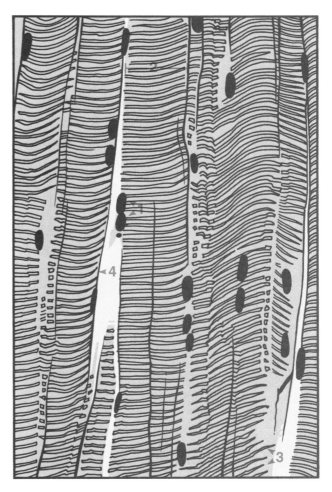

Drawing:

Figure 4.17

Cross section of skeletal muscle; patchlike areas of muscle fibers contain myofibrils; nuclei are located at periphery of cells (100×)

1. perimysium surrounding a fascicle (bundle of muscle fibers)
2. endomysium surrounding an individual muscle fiber
3. muscle fibers
4. nuclei
5. muscle fibers (longitudinal section)

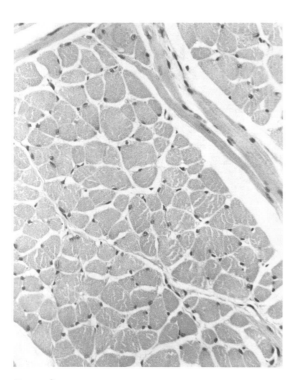
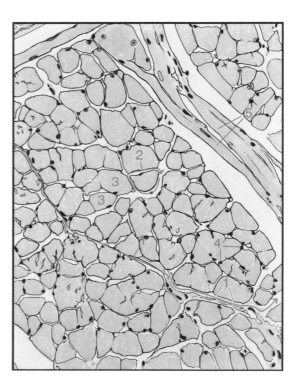

Drawing:

Figure 4.18

Smooth muscle; this tissue type is found in blood vessels and in the digestive tract and is under involuntary nervous control (430×) (Color Plate 29)

1. nuclei of smooth muscle fibers
2. capillary

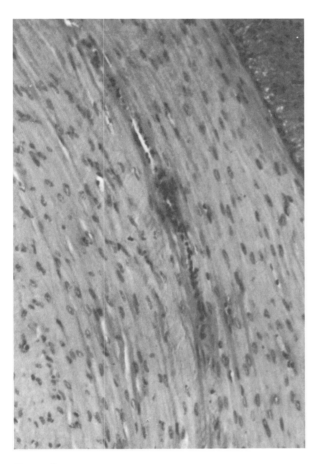
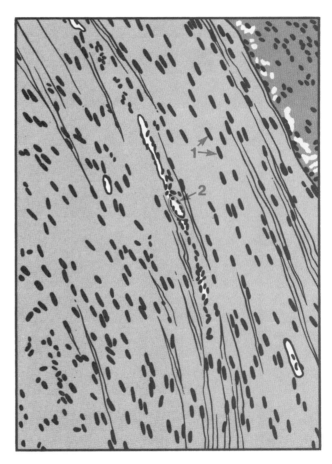

Drawing:

Figure 4.19

Smooth muscle; cells are elongated, contain oval nuclei, and "dovetail" into each other (1000×) (Color Plate 29)

1. smooth muscle fibers
2. nucleus of muscle fiber
3. sarcolemma

Question 4.7

Is smooth muscle considered to be voluntary or involuntary?

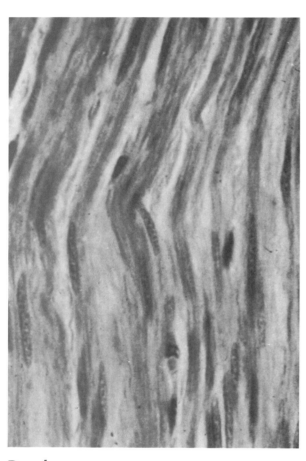

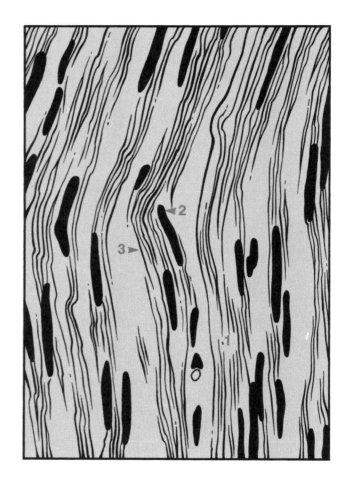

Drawing:

Figure 4.20

Cardiac muscle fibers are connected end to end by intercalated disks (40×) (Color Plate 28)

1. nucleus of cardiac muscle fiber
2. intercalated disks

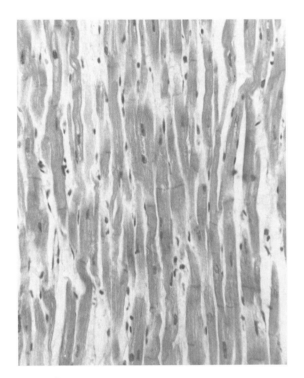

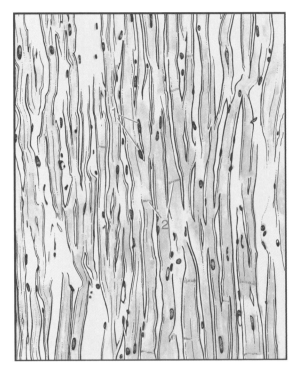

Drawing:

Figure 4.21

Cardiac muscle is a type of striated muscle that contains bifurcations and centrally located nuclei (100×) (Color Plate 27)

1. bifurcations
2. nuclei
3. striations
4. intercalated disks

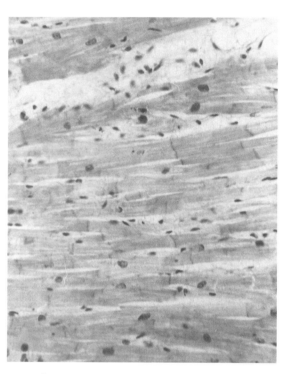
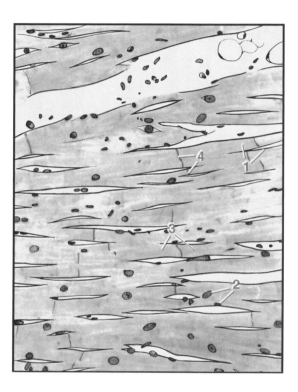

Drawing:

EXERCISE 2 Microscopic Identification of the Skin (Color Plates 34 and 35)

Now that you have examined and identified various tissues, you will be able to observe some of them in the microscopic examination of the skin. The skin is the largest organ in the body. An organ, remember, is a group of tissues that performs a specific function.

The skin is composed of an outer **epidermis**, a deeper **dermis**, or **corium**, and an underlying **hypodermis**. The epidermis is made up of stratified squamous epithelium and consists of five layers:

1. **Stratum corneum:** This is the outermost layer of the epidermis, consisting of flattened, dead, keratinized cells that are continuously shed. The stratum corneum helps serve as a barrier to light, heat, bacteria, and certain chemicals.
2. **Stratum lucidum:** This layer is directly beneath the stratum corneum and is one or two cell layers thick. It is most easily seen in the epidermis of the palms and soles.
3. **Stratum granulosum:** This is a thin layer lying beneath the stratum lucidum. It is thought to be the layer in which keratinization takes place. In this layer granules are numerous and tend to stain heavily.
4. **Stratum spinosum:** This layer may be difficult to see. It is beneath the stratum granulosum and consists of "prickly" cells.
5. **Stratum basale:** This layer is the deepest layer of the epidermis. Cells in this layer actively undergo mitotic division and give rise to the four outer epidermal layers. **Melanin,** the principal pigment of the skin, is formed in this layer by **melanocytes.** Increased melanocyte activity results in darker skin.

The dermis lies beneath the epidermis. It contains connective tissue fibers, blood vessels, nerves, sweat glands, sebaceous glands, and hair follicles.

The hypodermis or subcutaneous layer is the deepest layer of skin. It binds the dermis to underlying muscle. In this layer you will find various tissues, such as loose areolar connective tissue, elastic and adipose connective tissues, and nerve endings. The hypodermis also contains blood vessels.

Examine slides of pigmented and nonpigmented skin, identifying the above structures (**Figures 4.22–4.28**). Draw and label various tissue types that you observe in the skin.

Figure 4.22

Thick skin, palm, showing epidermal layers (430×)

1. stratum corneum
2. stratum lucidum
3. stratum granulosum
4. stratum basale
5. dermal papilla
6. epidermis
7. dermis
8. melanocytes

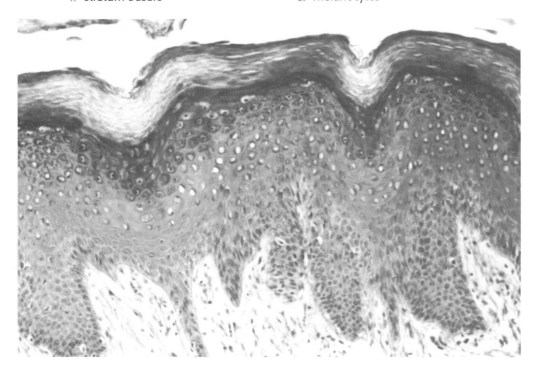

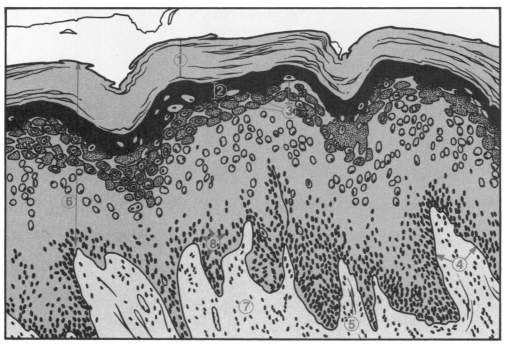

Figure 4.23

Skin, human scalp, showing general view of hair follicles (100×)

1. epidermis
2. dermis
3. hypodermis
4. adipose cells
5. hair follicle
6. hair shaft

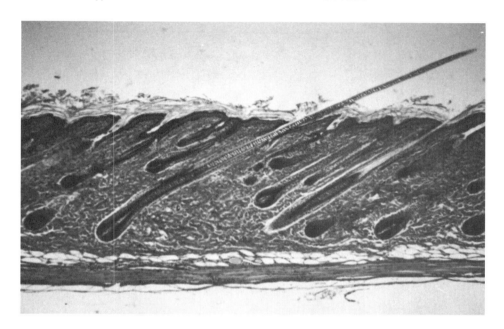

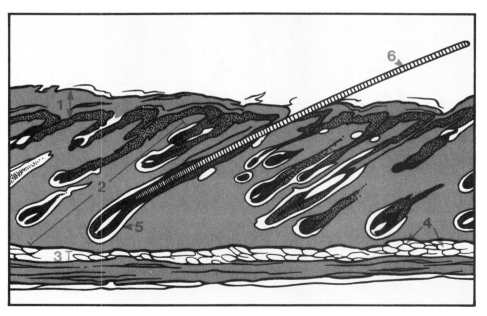

Figure 4.24

Skin showing sebaceous and sweat glands (100×)

1. sebaceous glands
2. ducts of sweat glands
3. adipose cells

4. arrector pili muscle
5. hair shaft

Drawing:

Figure 4.25

Skin showing detail of a sebaceous gland (1000×) (Color Plate 35)

1. sebaceous gland
2. duct of sebaceous gland
3. squamous cells

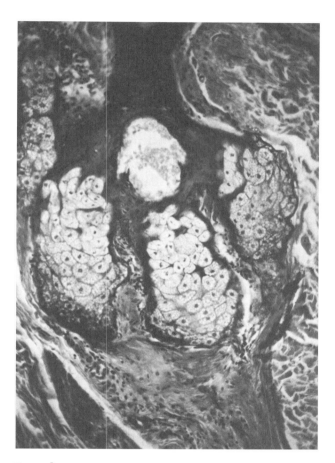 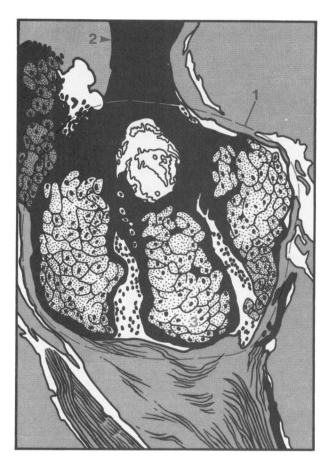

Drawing:

Figure 4.26

Detail of sweat gland ducts (430×) (Color Plate 4)

1. ducts of sweat gland
2. cuboidal cells
3. lumen

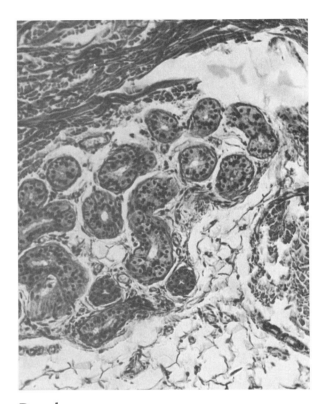
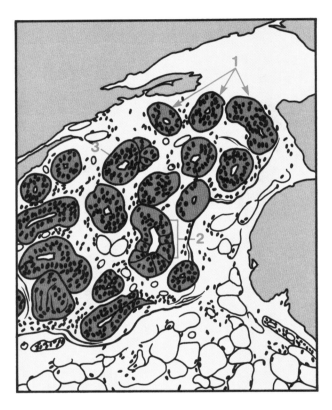

Drawing:

Figure 4.27

Pacinian (lamellated) corpuscles (sensory neurons specialized for pressure) in dermis of skin (100×)

1. Pacinian corpuscles
2. dermis
3. hypodermis
4. ducts of sweat glands
5. adipose tissue
6. artery

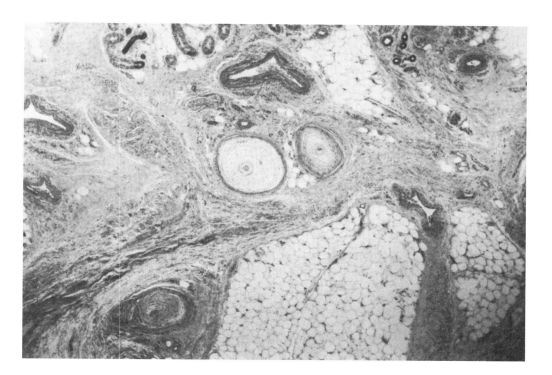

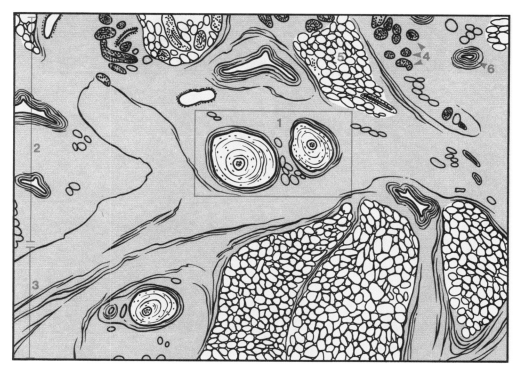

Figure 4.28

Detail of Pacinian corpuscle in dermis of skin (430×)

1. Pacinian corpuscle
2. Arteriole
3. Capillary

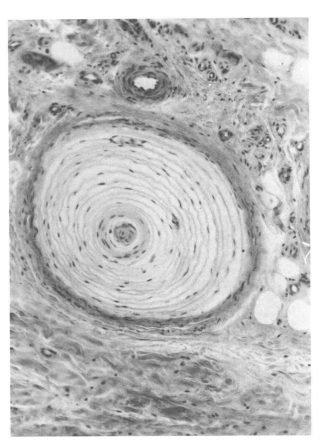

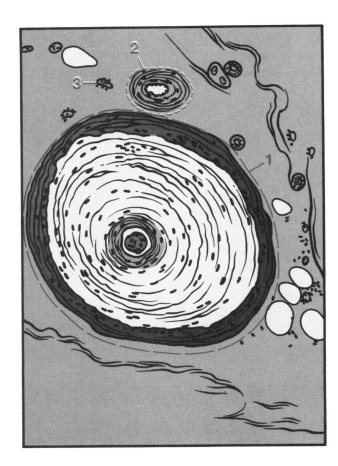

Drawing:

EXERCISE 3   Human Skin

On the skin model identify the **epidermis**, **dermis**, and **hypodermis**. In the epidermis find the **stratum corneum**, **stratum granulosum**, and **stratum basale**. In the dermis locate nerve endings, blood vessels, **sebaceous glands**, **sweat glands**, and **hair follicles**. In the hypodermis locate **adipose tissue**.

EXERCISE 4   Human Fingerprints*

The skin of the fingertips is not smooth, but contains dermal ridges that form a pattern, commonly known as fingerprints, which are unique to each individual. These patterns of ridges are inherited and may be categorized as arches, loops, or whorls. In this exercise, you will observe your dermal ridges, categorize their patterns, determine total ridge counts (TRC) and compare your data with other members of the class.

MATERIALS

Sheet of white paper
No. 2 pencil
Roll of ¾″ Scotch Brand Magic Tape
Dissecting microscope or magnifying glass

PROCEDURE

Blacken a 1.5″ square area on the paper by briskly and firmly rubbing it with the edge of the lead pencil.

Remove a 1.5″ strip of tape from the roll and attach one end loosely to a fingertip of your right hand. Be careful not to touch the surface of the tape.

Press the pad of your left thumb firmly against the blackened square of the paper and rub it back and forth. Fasten the tape lengthwise along the entire blackened area of your thumb. Avoid wrinkling or touching the sticky surface of the tape with your right hand as you apply it firmly to cover the entire thumb print. Remove the tape and place it in the portion of the following table marked for the left-hand thumb. Repeat this procedure for your remaining nine digits, using freshly blackened squares of paper when necessary to yield dark, discernible prints. You may also find it helpful to have your lab partner apply and remove the tape from your fingers.

The **arch** pattern is the simplest, in that the ridges rise over the middle of the finger. Arches have a ridge count of zero and do not contain a triradius. The **loop** pattern contains a triradius and a core. A *triradius* is the point at which three sets of ridges meet at angles of about 120°. (An example is screened in red in **Figure 4.29b**). A *core* is a blind-ended ridge that is surrounded by ridges that turn back on themselves at 180°. The **whorl** pattern appears as a series of concentric circles in the center, then broadens out to

---

*Adapted from G. Mendenhall, G. Mertens, and J. Hendrix, 1989. Fingerprint ridge count—a polygenic trait useful in classroom instruction. *The American Biology Teacher.* 51(4):203–207.

meet two triradii. According to the averages reported by Holt (The genetics of dermal ridges, 1968), the arch pattern in **Figure 4.29a** occurs in about 5% of the population, the loop pattern in **Figure 4.29b** in 68.9%, and the whorl pattern in **Figure 4.29c** in 26.1%.

Left Hand	Thumb	Index Finger	Third Finger	Fourth Finger	Fifth Finger
Pattern	_____	_____	_____	_____	_____
Ridge Count	_____	_____	_____	_____	_____

Right Hand	Thumb	Index Finger	Third Finger	Fourth Finger	Fifth Finger
Pattern	_____	_____	_____	_____	_____
Ridge Count	_____	_____	_____	_____	_____
			Total Ridge Count (left and right hands)		_____

Using **Figure 4.29** as a guide, determine the pattern of dermal ridges for each of your digits and enter it under each respective fingerprint. Then count the ridges for each pattern and write them in the corresponding spaces, remembering that the arch pattern has a ridge count of zero. You may find it helpful to use a dissecting microscope or magnifying glass to determine patterns and count ridges. After you have determined the ridge count for each digit, enter the total ridge count (TRC) in the corresponding space at the bottom of the table.

Figure 4.29

Three principal types of fingerprint patterns: (a) arch with no triradius and a ridge count of 0; (b) loop with one triradius and a ridge count of 12; and (c) whorl with two triradii and a ridge count of 15 (the higher of the two possible counts).

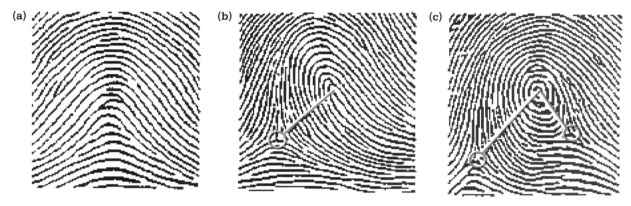

Adapted with permission of the Biological Sciences Curriculum Study from *Basic genetics: A human approach.* (1983). Dubuque, IA: Kendall/Hunt Publishing Co. Adaptation with permission from NABT, Reston, VA.

How do the fingerprint patterns (of digits) of the class compare with those of the general population, according to Holt (1968)?

	Percentage for Class	*Percentage for General Population*
Arch	_____%	5.0%
Loop	_____%	68.9%
Whorl	_____%	26.1%
	100%	100%

Total Ridge Count (TRC)

Individual	Class	General Population
Your TRC _____	*Males* _____	145
Your Gender _____	*Females* _____	126

Question 4.8

How does your individual TRC and those of the average of the class compare with the average TRC of the general population as reported by Holt (1968) as 145 for males and 126 for females?

Question 4.9

AABBCCDD* represents the four gene loci for the genotype of an individual with maximum TRC. The baseline or minimum TRC for an individual who is genotype aabbccdd is 40 for males and 20 for females. If each active, dominant allele of a gene pair adds 30 ridges to the TRC, determine the TRC for each of the following individuals. (Capital letters represent dominant alleles; lowercase letters represent recessive alleles; for example, AaBbCc consists of three gene pairs with six alleles.)

AaBbCcDd male _____   AABbccDD male _____

AaBbCcDd female _____   aaBBCcdd female _____

Using the same technique, prepare fingerprints for a relative in the following tables:

Left Hand

	Thumb	Index Finger	Third Finger	Fourth Finger	Fifth Finger
Pattern	_____	_____	_____	_____	_____
Ridge Count	_____	_____	_____	_____	_____

Right Hand

	Thumb	Index Finger	Third Finger	Fourth Finger	Fifth Finger
Pattern	_____	_____	_____	_____	_____
Ridge Count	_____	_____	_____	_____	_____

Total Ridge Count (left and right hands) _____

Question 4.11

How does the TRC of your relative compare with your own and that of the general population (Males, 145; Females, 126)?

	Relative	You	General Population
TRC	_____	_____	Males 145
Sex	_____	_____	Females 126
	_____		
	Relationship		

Question 4.10

Is there a similarity between your relative's dermal pattern and your own? Explain.

*Penrose (1969) proposed that a minimum of seven gene loci contributed to the TRC. This problem uses a four-locus model.

Tissues

MATCHING

Name

Section

Date

Column A

_____ 1. This tissue type functions in impulse conduction.

_____ 2. This tissue type lines the stomach and small intestine.

_____ 3. This tissue type lines the mouth.

_____ 4. This tissue type functions in contraction.

_____ 5. Blood is classified as this type of tissue.

_____ 6. This tissue functions in protection, absorption, and secretion.

_____ 7. Cartilage is of this tissue type.

_____ 8. This tissue includes striated and smooth types.

_____ 9. A glomerular capsule is composed of this tissue type.

_____ 10. Adipose tissue is of this type.

Column B

a. epithelial

b. connective

c. muscular

d. nervous

MULTIPLE CHOICE

_____ 11. Which of the following is *not* characteristic of epithelial tissue?
- a. has no blood vessels
- b. covers free surfaces of the body
- c. typically involved in secretion, excretion, or absorption
- d. contains an extensive nonliving matrix

_____ 12. Which is *not* a function of the skin?
- a. protection against mechanical injury
- b. protection against foreign invaders
- c. regulation of body heat
- d. all of the above are functions

_____ 13. The layer of skin that lacks blood vessels is
- a. subcutaneous.
- b. dermis.
- c. integument.
- d. epidermis.

_____ 14. The dermis does *not* contain
- a. sebaceous glands.
- b. hair follicles.
- c. mucous glands.
- d. nerves.

_____ 15. The color of human skin depends upon
- a. the yellowish tinge of epidermal cells.
- b. the number of underlying blood vessels.
- c. the kind and amount of pigment.
- d. all of the above.

_____ 16. Through which epithelial type would diffusion most easily take place?
- a. simple squamous
- b. transitional
- c. cuboidal
- d. pseudostratified columnar

_____ 17. In which of these organs would striated involuntary muscle be found?
- a. small intestine
- b. lung
- c. liver
- d. heart

_____ 18. Which layer of epidermis would be gradually shed through bathing?
- a. stratum granulosum
- b. stratum corneum
- c. stratum basale
- d. stratum lucidum

_____ 19. In which tissue type are concentric rings of cells found?
- a. osseous
- b. hyaline cartilage
- c. adipose
- d. epithelial

_____ 20. The dermis is primarily composed of which tissue type?
- a. nervous
- b. muscle
- c. connective
- d. epithelial

_____ 21. The amount of melanin produced in the skin is determined by the
    a. number of melanocytes.
    b. activity of melanocytes.
    c. diet.
    d. proximity of blood vessels to the skin

_____ 22. Most of the volume of adipose tissue cells contain
    a. cytoplasm.
    b. lipids.
    c. elastic tissue.
    d. chromatin.

_____ 23. What tissue types are found in the intestinal tract?
    a. cuboidal epithelium, smooth muscle
    b. stratified squamous epithelium, striated muscle
    c. simple columnar epithelium, smooth muscle
    d. pseudostratified columnar epithelium, reticular tissue

_____ 24. The criterion for categorizing epithelia as simple, stratified, or pseudostratified is
    a. cell size.
    b. secretions of cells.
    c. cell shape.
    d. arrangement of cells.

_____ 25. Which epidermal layer is closest to a blood supply?
    a. stratum basale
    b. stratum spinosum
    c. stratum granulosum
    d. stratum corneum

26. For review, draw the following tissue types in the spaces provided.

Stratified Squamous Epithelium	Simple Columnar Epithelium	Adipose	Dense Fibrous Connective Tissue
Hyaline Cartilage	Osseous	Skeletal Muscle	Smooth Muscle

27. a. In which epidermal layer are melanocytes found?

b. Which layer(s) of epidermis is (are) composed of dead cells?

c. Which epidermal layer(s) is (are) keratinized?

28. Which specialized tissue type(s) would you find in each of the following locations?

a. collecting tubules of the kidney

b. lining the urinary tract

c. trachea

d. muscle layer of the small and large intestines

e. mesenteries

f. biceps muscle

g. tendon

h. nasal septum

i. lining the intestinal tract

j. fat

CASE STUDY

HISTOLOGY CASE STUDY

John Moss, a first semester anatomy student was studying a cross-section of the ureter and was asked in a test to identify and give the functions of tissues from the epithelial, connective, and muscular histological groups. John thought that he could identify stratified squamous epithelium lining the lumen of the ureter. Smooth muscle was located in the middle region. He found dense fibrous connective tissue as an outer protective cover of the ureter.

Do you think that John is correct in his identification of the three histological samples and their respective functions? Explain your answer.

The Skeletal System

A. Skeletal Anatomy

PURPOSE

Unit 5A will familiarize you with the various divisions of the human skeleton and the names and numbers of bones included in each division.

OBJECTIVES

After completing Unit 5A, you will be able to identify
- major components of a Haversian system, or osteon.
- major anatomical structures of a long bone.
- bones of both axial and appendicular skeletons.
- major bone markings.

MATERIALS

slides of dry ground bone
  (osseous tissue)
articulated human skeleton
disarticulated human skeletons
articulated vertebral columns

models: sagittal section of a long
  bone, labeled adult skull,
Beauchene disarticulated skull
compound microscope
colored pencils (optional)

PROCEDURE

In this unit, you will study slides of osseous tissue and use articulated and disarticulated skull models and human skeletons to identify bones and bone markings.

EXERCISE 1 **Microscopic Examination of Osseous Tissue**

Since you studied osseous tissue in Unit 4, this will be a review. Examine a prepared slide of dry ground bone and identify: **Haversian** (or **central**) **canal**, **lamella, lacunae** containing **osteocytes, canaliculi**, and **interstitial material** (**Figure 4.12** and **Color Plate 19**).

Draw a Haversian system (osteon) and label the structures listed above to the right of your diagram.

Question 5A.1

Which end is proximal?

EXERCISE 2 **Gross Anatomy of a Long Bone**

You are to identify the gross anatomical features of a sagittal section of a long bone. Examine a long bone (preferably a femur) and identify the following: **articular cartilage(s), compact bone (dry ground bone), cancellous bone, diaphysis, endosteum, epiphysis, medullary canal,** and **periosteum** (Figure 5.1).

Figure 5.1

Femur, longitudinal section, showing gross structure of the bone

1. articular surface (cartilage)
2. spongy (cancellous) bone
3. compact bone
4. medullary canal
5. diaphysis
6. epiphysis

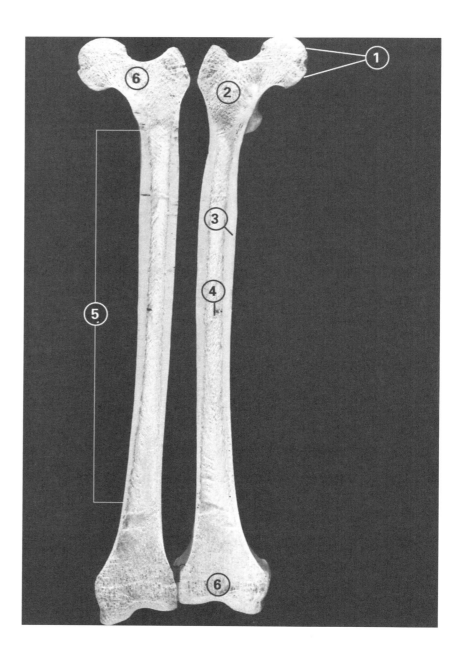

Question 5A.2

What is the purpose of cartilage on the epiphysis?

EXERCISE 3 Identification of Bones of the Human Skeleton

Question 5A.3
Which appendicular division has a greater number of bones? Why?

You are to identify the bones of the axial skeleton first and then the appendicular skeleton. If you are working with a disarticulated (bones disjointed) skeleton, arrange the bones in anatomical position on the laboratory table (**Figure 5.2**).

There are more than 200 bones in the human adult; some are paired, some are single. Other bones, such as the skull, sternum, and hip, are fusions of a number of bones. Be sure to identify the separate bones that make up each fusion. In addition to the bones listed in these tables, a variable number of **Wormian bones** are embedded in the skull.

In infancy and early childhood, ossification has not been completed; therefore bones are more cartilaginous. If you have a young skeleton, observe the delicate nature of the bones and obvious cartilage.

Identify all the bones in the following tables.

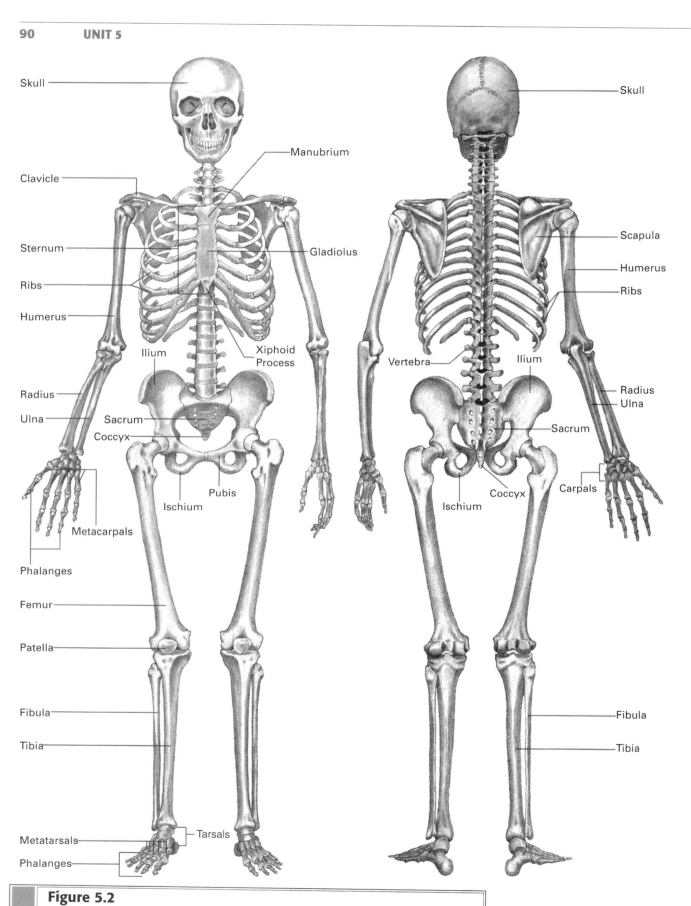

Skull

Clavicle

Sternum

Ribs

Humerus

Radius

Ulna

Metacarpals

Phalanges

Femur

Patella

Fibula

Tibia

Metatarsals

Phalanges

Manubrium

Gladiolus

Xiphoid Process

Ilium

Sacrum

Coccyx

Ischium

Pubis

Tarsals

Skull

Scapula

Humerus

Ribs

Vertebra

Ilium

Radius

Ulna

Sacrum

Coccyx

Ischium

Carpals

Fibula

Tibia

Figure 5.2

Anterior and posterior views of the human skeleton showing the normal position of each bone

Bones of the Axial Skeleton (80 total)

Skeletal Division	Name(s) of Bone(s)	Number of Bones
Skull (Figures 5.4–5.19)		28 (total)
Cranium		8 (total)
	Frontal	1
	Parietal	2
	Temporal	2
	Occipital	1
	Sphenoid	1
	Ethmoid	1
Face		14 (total)
	Mandible	1
	Nasal	2
	Lacrimal	2
	Vomer	1
	Inferior concha *or* turbinate	2
	Zygomatic *or* malar	2
	Palatine	2
	Maxilla	2
Ear ossicles (Figure 8.9)		6 (total)
	Malleus	2
	Incus	2
	Stapes	2
Hyoid (Figure 5.20)		1
Vertebrae (Figures 5.21–5.29)		26 (total)
	Cervical	7
	Thoracic	12
	Lumbar	5
	Sacrum	1 (fusion of 5)
	Coccyx	1 (fusion of 3–5)
Sternum (Figure 5.30)		1 (fusion of 3)
	Manubrium	1
	Gladiolus *or* body	1
	Xiphoid process	1
Ribs (Figure 5.31)		12 pairs (7 pairs true, 5 pairs false)

Bones of the Appendicular Skeleton (126 total)

Skeletal Division	Name(s) of Bone(s)	Number of Bones
Upper Division		64 (total)
(Figures 5.32–5.40)	Clavicle	2
	Scapula	2
	Humerus	2
	Ulna	2
	Radius	2
	Carpals (Figure 5.39)	16 (8 per hand)
	Scaphoid (SKAF-oyd)	1
	Lunate (LOO-nāt)	1
	Triquetrum (trī-KWĒ-trum)	1
	Pisiform (PĪ-si-form)	1
	Trapezium (tra-PĒ-zē-um)	1
	Trapezoid (TRAP-e-zoyd)	1
	Capitate (KAP-i-tāt)	1
	Hamate (HAM-āt)	1
	Metacarpals	10 (5 per hand)
	Phalanges	28 (14 per hand)
Lower Division		62 (total)
(Figures 5.41–5.48)	Os coxa *or* Innominate	2 (each a fusion of 3)
	Ilium	
	Ischium	
	Pubis	
	Femur	2
	Patella	2
	Tibia	2
	Fibula	2
	Tarsals (Figure 5.48)	14 (7 per foot)
	Calcaneus	1
	Talus	1
	Cuboid (KYOO-boyd)	1
	Navicular (na-VIK-yoo-lar)	1
	Cuneiforms (kyoo-NĒ-i-formz)	3
	Metatarsals	10 (5 per foot)
	Phalanges	28 (14 per foot)

EXERCISE 4   Identification of Bone Markings

The skeletal system serves as a basis for body topography. Now that you have learned the names of the bones, their numbers, and to which skeletal division they belong, you can learn greater detail about each bone. A thorough knowledge of **bone markings**—specific identifiable projections or depressions on bones—is of great importance, because it facilitates identification of other body structures, such as muscles, blood vessels, and nerves.

Listings of the major bone markings follow. Those classified as **projections** (also called **processes**) grow out from the bone, whereas those classified as **depressions** (also called **fossae**) are indentations in the bone. Locating specific bone markings will be easier if you first know the general names of the different types of markings (**Figure 5.3**).

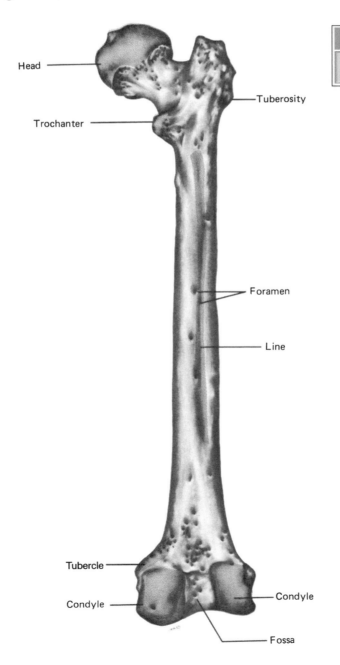

Figure 5.3

Femur showing general types of bone markings

BONE MARKINGS—GENERAL TYPES

Projections

1. **Trochanter** (tro-KAN-ter): a large irregularly shaped projection to which muscles attach.
2. **Tuberosity** (toó-ber-OS-i-tē): a large rounded projection to which muscles attach.
3. **Tubercle** (TOO-ber-kul): a small rounded projection to which muscles attach.
4. **Condyle** (KON-dīl): a rounded convex projection.
5. **Head:** a rounded extension projecting from a tapering neck.
6. **Ramus** (RĀ-mus): an armlike branch extending from the body of a bone.
7. **Spine:** a sharp projection that serves as a point for muscle attachment.
8. **Crest:** an elevation or ridge that serves as a point of muscle attachment.
9. **Line:** a lesser elevation or ridge that serves as a point of muscle attachment.
10. **Process:** a term used to designate a general outgrowth of varying shape and size.

Depressions

1. **Fossa:** a hollow indentation often found in joint formations.
2. **Foramen** (fo-RĀ-men): a hollow opening that serves as a passageway for nerves and blood vessels.
3. **Sinus** (SĪ-nus): an irregularly shaped space often filled with air and lined with mucosal tissue.
4. **Fissure** (FISH-er): a narrow slitlike depression often serving as a passageway for nerves.
5. **Meatus** (mē-Ā-tus): a canal-like opening that sometimes serves as a passageway for nerves.
6. **Sulcus** (SUL-kus): a furrow or groove.

Question 5A.4

How does a tuberosity differ from a tubercle?

Read through the tables on the next several pages. Following the tables sequentially, identify the bone markings in the accompanying figures. Use a set of colored pencils to color each of the bones and bone markings. When you have finished your identification of the skull, review your work. Then go on to identify and color the next division of the skeleton, the vertebrae.

Question 5A.5

Draw a fissure and a fossa in the space below.

Identification of Bone Markings—Axial Skeleton

Use the following tables, illustrations, and photographs to identify bones and bone markings on articulated or disarticulated skeletons. *Boldface figure labels denote the first time that the bones and bone markings listed in the tables appear in the illustrations.* After identifying the bone markings on a skull, you might find it helpful to quiz yourself by covering the legends that accompany the photographs.

Bone Markings of the Skull

Bone	Bone Marking	Description
CRANIUM (KRĀ-nē-um) **Frontal**		A flat convex bone in the adult, formed by the union of two flat bones joined at frontal suture (suture partly visible in adult)
(Figures 5.4–5.6, and 5.19; also Figures) 5.7 and 5.8)	Frontal eminences *or* protuberances	Protrusions above each optic orbit
	Supraorbital (soo'-pra-OR-bi-tal) margins	Flat arched regions immediately inferior to eyebrows
	Supraorbital notches *or* foramina (fo-RĀM-i-na)	Irregularly shaped small openings in supraorbital margins toward nasal bones, serving as openings for nerves and blood vessels
	Superciliary (soo'-per-SIL-ē-ar-ē) arches	Anterior ridges superior to optic orbits that form base for eyebrows; formed by frontal sinuses within them
	Glabella (gla-BEL-a)	Smooth area formed by the union of the superciliary arches above nasal bones
Parietal (pa-RĪ-e-tal) (Figure 5.6; also Figures 5.7 and 5.8)		Large irregularly shaped bones lateral and posterior to frontal bones; parietals merge at superior medial portion of the cranium to form sagittal suture
	Parietal foramen	Small rounded opening toward the posterior of the bone; immediately lateral to the sagittal suture; may be closed in some skeletons
	Parietal tuberosity	Slightly rounded protrusion on lateral surface of bone
Temporal (TEM-po-ral)		Bones form lateral inferior sides of cranium; house structures of middle and inner ear
(Figures 5.4, 5.6, 5.15–5.18; also Figures 5.7–5.9)	Mastoid process	Rough projection immediately posterior to ear or external auditory meatus
	Squamous (SKWĀ-mus) portion	Flat, thin superior portion of bone; translucent when held to light
	Zygomatic (zī-gō-MAT-ik) process *or* arch	Narrow bridgelike extension that articulates anteriorly with zygomatic bone
	External auditory *or* acoustic meatus	Canal into ear running from exterior to interior of temporal bone
	Mandibular (man-DIB-yoo-lar) fossa	Rounded depression inferior to and partially formed by zygomatic process; immediately anterior to external auditory meatus; forms socket for mandibular condyle
	Styloid (STĪ-loyd) process	Long needlelike projection anterior to mastoid process; serves as attachment for muscles and ligaments
	Stylomastoid foramen	Small opening between styloid process and mastoid process; serves as a passageway for facial nerve
	Jugular (JUG-yoo-lar) fossa	Irregularly shaped depression lateral and anterior to occipital condyles for internal jugular vein
	Jugular foramen	Fairly large oval-shaped opening at posterior end of jugular fossa; lateral to occipital condyle; serves as passageway for lateral sinus and cranial nerves IX, X, and XI
	Carotid (ka-ROT-id) foramen *or* canal	Round opening medial and anterior to jugular foramen; serves as passageway for internal carotid artery

Figure 5.4

Anterior aspect of human skull

Frontal Bone

Temporal Bone

Greater Wing of Sphenoid Bone

Supraorbital Foramen

Supraorbital Margin

Superior Orbital Fissure

Optic Foramen

Lacrimal Bone

Inferior Orbital Fissure

Inferior Nasal Concha

Vomer

Anterior Inferior Nasal Spine

Maxilla

Alveolar Process

Mental Foramen

Body of Mandible

Mental Tubercle

Glabella

Frontal Eminence (Protuberance)

Superciliary Arch

Nasal Bone

Sphenoid Bone

Middle Nasal Concha

Infraorbital Foramen

Zygomatic Bone

Styloid Process

Mastoid Process

Ramus of Mandible

Angle of Mandible

Mandible

Alveolar Process

Figure 5.5

Human skull, anterior view

1. frontal bone
2. supraorbital foramen
3. sphenoid bone
4. supra orbital margin
5. zygomatic bone
6. inferior orbital fissure
7. inferior concha
8. glabella
9. supraorbital fissure
10. frontal eminence
11. superciliary arch
12. nasal bones
13. middle conchae
14. ramus of mandible
15. infraorbital foramen
16. angle of mandible
17. perpendicular plate of ethmoid
18. maxilla
19. mandible
20. mental foramen
21. alveolar processes
22. lateral mass

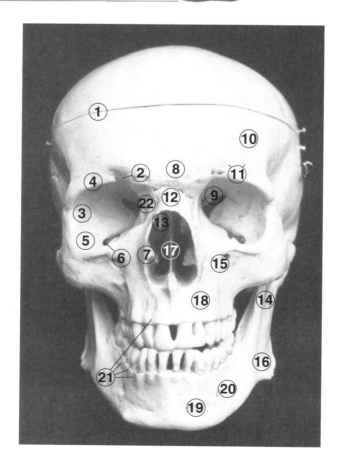

Bone Markings of the Skull

Bone	Bone Marking	Description
Temporal (Continued)	Petrous (PET-rus) portion	Rocklike protrusions in center of cranial floor; flare from medial to lateral region; middle and inner ear structures are contained within
	Internal auditory meatus	Opening in medial side of petrous portion; is the internal opening of canal running from external auditory meatus through which the acoustic nerve passes
Occipital (ok-SIP-i-tal)		Large flaring bone forming inferior posterior section of cranium; articulates with parietal bones to form lambdoidal suture
(Figures 5.6, 5.15–5.18; also Figure 5.7)	External occipital protuberance	Protrusion or eminence that extends most posteriorly in medial portion
	Superior nuchal (NOO-kul) line	Elevation projecting laterally from external occipital protuberance
	Inferior nuchal line	A lesser elevation inferior to superior nuchal line
	Foramen magnum	Large round opening that serves as passageway for spinal cord in its connection with brain; medial to condyles
	Condyles	Oval-shaped protrusions with flat surfaces lateral to foramen magnum; articulate with atlas
	Internal occipital protuberance	Linear projection in internal medial region of bone
	Internal occipital crest	A midline protrusion that extends vertically from inferior to superior
	Fossa for cerebrum	Superior lateral depressions supporting cerebrum on internal surface
	Fossa for cerebellum	Large inferior depressions supporting cerebellum on internal surface
Sphenoid (SFĒ-noyd)		Bat-shaped bone forming part of the floor of the cranial cavity; seen laterally anterior to the temporal bone
	Body	Medial cubelike region; hollow centrally
	Greater wings	Laterally flaring portion forming floor of cranial cavity and outer wall of optic orbit
	Lesser wings	Thin flaring portions superior to greater wings and body; form superior posterior section of optic orbit
(Figures 5.4–5.6, 5.15–5.19; also Figures 5.7, 5.8, 5.10 and 5.11)	Sella turcica (TER-si-ka)	Saddle-shaped indentation medial to wings; houses pituitary gland
	Anterior clinoid process	Posterior tip of the lesser wing of sphenoid
	Posterior clinoid process	Superiorly projecting posterior wall of the sella turcica
	Optic foramen	Rounded opening inferior to lesser wings at medial point; can be seen anteriorly through optic orbit; transmits cranial nerve II

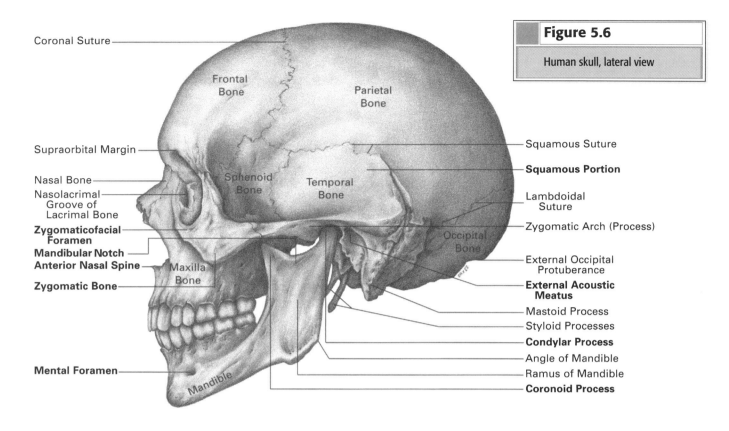

Coronal Suture

Frontal Bone

Parietal Bone

Figure 5.6

Human skull, lateral view

Supraorbital Margin

Squamous Suture

Squamous Portion

Nasal Bone

Sphenoid Bone

Temporal Bone

Nasolacrimal Groove of Lacrimal Bone

Lambdoidal Suture

Zygomatic Arch (Process)

Zygomaticofacial Foramen

Occipital Bone

Mandibular Notch

External Occipital Protuberance

Anterior Nasal Spine

Maxilla Bone

External Acoustic Meatus

Zygomatic Bone

Mastoid Process

Styloid Processes

Condylar Process

Angle of Mandible

Mental Foramen

Ramus of Mandible

Mandible

Coronoid Process

Identification 5.1

 5.1

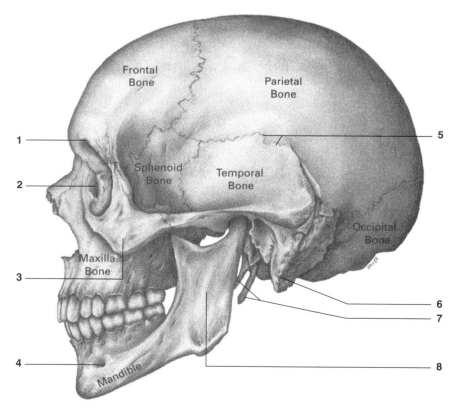

Frontal Bone

Parietal Bone

1

5

Sphenoid Bone

Temporal Bone

2

Occipital Bone

Maxilla Bone

3

6

7

4   Mandible

8

Figure 5.7

Human skull, lateral view—Beauchene skull

1. parietal bone
2. frontal bone
3. zygomatic bone
4. sphenoid bone
5. maxilla
6. coronoid process of mandible
7. mandible
8. temporal bone
9. pterygoid process of sphenoid
10. occipital bone
11. condylar process of mandible
12. nasal bone

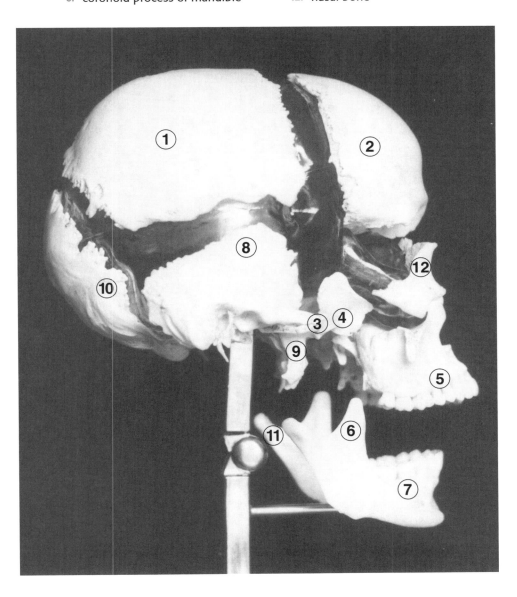

Figure 5.8

Human skull bones, anterolateral view

1. frontal
2. parietal
3. temporal
4. zygomatic
5. maxilla
6. sphenoid
7. mandible
8. inferior concha
9. ethmoid
10. nasal

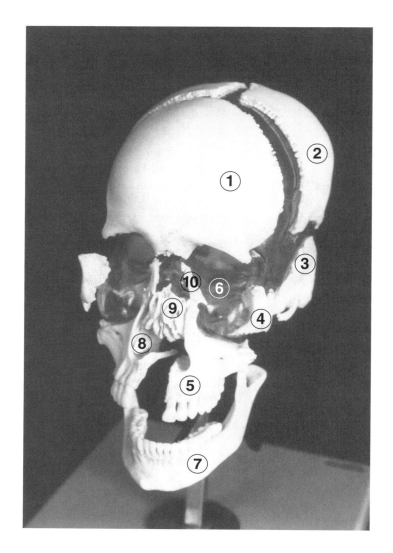

Figure 5.9

Temporal bone, interior view

1. squamous portion
2. external auditory meatus
3. styloid process
4. zygomatic process
5. petrous portion
6. internal auditory meatus

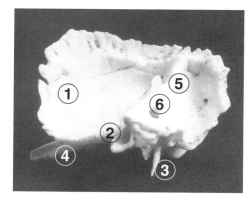

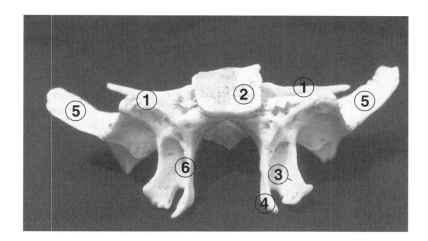

Figure 5.10

Human sphenoid bone, superior view

1. lesser wings
2. body of sphenoid
3. lateral lamina of pterygoid process
4. medial lamina of pterygoid process
5. greater wings
6. foramen ovale

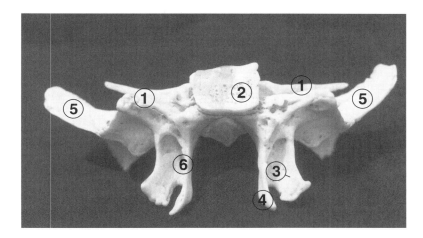

Figure 5.11

Sphenoid bone, frontal view

1. lesser wings
2. optic foramen
3. body of sphenoid
4. superior orbital fissure
5. lesser wing
6. greater wings
7. foramen ovale
8. foramen lacerum
9. pterygoid process

Figure 5.12

Human ethmoid bone, superior view

1. ethmoid sinuses
2. lateral masses
3. horizontal (cribriform) plate
4. crista galli
5. perpendicular plate

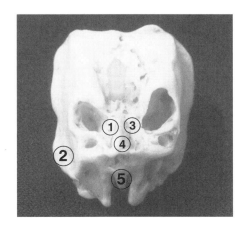

Bone Markings of the Skull

Bone	Bone Marking	Description
Sphenoid (*Continued*)	Superior orbital fissure	Irregularly shaped furrow, immediately inferior to and covered by lesser wings; can be seen anteriorly through optic orbit; transmits cranial nerves III, IV, and V
	Inferior orbital fissure	Irregularly shaped furrow running medially to laterally; can be seen anteriorly through optic orbit; transmits cranial nerve V (trigeminal), maxillary nerve, and infraorbital vessels
	Foramen rotundum	Small rounded hole in greater wings; lateral to sella turcica; transmits maxillary division of trigeminal cranial nerve; interior view
	Foramen ovale (ō-VA-lē)	Oval-shaped hole posterior to foramen rotundum; transmits mandibular division of trigeminal nerve; interior view
	Foramen lacerum (LAS-er-um)	Irregular oval-shaped opening in greater wing toward body; lateral to sella turcica at apex of petrous portion; transmits internal carotid arteries and branch of ascending pharyngeal artery; interior view
	Foramen spinosum (spī-NŌ-sum)	Small, irregularly shaped opening posterior and lateral to foramen ovale; transmits mandibular nerves and meningeal arteries; interior view
	Pterygoid (TER-i-goyd) process	Wing-shaped downward projection posterior to maxilla and molar teeth; consists of medial and lateral lamina; inferior view
Ethmoid (ETH-moyd) (Figures 5.4, 5.5, 5.15–5.19; also Figures 5.12 and 5.13)		Irregularly shaped bone that forms anterior superior section of cranial floor, medial posterior walls of optic orbits, lateral walls, posterior septum, and roof of nasal cavity
	Crista galli (KRIS-ta GAL-ē)	Anterior flaglike structure pointing anteriorly; serves as point of attachment for meninges; interior view
	Horizontal plate *or* cribriform (KRIB-ri-form) plate	Porous flat section of bone lateral to crista galli; transmits olfactory nerves
	Perpendicular plate	Linear projection extending inferiorly from crista galli; best seen from anterior view
	Lateral masses	Lateral portions of ethmoid; inner sections form walls of nasal cavities and conchae; contain many air cavities
	Ethmoid sinus	Air cavity within lateral masses
	Superior and middle conchae (KON-kā) *or* turbinates	Extensions from lateral sides of nasal cavity pointing medially toward septum; contain spongy, air-filled cavities
FACE BONES **Mandible** (MAN-di-bul) (Figures 5.4–5.6; also Figures 5.7 and 5.8)		Lower jaw; articulates with temporal bone; forms the only diarthrodial (movable) joint in cranium
	Body	Rounded anterior section; posteriorly convex, forms chin
	Mental tubercles	Small rounded protrusions on external surface of inferior rim of body
	Mental foramen	Small rounded opening on external surface below bicuspid teeth; transmits nerves and blood vessels

Figure 5.13

Human ethmoid bone, frontal view

1. superior nasal concha
2. crista galli
3. horizontal (cribriform) plate
4. lateral masses
5. perpendicular plate
6. middle nasal concha

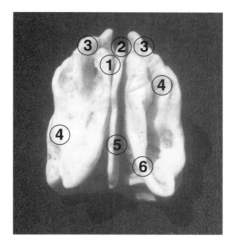

Figure 5.14

Human, lacrimal, palatine, maxilla, and inferior nasal concha, internal view

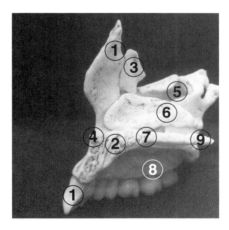

1. maxilla
2. incisive foramen
3. lacrimal bone
4. anterior nasal spine
5. orbit surface
6. inferior nasal concha
7. palatine process of maxilla
8. alveolar process of maxilla
9. horizontal plate of palatine bone

Bone Markings of the Skull

Bone	Bone Marking	Description
Mandible (*Continued*)		
	Angle	Junction of body and ramus on inferior side
	Ramus	Superior armlike projections on posterior part of either side of body
	Mandibular foramen	Opening on internal surface of ramus; passageway for nerves and blood vessels to teeth of mandible
	Mandibular condyle *or* condylar process	Posterior rounded head that fits into mandibular fossa of temporal bone
	Mandibular notch	Depression between condylar process and coronoid process
	Neck	Constriction at base of condylar process
	Coronoid process	Thin pointed anterior projection that serves as point for muscle attachment
	Alveolar (al-VĒ-ō-ler) processes	Sockets; processes of bone into which teeth are anchored centrally

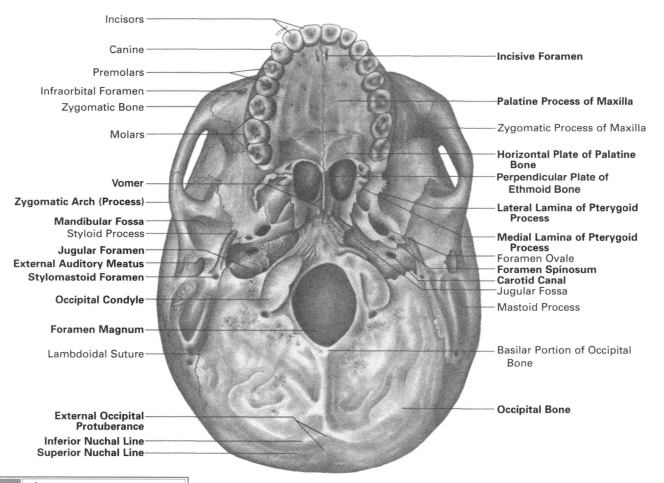

Figure 5.15

Inferior aspect of human skull

Figure 5.16

Human skull, inferior aspect

1. incisive foramen
2. alveolar process of maxilla
3. zygomatic bone
4. zygomatic process of temporal bone
5. perpendicular plate of ethmoid bone
6. vomer bone
7. jugular foramen
8. jugular fossa
9. superior nuchal line
10. inferior nuchal line
11. occipital bone
12. occipital condyle
13. foramen magnum
14. styloid process
15. pterygoid process
16. horizontal plate of palatine bone
17. palatine process of maxilla
18. mandibular fossa

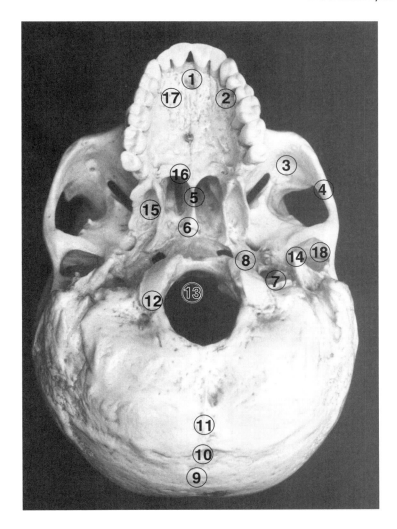

Question 5A.6

What is the function of the following foramina?

jugular:

carotid:

Question 5A.7

What is the function of the following bone markings?

stylomastoid foramen:

jugular fossa:

Identification 5.2

5.2

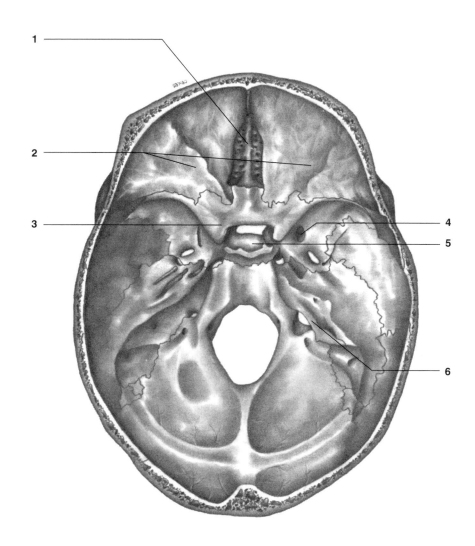

Figure 5.17

Human skull, interior aspect

1. optic foramen
2. cribriform plate of ethmoid bone
3. crista galli of ethmoid bone
4. frontal bone
5. lesser wing of sphenoid bone
6. sphenoid bone
7. temporal bone
8. foramen ovale
9. petrous portion of temporal bone
10. occipital bone
11. jugular foramen
12. jugular fossa
13. anterior clinoid process
14. greater wing of sphenoid
15. foramen spinosum

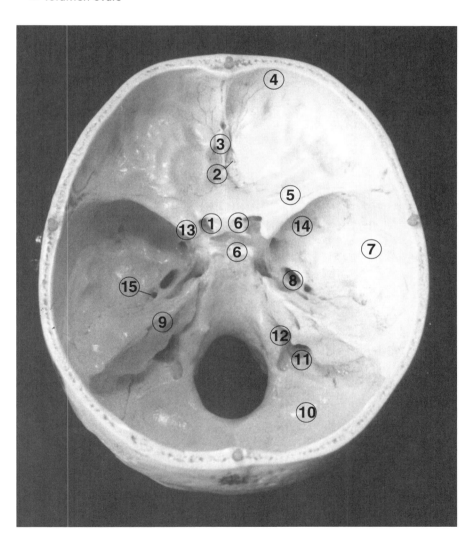

Bone Markings of the Skull

Bone	Bone Marking	Description
Nasal (Figures 5.4–5.6 and 5.19; also Figures 5.7 and 5.8)		Small irregularly shaped bones inferior to glabella; form upper division of nasal bridge
Lacrimal (LAK-ri-mal) (Figures 5.4, 5.6, and 5.19; also Figure 5.14)	Groove *or* foramen	Thin delicate bones lateral to nasal bones; form medial wall of optic orbit and lateral wall of nasal cavity Slight furrow running in a superior to inferior direction; serves as a passageway for tears
Vomer (VŌ-mer) (Figures 5.4, 5.15, and 5.16) **Inferior conchae or turbinates** (Figures 5.4, 5.5, and 5.19; also Figures 5.8 and 5.14)		Single irregularly shaped bone found medially and posteriorly in the nasal cavity Thin delicate bones flaring from lateral walls of nasal cavity; most posterior projections
Zygomatic or **malar** (MA-lar) (Figures 5.4–5.6, 5.15, 5.16, and 5.19; also Figures 5.7 and 5.8)	Zygomaticofacial (zī'-gō-MAT-i-kō-fā'-shul) foramen	Hard, heavy cheekbone Small round opening anterior to zygomatic arch; inferior and slightly lateral to optic orbit

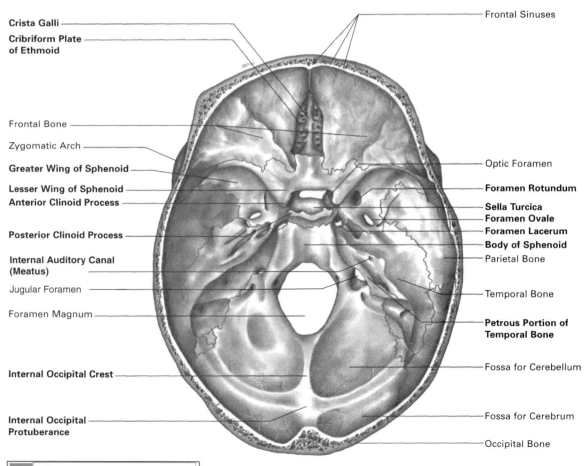

Crista Galli
Cribriform Plate of Ethmoid
Frontal Bone
Zygomatic Arch
Greater Wing of Sphenoid
Lesser Wing of Sphenoid
Anterior Clinoid Process
Posterior Clinoid Process
Internal Auditory Canal (Meatus)
Jugular Foramen
Foramen Magnum
Internal Occipital Crest
Internal Occipital Protuberance

Frontal Sinuses
Optic Foramen
Foramen Rotundum
Sella Turcica
Foramen Ovale
Foramen Lacerum
Body of Sphenoid
Parietal Bone
Temporal Bone
Petrous Portion of Temporal Bone
Fossa for Cerebellum
Fossa for Cerebrum
Occipital Bone

Figure 5.18

Interior aspect of human skull

Bone Markings of the Skull

Bone	Bone Marking	Description
Palatine (PAL-a-tīn) (Figures 5.15 and 5.16; also Figure 5.14)		Forms the posterior part of the hard palate; also forms lateral and inferior walls of posterior division of nasal cavity
	Horizontal plate	Horizontal junction with maxilla; forms most posterior section of hard palate
Maxilla (mak-SIL-a) (Figures 5.4–5.6, 5.15, 5.16, and 5.19; also Figures 5.7, 5.8, and 5.14)		Upper jawbone, main bone of face; all other facial bones articulate with maxilla
	Antrum of Highmore *or* maxillary sinus	Large sinus within central cavity of maxilla; mucosa lined and air-filled
	Incisive (in-SĪ-siv) foramen	Triangular opening between central incisors; located at midline of maxilla
	Palatine process	Forms most of hard palate; curves downward to form alveoli
	Alveolar processes	Archlike sockets containing teeth
	Anterior nasal spine	Projection inferior and anterior to nasal septum
	Infraorbital (in'-fra-OR-bi-tal) foramen	Opening inferior to optic orbit
EAR OSSICLES (OS-i-kulz) (Figures 8.8 and 8.9)		Small bones contained within the middle ear; conduct vibrations from eardrum to inner ear (not visible in your specimen)
	Malleus (MAL-ē-us)	Largest of three ear ossicles; attached to eardrum; articulates with incus; also known as "hammer"
	Incus	Middle of three ossicles in middle ear; also known as "anvil"
	Stapes (STĀ-pēz)	Most medial of three ossicles; vibrates against oval window of inner ear; also known as "stirrup"

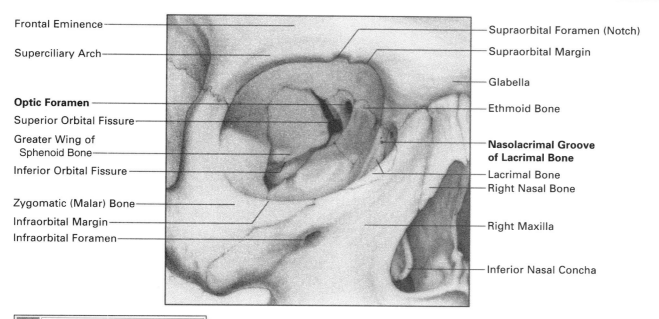

Frontal Eminence
Superciliary Arch
Optic Foramen
Superior Orbital Fissure
Greater Wing of Sphenoid Bone
Inferior Orbital Fissure
Zygomatic (Malar) Bone
Infraorbital Margin
Infraorbital Foramen

Supraorbital Foramen (Notch)
Supraorbital Margin
Glabella
Ethmoid Bone
Nasolacrimal Groove of Lacrimal Bone
Lacrimal Bone
Right Nasal Bone
Right Maxilla
Inferior Nasal Concha

Figure 5.19

The orbit of right eye, anterior view

Question 5A.8

What is the function of the following bone markings?

lacrimal groove:

alveolar process of the maxilla:

Bone Markings of the Hyoid   (Figure 5.20)

Bone	Bone Marking	Description
Hyoid (HĪ-oyd) (Figure 5.20)		U-shaped bone anteriorly located between larynx and mandible; serves as point of muscle attachment for some muscles of mouth and tongue
	Transverse body	Rounded anterior portion of bone
	Greater cornu (KOR-noo)	Posterior projection extending from transverse body
	Lesser cornu	Anterior projection on either side of anterior portion of transverse body

Figure 5.20

Hyoid bone, anterior view

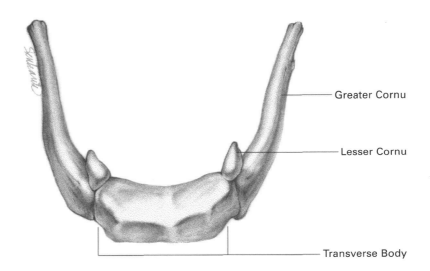

Greater Cornu

Lesser Cornu

Transverse Body

Vertebrae are irregularly shaped bones. The **vertebral column** is composed of blocks of vertebrae of various sizes and shapes. Their major functions are to carry the weight of the head and trunk, to permit various movements by the trunk, and to protect the spinal cord. The vertebral column is made up of five groups of vertebrae: (1) 7 **cervical vertebrae** (C1, C2, C3, C4, C5, C6, C7) in the neck region; (2) 12 **thoracic vertebrae** (T1, T2, T3, T4, T5, T6, T7, T8, T9, T10, T11, T12) in the thoracic region, to which the ribs are attached; (3) 5 **lumbar vertebrae** (L1, L2, L3, L4, L5) that form the lower back; (4) 5 **sacral vertebrae** that fuse by age 26 to form one bone, the **sacrum**; and (5) **coccygeal vertebrae**, including 3–5 vertebrae making up the "tailbone" or **coccyx**, which also fuses by age 26. Each group of vertebrae is illustrated in **Figures 5.21–5.29**. Structures common to all types of vertebrae are described in the table entitled Typical Structures of All Vertebrae.

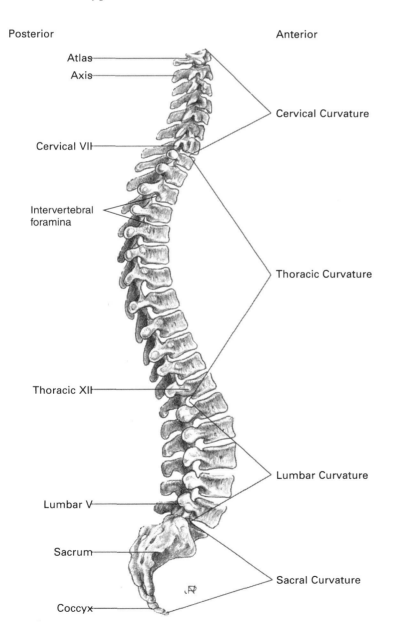

Posterior

Anterior

Atlas
Axis

Cervical VII

Intervertebral foramina

Thoracic XII

Lumbar V

Sacrum

Coccyx

Cervical Curvature

Thoracic Curvature

Lumbar Curvature

Sacral Curvature

Figure 5.21

Normal curvatures of the human spinal column

Question 5A.9

How many vertebrae are there of each type?

Cervical:

Thoracic:

Lumbar:

Sacral:

Coccygeal:

Identification 5.3

Posterior   Anterior

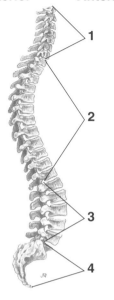

From the lateral view, the normal adult vertebral column presents four curvatures: (1) **cervical curvature**, which presents an anterior convex curvature; (2) **thoracic curvature**, which presents an anterior concave curvature; (3) **lumbar curvature**, which presents an anterior convex curvature, and (4) **sacral curvature**, which presents an anterior concave curvature. See **Figure 5.21** to identify the four curvatures of the vertebral column. Use an articulated human skeleton to identify the five groups of vertebrae and the curvatures. Also observe your partner's vertebral column, if possible. These curvatures help to bear the weight of the trunk and to permit various types of positions and movements, such as standing, sitting, or running. Factors such as poor nutrition, posture, disease, or injury may cause an abnormal curvature to develop: The term *kyphosis* (kī-FŌ-sis) is used to describe a posterior exaggeration of the thoracic curvature in an individual, resulting in a "hunchback" condition. The term *lordosis* (lor-DŌ-sis) is used to describe a marked exaggeration of the lumbar curvature, creating a "swayback" appearance. The term *scoliosis* (skō'-lē-Ō-sis) is applied to a lateral deviation of the thoracic curvature.

Typical Structures of All Vertebrae   (Figures 5.22–5.28)

Structure	Description
Body	Rounded central bony portion facing anteriorly in column; point of attachment for intervertebral discs (not present in C-1, the atlas)
Vertebral arch	Junction of all posterior extensions from body
Vertebral foramen *or* neural canal	Central opening through which spinal cord passes
Pedicle (PED-i-kul)	Thick, bony lateral extension posterior to body; forms anterior lateral portion of vertebral foramen
Lamina (LAM-i-na)	Flat, thinner bony extension; extends pedicle toward midline posteriorly; forms posterior portion of vertebral foramen
Intervertebral (in'-ter-VER-te-bral) foramen	Opening between vertebrae for emergence of spinal nerves
Processes	
Transverse processes	Lateral extensions from junction of pedicle and lamina
Superior articular facets or surfaces	Paired projections on superior surface lateral to vertebral foramen; slightly indented centrally; form articulation with immediately superior vertebra (or, in the atlas, with occipital condyles)
Inferior articular facets or surfaces	Paired projections on inferior surface lateral to vertebral foramen; slightly indented centrally; form articulation with immediately inferior vertebra
Spinous (SPĪ-nus) process	Single posterior process extending inferiorly at midline of vertebra

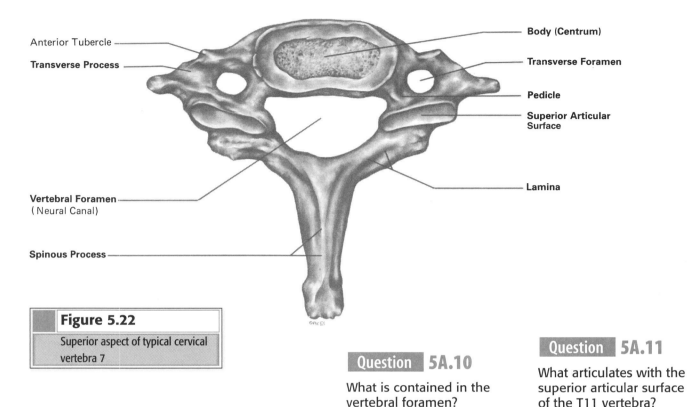

Anterior Tubercle

Transverse Process

Vertebral Foramen
(Neural Canal)

Spinous Process

Body (Centrum)

Transverse Foramen

Pedicle

Superior Articular
Surface

Lamina

Figure 5.22

Superior aspect of typical cervical
vertebra 7

Question 5A.10

What is contained in the
vertebral foramen?

Question 5A.11

What articulates with the
superior articular surface
of the T11 vertebra?

Bone Markings of Specific Vertebral Types

Bone	Description
(Structures unique to each vertebral group are labeled in Figures 5.22–5.29)	
Cervical (SER-vi-kal) (Figure 5.22)	Smallest, lightest vertebrae; has all of the typical structures; slightly larger, triangular vertebral foramen; spinous processes bifurcated and short; transverse processes contain **transverse foramen**
Atlas (first cervical vertebra) (Figures 5.23 and 5.24)	Has no body; contains large concave articulating surfaces which support occipital condyles
Axis (second cervical vertebra) (Figure 5.25)	Acts as a pivot for rotation of atlas and cranium; **odontoid process** is a superior vertical projection serving as pivot point and body
Thoracic (thō-RAS-ik) (Figures 5.26 and 5.27)	Has all of the typical structures; larger body than cervical; oval vertebral foramen; sharp inferiorly projecting spinous process; contains lateral articulating surfaces, superior and inferior for rib articulation
Lumbar (Figure 5.28)	Has all of the typical structures; large; thick blocklike body with superior articulating surfaces projecting inward and inferior articulating surfaces projecting outward; short, thick spinous process projecting posteriorly
Sacrum (SĀ-krum) (Figure 5.29)	A fusion in the adult of the five sacral vertebrae that, with the coccyx, forms the dorsal portion of the pelvic girdle
Coccyx (KOK-six) (Figure 5.29)	Fusion of three to five small, irregularly shaped vertebrae forming tail-like projection inferiorly

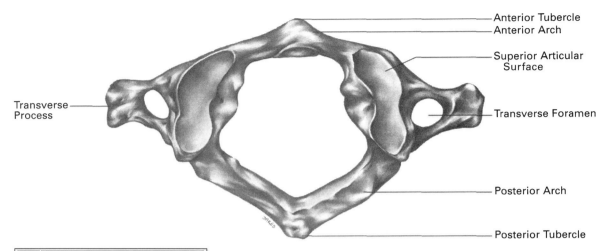

Figure 5.23

Superior view of the atlas

Question 5A.12

How would you distinguish between T12 and L1 vertebrae?

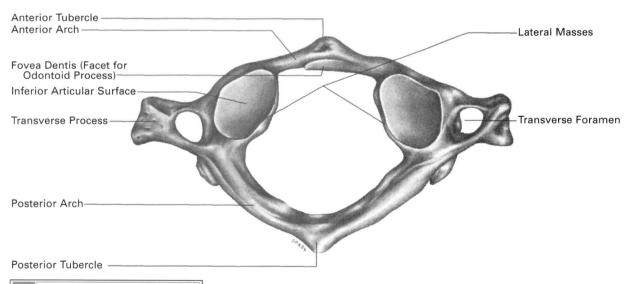

Figure 5.24

Inferior aspect of the atlas

Question 5A.13

What common feature is shared by all cervical vertebrae?

Question 5A.14

How does the atlas differ from all other cervical vertebrae?

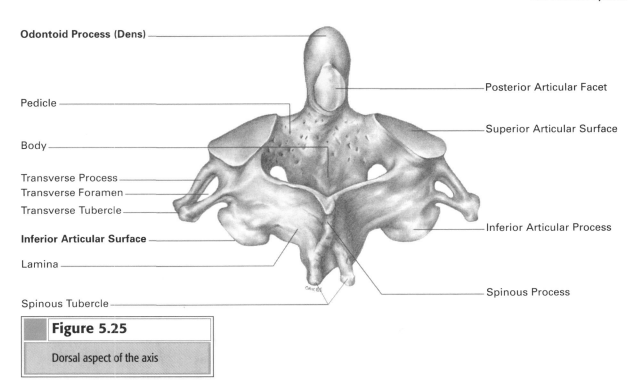

Odontoid Process (Dens)

Pedicle

Body

Transverse Process
Transverse Foramen
Transverse Tubercle

Inferior Articular Surface

Lamina

Spinous Tubercle

Posterior Articular Facet

Superior Articular Surface

Inferior Articular Process

Spinous Process

Figure 5.25

Dorsal aspect of the axis

Question 5A.15

What is the purpose of the dens?

Question 5A.16

With what two vertebrae does the axis articulate?

Question 5A.17

How many thoracic vertebrae articulate with ribs?

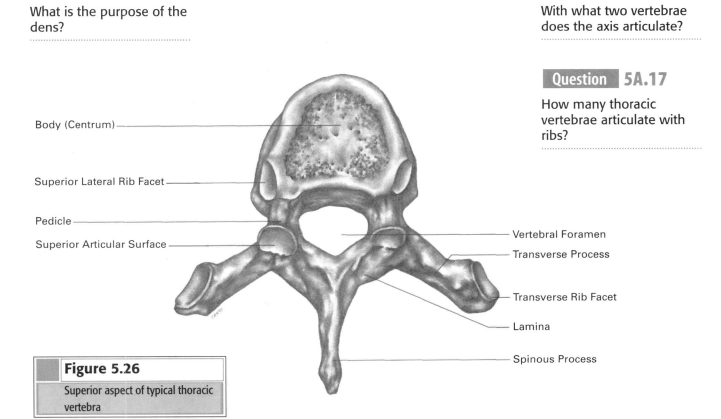

Body (Centrum)

Superior Lateral Rib Facet

Pedicle

Superior Articular Surface

Vertebral Foramen

Transverse Process

Transverse Rib Facet

Lamina

Spinous Process

Figure 5.26

Superior aspect of typical thoracic vertebra

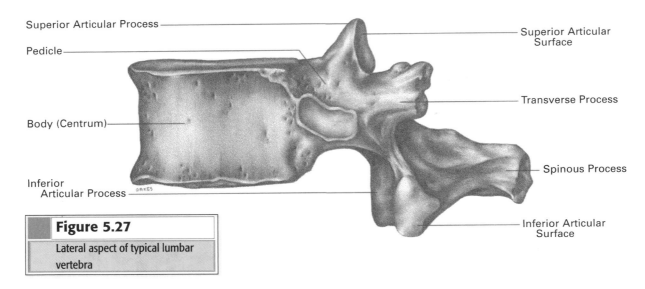

Superior Articular Process

Pedicle

Body (Centrum)

Inferior Articular Process

Superior Articular Surface

Transverse Process

Spinous Process

Inferior Articular Surface

Figure 5.27
Lateral aspect of typical lumbar vertebra

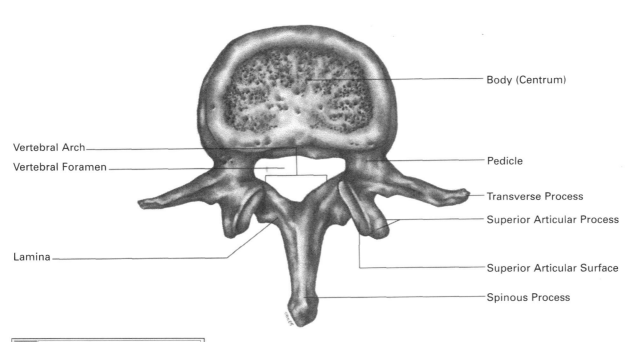

Body (Centrum)

Vertebral Arch

Vertebral Foramen

Lamina

Pedicle

Transverse Process

Superior Articular Process

Superior Articular Surface

Spinous Process

Figure 5.28
Superior aspect of typical lumbar vertebra

Question 5A.18

Why are lumbar vertebrae thick and dense?

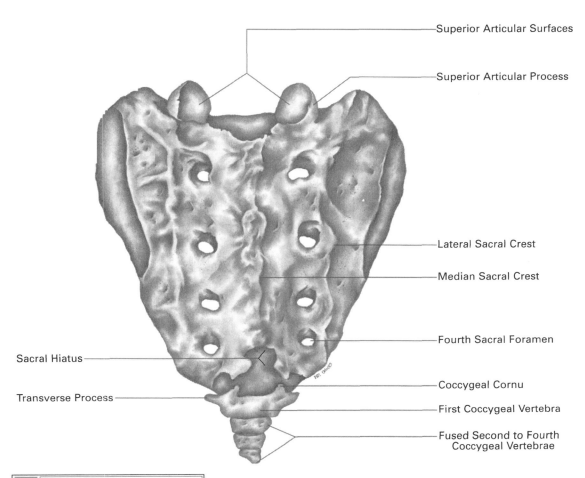

Superior Articular Surfaces

Superior Articular Process

Lateral Sacral Crest

Median Sacral Crest

Fourth Sacral Foramen

Sacral Hiatus

Coccygeal Cornu

Transverse Process

First Coccygeal Vertebra

Fused Second to Fourth Coccygeal Vertebrae

Figure 5.29

Sacrum and coccyx, posterior aspect

Bones of the Sternum   (Figure 5.30)

Bone	Description
Sternum	Breastbone; fusion of three bones; superiorly flared bone articulating with true ribs
Manubrium (ma-NOO-brē-um)	Superior horizontal portion
Gladiolus (glad'-ē-Ō-lus) or **body**	Central vertical portion
Xiphoid (ZĪ-foyd) **process**	Cartilaginous inferior projection

Figure 5.30

Sternum, anterior view

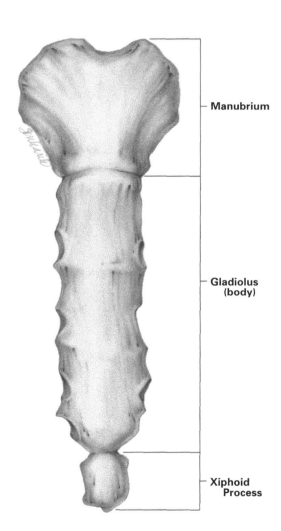

— Manubrium

— Gladiolus (body)

— Xiphoid Process

Question 5A.19

What is a bone fusion?

Question 5A.20

What bones are fused in the sternum?

Bone Markings of the Ribs   (Figure 5.31)

Bone	Bone Marking	Description
Ribs (seven true pairs, five false pairs)		Form framework of thoracic cavity; true ribs articulate with both vertebrae and sternum; false ribs articulate with vertebrae and have cartilaginous connections to sternum, except for last two pairs of false ribs ("floating" ribs with no anterior connection to sternum)
	Head	Posterior rounded projection, articulating with a thoracic vertebra
	Neck	Narrow portion below head and connecting head with shaft
	Shaft *or* body	Long, thin, curved portion
	Tubercle	Small rounded protrusion below neck on shaft
	Costal cartilage	Cartilaginous connection of ribs to sternum (not visible on disarticulated bones)

Question 5A.21

With which bones do the true ribs articulate?

Question 5A.22

With which bones do the false ribs articulate?

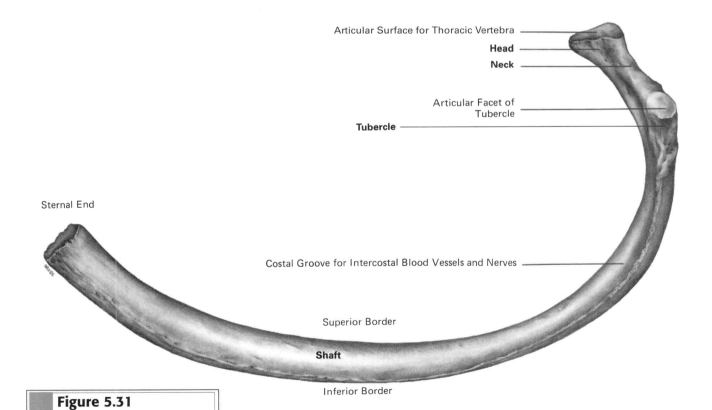

Articular Surface for Thoracic Vertebra

Head

Neck

Articular Facet of Tubercle

Tubercle

Sternal End

Costal Groove for Intercostal Blood Vessels and Nerves

Superior Border

Shaft

Inferior Border

Figure 5.31

Sixth rib, inferior view

Bone Markings of the Upper Division

Bone	Bone Marking	Description
Clavicle (KLAV-i-kul) (Figure 5.34)		Collarbone; S-shaped body articulating posteriorly with scapula and anteriorly with sternum
	Sternal end	Blunted end that articulates with sternum
	Acromial (a-KRŌ-mē-al) end	Flat, thinner end that articulates with acromion process of scapula
		Two wing-shaped shoulder bones projecting posteriorly on either side of body midline
Scapula (SKAP-yoo-la) (Figures 5.32 and 5.33)	Body	Flat, thin, central portion
	Superior border	Upper margin of body
	Vertebral (VER-te-bral) or medial border	Medial margin that runs parallel to vertebral column
	Axillary (AK-si-lar-ē) or lateral border	Lateral margin
	Spine	Posterior ridgelike projection extending from vertebrae toward axillary border
	Acromial process *or* acromion (a-KRŌ-mē-on)	Articulation with clavicle; rounded end of spine
	Coracoid (KOR-a-koyd) process	Lateral rounded projection on superior border; anterior to acromion
	Glenoid (GLĒN-oyd) cavity	Articulation point for humerus; rounded fossa on superior end of axillary border
Humerus (HYOO-mer-us) (Figures 5.35 and 5.36)		Upper arm bone articulating with scapula proximally and with ulna and radius distally
	Head	Rounded posteriomedial projection on proximal end
	Anatomical neck	Oblique constriction on anterior surface below head
	Greater tubercle	Larger anteriolateral projection from head
	Lesser tubercle	Smaller anterior projection below anatomical neck
	Intertubercular (in'-ter-too-BER-kyoo-lar) groove	Indentation between greater and lesser tubercles
	Surgical neck	Posterior constriction below tubercles
	Deltoid tuberosity	Rough lateral protrusion on mid-diaphysis
	Medial epicondyle (ep'-i-KON-dīl)	Rough rounded projection on medial surface at distal end
	Lateral epicondyle	Rough smaller protrusion on lateral surface at distal end
	Capitulum (ka-PIT-yoo-lum) *or* radial head	Lateral rounded knob immediately anterior to lateral epicondyle; articulates with radius
	Trochlea (TRŌK-lē-a)	Pulley-shaped structure medial and inferior to medial epicondyle; articulates with ulna
	Olecranon (ō-LEK-ra-non) fossa	Deep triangular depression on posterior surface between epicondyles; articulates with olecranon process of ulna during extension of forearm
	Coronoid (KOR-o-noyd) fossa	Depression on anterior surface immediately superior to trochlea; articulates with coronoid process of ulna during flexion of forearm

Bone Markings of the Upper Division

Bone	Bone Marking	Description
Ulna (UL-na) (Figures 5.37 and 5.38)		Medial bone of forearm; longer than radius
	Olecranon process	Large round projection at proximal end; concave center (semilunar notch) resembling open mouth; forms elbow
	Coronoid process	Protrusion on anterior surface below olecranon process; lower half of "open mouth"; articulates with humerus
	Semilunar (sem'-ē-LOO-ner) notch	Concave notch that articulates with trochlea of humerus
	Radial notch	Lateral concave indentation that articulates with medial aspect of radial head
	Styloid (STĪ-loyd) process	Small V-shaped projection at distal end of bone
Radius (Figures 5.37 and 5.38)		Shorter lateral bone of forearm
	Radial head	Rounded knoblike projection at proximal end of bone
	Radial tuberosity	Raised rough protrusion on medial aspect near proximal end
	Styloid process	Needlelike projection on lateral aspect of distal end
Hand (Figures 5.39 and 5.40)		
Carpals (Table, page 92)		Sixteen irregularly shaped bones; eight in each wrist
Metacarpals (met'-a-KAR-pulz)		Bones of the palm of the hand (five per hand)
Phalanges (fa-LAN-jēz)		Bones in fingers and thumbs

Question 5A.23

Which bones form the shoulder joint?

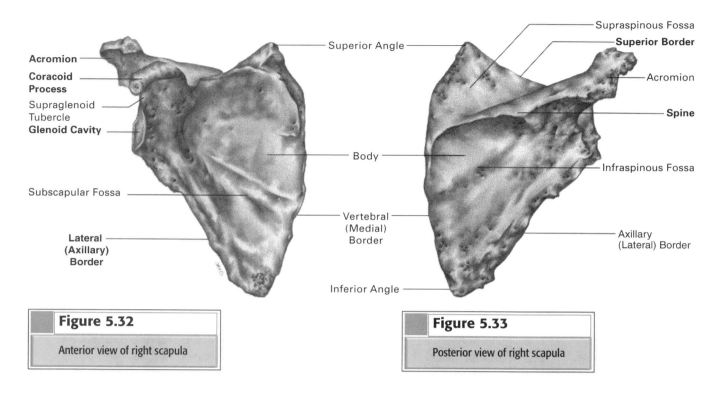

Acromion
Coracoid Process
Supraglenoid Tubercle
Glenoid Cavity
Subscapular Fossa
Lateral (Axillary) Border

Superior Angle
Body
Vertebral (Medial) Border
Inferior Angle

Supraspinous Fossa
Superior Border
Acromion
Spine
Infraspinous Fossa
Axillary (Lateral) Border

Figure 5.32

Anterior view of right scapula

Figure 5.33

Posterior view of right scapula

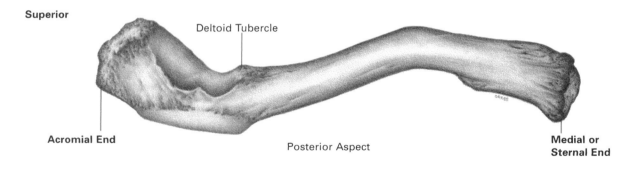

Superior

Deltoid Tubercle

Acromial End

Posterior Aspect

Medial or Sternal End

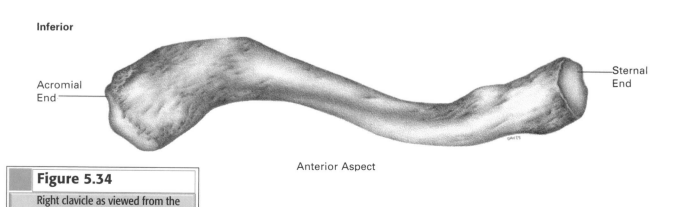

Inferior

Acromial End

Sternal End

Anterior Aspect

Figure 5.34

Right clavicle as viewed from the superior and inferior surfaces

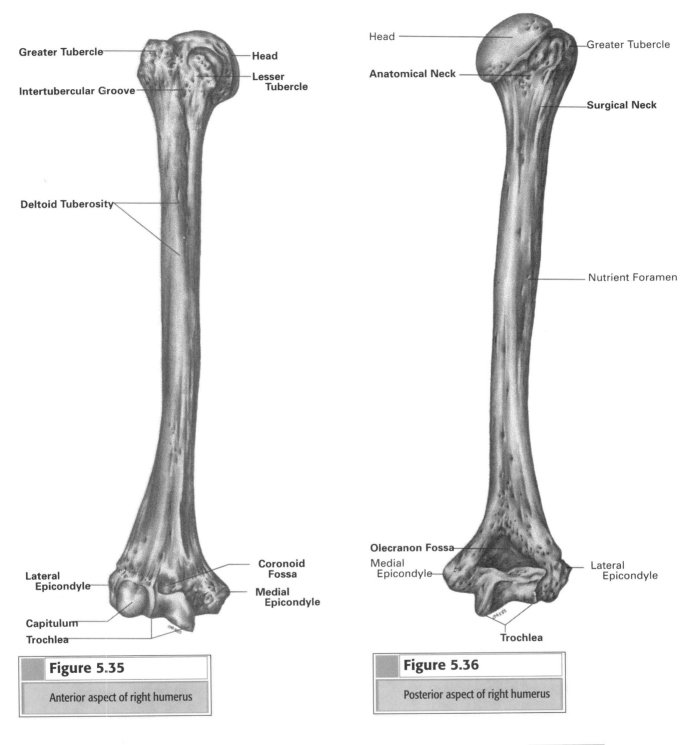

Greater Tubercle
Head
Intertubercular Groove
Lesser Tubercle
Deltoid Tuberosity
Lateral Epicondyle
Coronoid Fossa
Medial Epicondyle
Capitulum
Trochlea

Figure 5.35

Anterior aspect of right humerus

Head
Greater Tubercle
Anatomical Neck
Surgical Neck
Nutrient Foramen
Olecranon Fossa
Medial Epicondyle
Lateral Epicondyle
Trochlea

Figure 5.36

Posterior aspect of right humerus

Question **5A.24**

What is the difference between the surgical and anatomical necks?

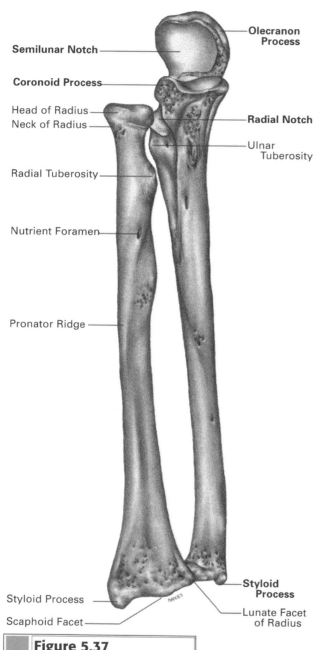

Semilunar Notch

Coronoid Process

Head of Radius
Neck of Radius

Radial Tuberosity

Nutrient Foramen

Pronator Ridge

Olecranon Process

Radial Notch

Ulnar Tuberosity

Styloid Process
Scaphoid Facet

Styloid Process
Lunate Facet of Radius

Figure 5.37

Anterior view of right radius and ulna

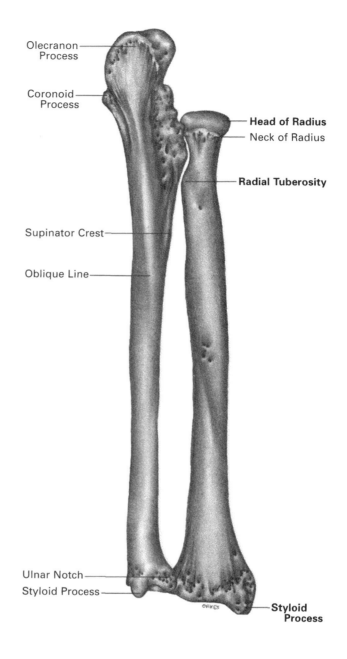

Olecranon Process

Coronoid Process

Head of Radius
Neck of Radius

Radial Tuberosity

Supinator Crest

Oblique Line

Ulnar Notch
Styloid Process

Styloid Process

Figure 5.38

Posterior view of right radius and ulna

Question 5A.25

Which is the shorter bone of the forearm?

....

Question 5A.26

Into which humeral fossa does the coronoid process fit when the forearm is flexed?

....

Question 5A.27

Into which humeral fossa does the olecranon process fit when the forearm is extended?

....

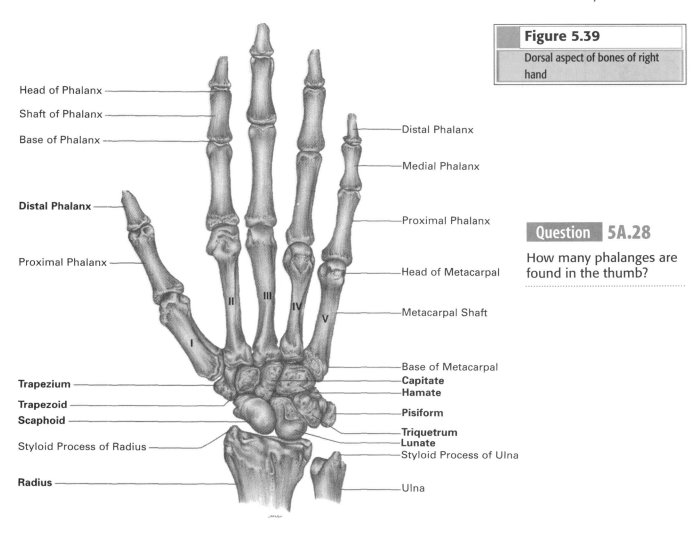

Head of Phalanx

Shaft of Phalanx

Base of Phalanx

Distal Phalanx

Proximal Phalanx

Trapezium

Trapezoid

Scaphoid

Styloid Process of Radius

Radius

Distal Phalanx

Medial Phalanx

Proximal Phalanx

Head of Metacarpal

Metacarpal Shaft

Base of Metacarpal

Capitate

Hamate

Pisiform

Triquetrum

Lunate

Styloid Process of Ulna

Ulna

Figure 5.39

Dorsal aspect of bones of right hand

Question 5A.28

How many phalanges are found in the thumb?

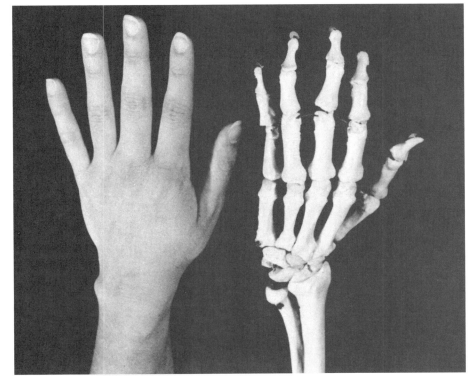

Figure 5.40

Left hand showing prone position and articulation

Identification 5.4

    5.4

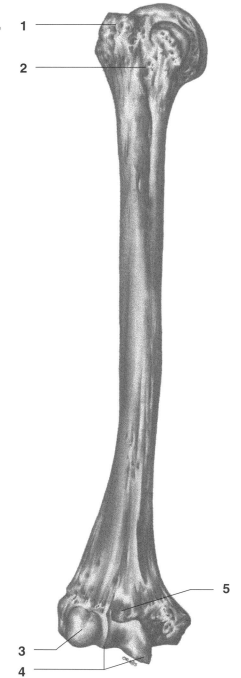

1

2

5

3

4

Identification 5.5

    5.5

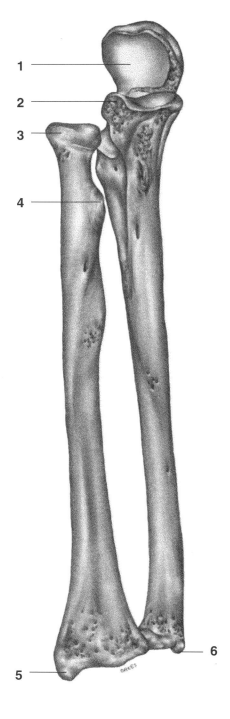

1

2

3

4

5

6

Bone Markings of the Lower Division

Bone	Bone Marking	Description
Os coxa (OS KOK-sa) or **innominate** (i-NOM-i-nāt) (Figures 5.41 and 5.42)		Fusion of three smaller bones to form right and left pelvic bones; commonly known as hip bone
	Acetabulum (as-e-TAB-yoo-lum)	Lateral cavity formed by fusion of ilium, ischium, and pubis; site of hip socket articulation with femur head
Ilium (IL-ē-um)		Superior flaring portion of bone
	Iliac (il-ē-ak) crest	Superior border of ilium
	Iliac spines	Projections on periphery of ilium (named by position)
	Anterior superior	Most anterior projection of hip
	Anterior inferior	Less prominent projection inferior to anterior superior spine
	Posterior superior	Most posterior projection on periphery of iliac crest
	Posterior inferior	Less prominent projection inferior to posterior superior spine
	Greater sciatic (sī-AT-ik) notch	Large irregularly shaped indentation inferior to posterior inferior spine
	Iliac fossa	Concave inner surface of ilium
Ischium (ISH-ē-um)		Inferior posterior bone that provides support when sitting
	Ischial (ISH-ē-al) tuberosity	Irregularly shaped projection on inferior surface that is in contact with chair when sitting
	Ischial spine	Slightly pointed projection immediately superior to tuberosity
Pubis (PYOO-bis)		Anterior irregularly shaped bones; two fuse to form anterior part of pelvic girdle
	Pubic (PYOO-bik) symphysis	Anterior medial cartilaginous articulation between 2 pubic bones
	Pubic tubercle	Small rounded projection on superior ramus; lateral and superior to pubic symphysis
	Obturator (OB-tyoo-rā'-ter) foramen	Very large rounded opening between pubis and ischium on anteriolateral surface
	Pubic arch	Anterior arch formed inferior to pubic symphysis by inferior rami of both pubic bones
	Pubic crest	Superior margin of superior rami above and lateral to pubic symphysis
	Pubic rami	Superior and inferior armlike portions of each pubic bone
Pelvis	Pelvic girdle	Articulation of two ossa coxae, sacrum, and coccyx; forms point of attachment for lower extremities
	Pelvic brim	Circumference of boundary formed by curved bony rim passing from anterior margin of superior sacrum to superior margin of pubic symphysis; oval-shaped in male, larger and rounded in female; of obstetrical importance in childbirth
	True pelvis	Space inferior to pelvic brim; contains pelvic organs
	False pelvis	Space superior to pelvic brim, medial to iliac bones; included in abdominal cavity

Question 5A.29

When sitting, what bone marking projects downward toward the chair?

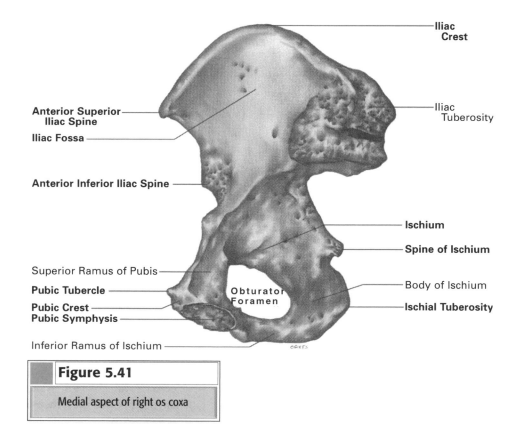

Figure 5.41

Medial aspect of right os coxa

Labels: Iliac Crest, Iliac Tuberosity, Anterior Superior Iliac Spine, Iliac Fossa, Anterior Inferior Iliac Spine, Ischium, Spine of Ischium, Superior Ramus of Pubis, Pubic Tubercle, Pubic Crest, Pubic Symphysis, Obturator Foramen, Body of Ischium, Ischial Tuberosity, Inferior Ramus of Ischium

Question 5A.30

Where does the fusion of the os coxa occur?

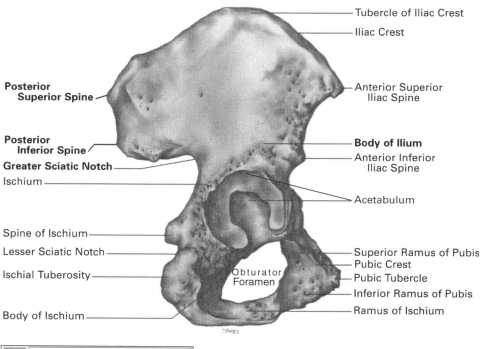

Figure 5.42

Lateral aspect of right os coxa

Labels: Tubercle of Iliac Crest, Iliac Crest, Posterior Superior Spine, Anterior Superior Iliac Spine, Posterior Inferior Spine, Body of Ilium, Greater Sciatic Notch, Anterior Inferior Iliac Spine, Ischium, Acetabulum, Spine of Ischium, Lesser Sciatic Notch, Superior Ramus of Pubis, Pubic Crest, Ischial Tuberosity, Obturator Foramen, Pubic Tubercle, Inferior Ramus of Pubis, Ramus of Ischium, Body of Ischium

Identification 5.6

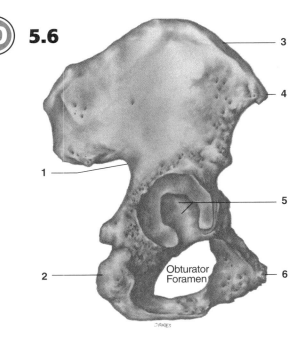

ID **5.6**

1

2

3

4

5

6

Obturator
Foramen

Identification 5.7

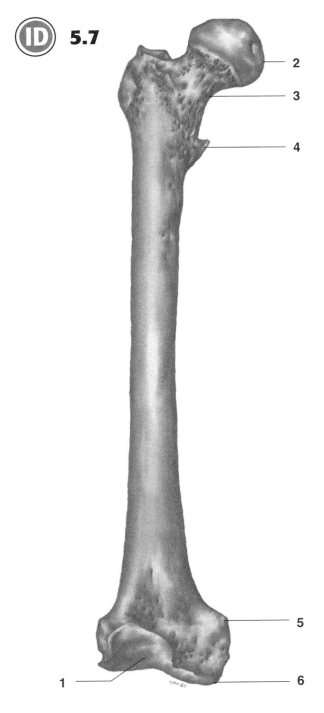

ID **5.7**

1

2

3

4

5

6

Bone Markings of the Lower Division

Bone	Bone Marking	Description
Femur (FĒ-mer) (Figures 5.43 and 5.44)		Large strong thigh bone
	Head	Rounded projection that articulates with acetabulum in ball-and-socket joint
	Greater trochanter	Large irregularly shaped process inferior and lateral to head
	Lesser trochanter	Small irregularly shaped process inferior and posterior to head
	Linea aspera (LIN-ē-a AS-per-a)	Ridge running approximately three-fourths the length of posterior bone surface
	Supracondylar (soo'-pra-KON-dī-lar) ridges	Inferior division of linea aspera into two ridges, medial and lateral
	Gluteal (GLOO-tē-al) tubercle	Small rounded projection inferior to greater trochanter
	Medial condyle	Large rounded protrusion on medial distal end
	Lateral condyle	Large rounded protrusion on lateral distal end
	Adductor tubercle	Very small rough projection above medial condyle marking termination of medial supracondylar ridge
	Patellar facet	Pulley-shaped structure on anterior surface between condyles
	Intercondyloid (in'-ter-KON-di-loyd) notch *or* fossa	Deep depression between condyles on posterior surface
Patella (pa-TEL-a) (Figures 5.45 and 5.46)		Irregularly shaped flat knee bone
Tibia (TIB-ē-a) (Figure 5.47)		Large strong medial bone of lower leg
	Lateral and medial condyles	Protuberances on lateral and medial surfaces at proximal end of bone
	Tibial (TIB-ē-al) tuberosity	Rough projection at midpoint on anterior surface at proximal end of bone
	Intercondylar (in'-ter-KON-dī-lar) eminence	Raised area with toothlike projections on superior surface between condyles
	Popliteal (pop-LIT-ē-al) line	Slight ridge on posterior surface running from superior to inferior; extends approximately one-third the length of the bone
	Medial malleolus (ma-LĒ-o-lus)	Medial protrusion on distal end
	Anterior crest	Vertical ridge along length of anterior tibia
Fibula (FIB-yoo-la) (Figure 5.47)		Long thin lateral bone of lower leg
	Lateral malleolus	Lateral protrusion on distal end
Foot (Figure 5.48)		
Tarsals (Table, page 92)		Fourteen irregularly shaped bones of foot (seven per foot)
Calcaneus (kal-KĀ-nē-us)		Large irregular bone that forms heel; one of the tarsals
Talus (TĀ-lus)		Rounded bone with smooth superior surface that articulates with tibia and fibula; one of the tarsals
Metatarsals (met'-a-TAR-salz)		Ten long bones articulating between the tarsals and proximal phalanges (five per foot)
Phalanges		Bones of toes

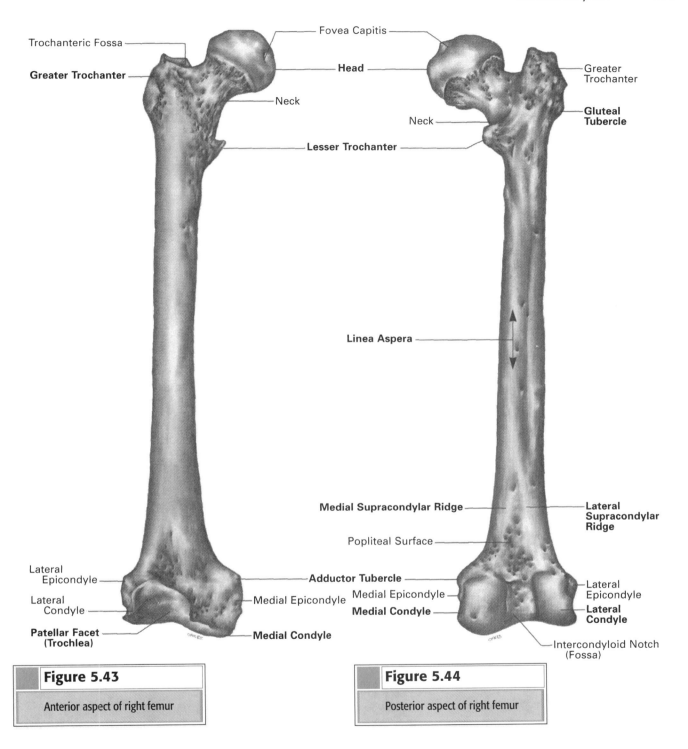

Trochanteric Fossa

Greater Trochanter

Fovea Capitis

Head

Neck

Lesser Trochanter

Greater Trochanter

Gluteal Tubercle

Neck

Linea Aspera

Medial Supracondylar Ridge

Popliteal Surface

Adductor Tubercle

Medial Epicondyle

Lateral Supracondylar Ridge

Lateral
Epicondyle

Lateral
Condyle

**Patellar Facet
(Trochlea)**

Medial Epicondyle

Medial Condyle

Medial Epicondyle

Medial Condyle

Lateral
Epicondyle

**Lateral
Condyle**

Intercondyloid Notch
(Fossa)

Figure 5.43

Anterior aspect of right femur

Figure 5.44

Posterior aspect of right femur

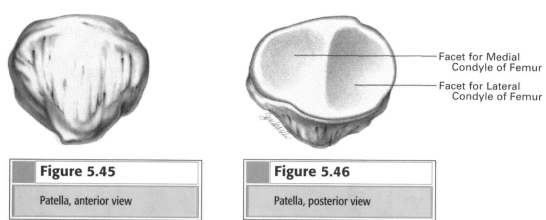

Figure 5.45

Patella, anterior view

Figure 5.46

Patella, posterior view

Facet for Medial Condyle of Femur

Facet for Lateral Condyle of Femur

Figure 5.47

Anterior aspect of right fibula and tibia

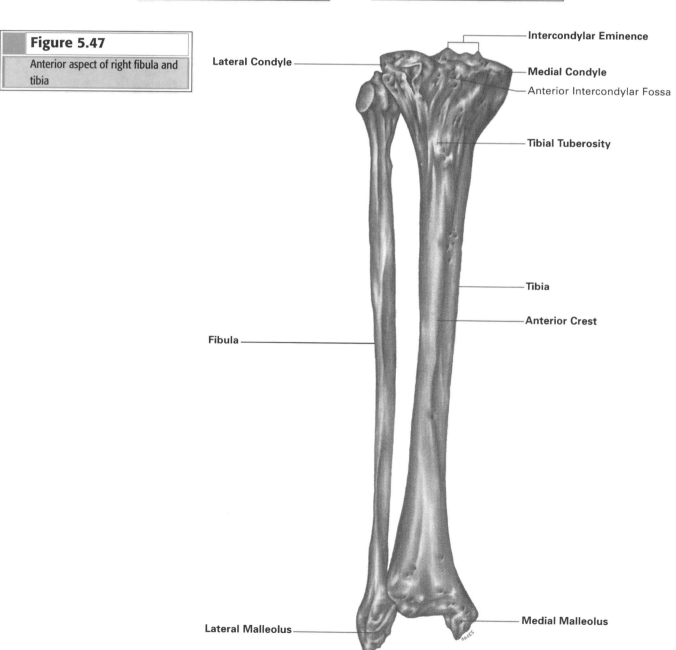

Lateral Condyle

Intercondylar Eminence

Medial Condyle

Anterior Intercondylar Fossa

Tibial Tuberosity

Tibia

Anterior Crest

Fibula

Lateral Malleolus

Medial Malleolus

Question 5A.31
What bone marking protrudes from your inner ankle? Your outer ankle?

Question 5A.32
How many phalanges are found in your fifth toe?

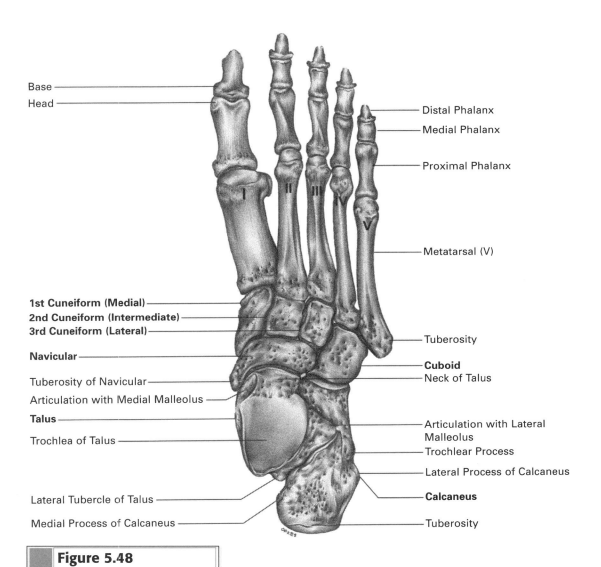

Base
Head

Distal Phalanx
Medial Phalanx
Proximal Phalanx

Metatarsal (V)

1st Cuneiform (Medial)
2nd Cuneiform (Intermediate)
3rd Cuneiform (Lateral)

Navicular

Tuberosity of Navicular
Articulation with Medial Malleolus

Talus

Trochlea of Talus

Lateral Tubercle of Talus

Medial Process of Calcaneus

Tuberosity

Cuboid
Neck of Talus

Articulation with Lateral Malleolus
Trochlear Process
Lateral Process of Calcaneus

Calcaneus

Tuberosity

Figure 5.48

Bones of right foot, dorsal surface

B. Identification of Joints

PURPOSE

Unit 5B will enable you to identify and understand, the various types of joints and selected examples of diarthrodial joint movements.

OBJECTIVES

After completing Unit 5B, you will be able to

- differentiate between types of synarthrodial and diarthrodial joints.
- locate major sutures and fontanels of adult and fetal skulls.
- demonstrate selected diarthrodial joint movements.

Most bones of the body join with one or more other bones. This union of bones is termed a **joint** or **articulation**. According to the amount of movement allowed by the anatomical structures of these unions, joints are divided into two major groups: **synarthroses** and **diarthroses**.

MATERIALS

articulated skeleton
mature human skulls
fetal skull

model of vertibral column
model of shoulder, elbow, hip,
    and knee joints

 **EXERCISE 1   Synarthroses**

Synarthrodial joints are those lacking a joint cavity and are therefore not freely movable. Instead, the bones themselves are united by cartilage or fibrous tissue. Synarthroses may be divided into two groups: those allowing *no* movement and those allowing *slight* movement. As you study the following synarthrodial joints, identify the structures written in boldface.

PROCEDURE

Immovable synarthrodial joints are located in the skull between cranial and facial bones, except for the mandible, which forms a movable joint. Immovable synarthrodial joints exist between mature, ossified bones and are referred to as **sutures**. Observe a mature skull. Note that the sutures appear to be jagged rather than straight. Find the **coronal suture**, which joins the frontal bone with the parietal bone (**Figure 5.6**). The **sagittal suture** joins the two parietal bones. The **lambdoidal suture** unites the parietals with the occipital bone. The **squamous suture** is formed by the union of the sphenoid, temporal, frontal, and parietal bones, and can be observed on the lateral aspect of the skull, also illustrated in **Figure 5.6**.

Prior to maturation, the four cranial bones are soft rather than ossified and do not come in contact with each other; nor are sutures of the skull completely formed. Areas of fibrous tissue, or **fontanels** ("soft spots") are found in the fetal skull. On a fetal skull find the **anterior** (or **frontal**) **fontanel**, which is located in the midline of the skull between the frontal and parietal bones. The **posterior** (or **occipital**) **fontanel**, also located in the midline, is between the parietal and occipital bones (**Figures 5.49** and **5.50**). If the fetal

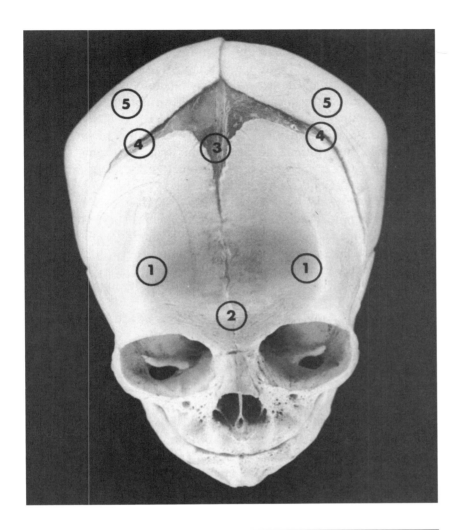

Figure 5.49

Fetal skull, full term, anterior view

1. frontal bone
2. frontal suture
3. anterior fontanel
4. coronal suture
5. parietal bone

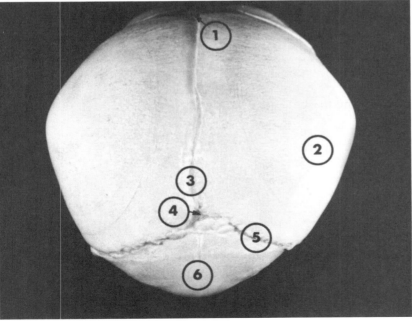

Figure 5.50

Fetal skull, full term, posterior view

1. anterior fontanel
2. parietal bone
3. sagittal suture
4. posterior fontanel
5. lambdoidal suture
6. occipital bone

Question 5B.1

Which of the fontanels is usually the last to ossify?

Question 5B.2

What is the advantage of sutures of the skull not fusing until early adulthood?

Question 5B.3

Of which tissue type are ligaments composed?

skull you are observing is from early pregnancy, you may also be able to find the paired **anterolateral** (or **sphenoidal**) **fontanels** at the junction of the frontal, parietal, and temporal bones, and the **posterolateral (mastoid) fontanels** at the junction of the parietal, temporal, and occipital bones.

Slightly movable synarthrodial (or **amphiarthrodial**) **joints** allow for limited motion in response to compression, twisting or stress. They are located between the ribs and sternum, where these bones are united by hyaline cartilage, and between the vertebrae and at the pubic symphysis, where the union is of fibrocartilage. Observe these amphiarthrodial joints on the articulated skeleton and the model of the vertebral column.

A particular type of amphiarthrodial joint is a **syndesmosis**, in which the distal ends of long bones are held together by ligaments. These are found at the distal ends of the tibia and fibula, and of the radius and ulna. Observe the locations of these syndesmoses on an articulated skeleton. You will probably not see the ligaments, which were removed in the process of preparing the skeleton.

EXERCISE 2 **Diarthroses**

Diarthrodial joints, the second major group of articulations, consist of those joints that have a **synovial cavity** and allow freedom of movement (**Figures 5.51–5.53**). The cavity, which is formed by a fibrous capsule, is lined with membranes that secrete a lubricating fluid called **synovial fluid.** Synovial fluid facilitates movement and reduces friction between articulating bones. Diarthrodial joints are further subdivided according to the shapes of the articulating bone surfaces or by the type of movement allowed by the joint (**Figure 5.51**). The major diarthrotic joint movements may be defined and illustrated as follows:

Movement	Definition	Illustration
Flexion	Bending or folding movements that decrease the angle of a joint	
Extension	The return from flexion; straightening movements that increase the angle of a joint	
Hyperextension	Movement that increases the angle of a joint beyond anatomical position	
Dorsiflexion	Backward flexion or bending of the hand or foot	

Movement	Definition	Illustration
Plantar flexion	Forward flexion or bending of the foot	
Abduction	Movement away from the median plane of the body	
Adduction	Movement toward the median plane of the body	
Rotation	Pivoting or movement of a bone upon its axis	
External rotation	Movement of a joint around its axis in a superior or lateral direction	
Internal rotation	Movement of a joint around its axis in an inferior or medial direction	
Circumduction	Movement that describes the surface of a cone	
Supination	Movement of bones of the forearm so the radius and ulna are parallel, palm up	
Pronation	Movement of bones of the forearm so the radius and ulna are not parallel; palm down	
Inversion	Movement of the sole of the foot medially at the ankle joint	
Eversion	Movement of the sole of the foot laterally at the ankle joint	

Figure 5.51

Left knee joint, anterior aspect

1. femur
2. patellar surface of femur
3. articular surface of patella
4. medial meniscus
5. lateral meniscus
6. anterior cruciate ligament
7. patellar ligament
8. tibia
9. fibula
10. interosseous membrane
11. fibular collateral ligament

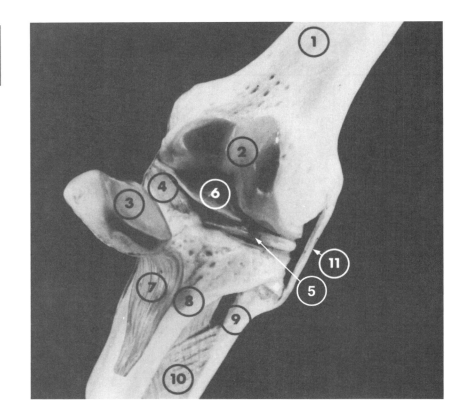

Figure 5.52

Left knee joint, posterior aspect

1. femur, posterior surface
2. lateral condyle
3. medial condyle
4. posterior cruciate ligament

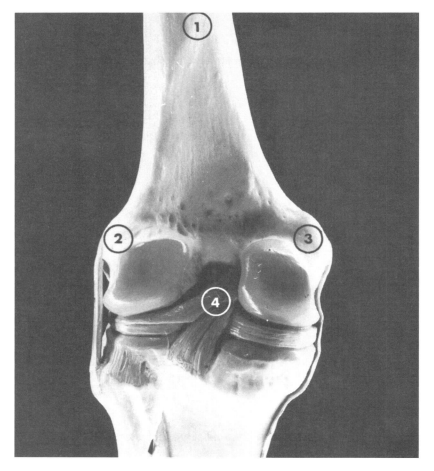

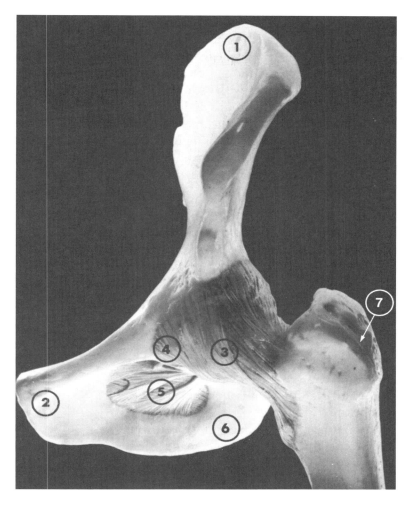

Figure 5.53

Anterior view of left pelvis and femur

1. iliac crest
2. pubis
3. iliofemoral ligament
4. ischiofemoral ligament
5. obturator membrane
6. ischium
7. greater trochanter

Question **5B.4**

Which type of diarthrodial joint is capable of the most extensive movement?

Observe models of shoulder, elbow, hip, and knee joints. The following table lists types of diarthrodial joints, movements allowed, and examples of diarthroses in the human body.

Diarthrodial Joints

Type of Joint	Movement	Example
Hinge (Figures 5.51 and 5.52)	Flexion/extension	Elbow Knee
Pivot	Rotation	Atlas/axis
Spheroidal (ball and socket) (Figure 5.53)	Flexion/extension Abduction/adduction External rotation Internal rotation Circumduction	Shoulder Hip
Saddle	Flexion/extension Abduction/adduction Circumduction	Thumb between carpals and metacarpals
Ellipsoidal	Flexion/extension Abduction/adduction Circumduction	Wrist Ankle

Movements of the head, torso, and limbs are dependent on various joints pulling on skeletal muscles of the body. When performed repeatedly and in sequence, these movements may be referred to as **range-of-motion exercises.** These exercises are often used to prevent joint immobility and to maintain muscle strength in individuals with orthopedic or neurological disorders.

Work in pairs or in small groups and perform each of these exercises three times. Using the definitions and illustrations of diarthrodial joint movements and examples of diarthrodial joints as a guide, identify the movement that each exercise demonstrates, using terms from the following list. Begin each exercise from anatomical position.

abduction	hyperextension
adduction	internal rotation
circumduction	inversion
dorsiflexion	plantar flexion
eversion	pronation
extension	rotation
external rotation	supination
flexion	

Description   *Movement*

1. Stand with both feet together. Without bending your knees, move your right leg laterally until the foot is about 12″ off the floor. _____

2. Without bending the knee, move your right leg medially, placing both feet together on the floor. _____

3. Bend your elbows. Without bending your wrists, turn the palms of your hands so they face superiorly, remaining parallel to the floor. _____

4. Keeping your elbows bent, turn the palms of your hands so they face inferiorly, remaining parallel to the floor. _____

5. Bend your neck anteriorly, moving your chin inferiorly until it meets resistance. _____

6. Elevate your chin until you are looking straight ahead. _____

7. Bend your neck posteriorly until you meet resistance. _____

8. Slowly turn your head to the right, then to the left, keeping your head perpendicular to your shoulders. _____

9. Keeping your elbows and fingers straight, extend your left arm to the side until it is parallel to the floor. Bend your elbow, allowing your forearm to hang, remaining perpendicular to the floor. _____

10. Keeping your left upper arm laterally extended and parallel to the floor, swing your left forearm, which is perpendicular to the floor, anteriorly and superiorly until your hand points toward the ceiling.

                           _____

11. Extend your left arm laterally, keeping the elbow straight. Move your arm, first in a clockwise motion, then counterclockwise.

                           _____

12. Bend your right foot inferiorly until the toes are pointing toward the floor.

                           _____

13. Keep your right heel on the floor. Bend your foot upward, off the floor.

                           _____

14. Place your right foot on the floor. Lift the medial edge of your foot so your weight is redistributed along its lateral edge.

                           _____

15. Place your right foot on the floor. Lift the lateral edge of your foot so your weight is redistributed along its medial edge.

                           _____

Question 5B.5

Draw an example of the following movements:

Flexion:

Extension:

The Skeletal System

SELF-TEST

MULTIPLE CHOICE

Name _____

Section _____

Date _____

_____ 1. Diarthrodial joints are characterized by
    a. a joint cavity.
    b. a synovial membrane.
    c. lubrication by a thick fluid.
    d. all of the above.

_____ 2. Which of these is a single, rather than a paired, bone?
    a. zygomatic
    b. vomer
    c. maxilla
    d. lacrimal

_____ 3. The articulation of the femur to the acetabulum is a good example of a _____ joint.
    a. pivot
    b. hinge
    c. ball-and-socket
    d. pivot-condyloid

_____ 4. Which of the following is *not* a part of the appendicular skeleton?
    a. skull
    b. pelvic girdle
    c. scapula
    d. pectoral girdle

_____ 5. The largest foramen in the skull is the
    a. foramen magnum.
    b. foramen lacerum.
    c. obturator foramen.
    d. mental foramen.

_____ 6. Intervertebral discs are composed of
    a. hyaline cartilage.
    b. bone.
    c. fibrocartilage.
    d. epithelial tissue.

_____ 7. Which of the following is *not* an example of a diarthrodial joint?
    a. knee
    b. hip joint
    c. atlas-axis
    d. sagittal suture

_____ 8. Vertebrae are classified as _____ bones.
    a. long
    b. sesamoid
    c. irregular
    d. short

_____ 9. An opening in a bone that serves as a passageway for nerves and blood vessels is a
    a. sinus.
    b. foramen.
    c. fissure.
    d. tubercle.

_____ 10. The odontoid process is found on which of the following bones?
    a. atlas
    b. axis
    c. sternum
    d. femur

_____ 11. Which of the following is a bone, *not* a bone marking?
    a. inferior concha   c. pedicle
    b. middle concha   d. acetabulum

_____ 12. An example of a synarthrosis is the
    a. elbow joint.   c. hip joint.
    b. coronal suture.   d. wrist joint.

_____ 13. An example of a pivotal joint is the
    a. wrist.   c. femur head-acetabulum.
    b. atlas-axis.   d. knee.

_____ 14. Another name for a ball-and-socket joint is
    a. saddle.   c. spheroidal.
    b. ellipsoidal.   d. hinge.

_____ 15. An example of a joint type that does *not* allow for rotation is the
    a. hinge.   c. spheroidal.
    b. pivot.   d. ball-and-socket.

_____ 16. Which of the following is a rounded extension that projects from a tapered neck and forms part of a joint?
    a. spine   c. tubercle
    b. head   d. trochanter

_____ 17. The shaft of a long bone is known as the
    a. hinge.   c. hypophysis.
    b. diaphysis.   d. metaphysis.

_____ 18. Which of these is most centrally located in an osteon?
    a. Haversian canal   c. lamella
    b. osteocyte   d. canaliculus

_____ 19. Which of these bones is the smallest?
    a. hyoid   c. calcaneus
    b. incus   d. nasal

_____ 20. Which of these bone markings would be found at a hinge joint?
    a. spine   c. head
    b. trochanter   d. trochlea

_____ 21. Which of these bones would have the highest ratio of compact bone to spongy bone?
    a. lunate   c. thoracic vertebra
    b. scapula   d. tibia

_____ 22. Which of these bones is *not* the result of fusion of smaller bones?
    a. sternum   c. scapula
    b. os coxa   d. coccyx

_____ 23. The ilium, ischium, and pubis meet
    a. at the pelvic brim.   c. in the acetabulum.
    b. at the pubic arch.   d. inferior to the sciatic notch.

—————— 24. If a patient is in a below-the-knee cast for a broken ankle, which of these movements could he *not* perform?
   a. inversion of the foot
   b. abduction of the leg
   c. hyperextension of the hip
   d. flexion of the knee

—————— 25. Which of these definitions is incorrect?
   a. spine: a sharp slender projection
   b. trochanter: a large terminal enlargement
   c. fissure: a narrow slit
   d. tubercle: a small rounded process

26. Define the following bone markings and give an example of each:
    a. trochlea _____
    b. tubercle _____
    c. trochanter _____
    d. spine _____
    e. process _____
    f. fossa _____
    g. foramen _____
    h. condyle _____
    i. groove _____

27. Which bones are included in (a) the axial skeleton and (b) the appendicular skeleton?

28. Give the function of each of the following bone markings:

    a. mandibular fossa _____

    b. jugular foramen _____

    c. carotid foramen _____

    d. sella turcica _____

    e. superior orbital fissure _____

    f. foramen rotundum _____

    g. crista galli _____

    h. foramen magnum _____

    i. superior articulating surface of a vertebra _____

    j. odontoid process _____

    k. rib tubercle _____

    l. acromion of scapula _____

    m. glenoid cavity _____

    n. coronoid fossa of humerus _____

    o. semilunar notch of ulna _____

    p. ischial tuberosity _____

    q. acetabulum _____

CASE STUDY

SKELETAL SYSTEM

While cycling in a 10K speed competition, Janet, 21 years old, hit a rough spot in the road and was thrown from her bike. She immediately felt severe stabbing pain in her right thigh. After x-raying the region, the doctor told her that she had an oblique fracture of the proximal femur. He said that she had no need to worry about the epiphyseal plate.

Draw and label Janet's bone(s) and fracture.

Explain the doctor's comment about the epiphyseal plate.

The Muscular System

A. Human Musculature

PURPOSE

Unit 6A will give you an understanding of the gross and microscopic anatomical relationships of human muscles.

OBJECTIVES

After completing Unit 6A, you will be able to
- microscopically recognize skeletal, smooth, and cardiac muscle.
- microscopically recognize a myoneural junction.
- identify the major muscles of the head, neck, trunk, and limbs.

MATERIALS

slides of skeletal, smooth, and
   cardiac muscle
slides of myoneural junction
prosected cadaver (if available)
dissection instruments

human torso models
human head, arm, and leg models
human anatomical charts
model of eyeball
disposable gloves

PROCEDURE

 EXERCISE 1 **Microscopic Identification of Muscle Types and Myoneural Junctions**

SKELETAL MUSCLE (COLOR PLATE 26)

Skeletal muscle is voluntary striated muscle. Obtain a prepared slide of a longitudinal section of skeletal muscle and observe the **muscle fibers** (**Figures 4.16** and **4.17**). Each fiber (muscle cell) is multinucleated. You should be able to distinguish these nuclei quite easily by their peripheral location. Each muscle fiber contains many **myofibrils** that are made up of **myofilaments.** Careful observation under high power will reveal striations in the myofibrils. The **A bands (anisotropic bands)** stain darkly and represent myosin and actin filaments. Between the A bands are light areas known as **I bands (isotropic bands)**, which represent actin filaments. Dividing the I bands are **Z bands** or **Z lines**, which can be seen only with an electron microscope. The segment of a fiber between successive Z bands is known as a **sarcomere**, the unit of structure and function of a muscle fiber.

An entire muscle is surrounded by connective tissue known as **epimysium.** The muscle is subdivided into **fascicles (fasciculi)** that are surrounded by **perimysium.** Observe a cross section of skeletal muscle. Look for muscle fibers that are contained in a fascicle. Under high power, you should be able to see the details of a muscle fiber, including nuclei and myofibrils. In the space provided, sketch a longitudinal section and a cross section of skeletal muscle. Label nucleus, A band, and I band on the longitudinal section. After viewing the cross section, draw and label nuclei and myofibrils.

SMOOTH MUSCLE (COLOR PLATE 29)

Smooth muscle appears nonstriated under the light microscope. It is usually considered involuntary and is present primarily in blood vessels and internal organs. Observe a slide of a longitudinal section of smooth muscle. Notice the "dovetailing" effect of the muscle fibers (**Figure 4.19**). In cross section, nuclei of smooth muscle fibers are located more centrally within the fiber rather than at the perimeter, as in skeletal muscle. Draw several smooth muscle fibers. Label nucleus and cytoplasm.

CARDIAC MUSCLE (COLOR PLATES 27 AND 28)

Cardiac muscle is involuntary striated muscle. The muscle fibers bifurcate and connect with adjacent fibers to form a three-dimensional network (**Figures 4.20 and 4.21**). Nuclei of cardiac muscle are situated deep within the fiber, as in smooth muscle. If your slide has been specially stained, you can recognize **intercalated disks** that join together the ends of cardiac muscle fibers. The pattern of cross-striations of myofibrils and the A and I bands and Z lines are the same as in skeletal muscle. Draw a longitudinal section of cardiac muscle. Label nucleus, A band, I band, and intercalated disks.

MYONEURAL JUNCTION (MOTOR END PLATE) (COLOR PLATE 30)

This slide represents the junction of a nerve ending and skeletal muscle fibers. Note the **motor end plate** detail under high power. **Figure 6.1** shows the motor end plate as seen under low power. Draw and label your high power observation.

Figure 6.1

Motor end plate details as viewed in skeletal muscle (430×)

1. motor nerve fibers
2. striated muscle fiber
3. motor end plate

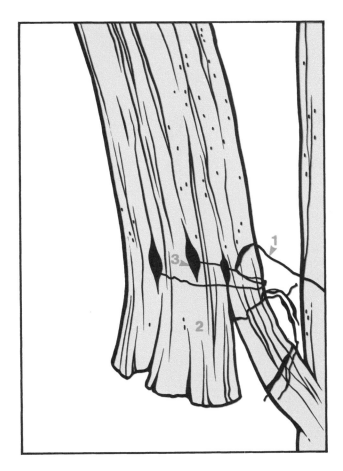

EXERCISE 2 Identification of Muscles

Using models and a prosected cadaver, identify these muscles. Muscles are listed from superior to inferior and lateral to medial within each of these body regions.

The muscle tables, accompanying photos, and illustrations will aid in the identification of muscles. The tables are organized to give you a description of each muscle, its **origin**, **insertion**, and **action**. The origin of a muscle is usually the proximal end, where it is attached to the immovable bone or bones. The insertion is usually the distal end of the muscle, where it is attached to a more movable bone or bones. See **Figure** 6.2.

Major Human Muscles   (Figures 6.2 and 6.13)

Muscles of the Head
Epicranius
Corrugator
Orbicularis oculi
Temporalis
Nasalis
Zygomaticus major
Zygomaticus minor
Orbicularis oris
Buccinator
Masseter
Mentalis

Muscles of the Neck
Platysma
Sternocleidomastoid
Sternohyoid
Sternothyroid
Mylohyoid
Scalenus group

Muscles of the Eyeball
Superior rectus
Inferior rectus
Medial rectus
Lateral rectus
Superior oblique
Inferior oblique

Muscles of the Chest
Pectoralis major
Pectoralis minor
External intercostals
Internal intercostals
Serratus anterior

Muscles of the Abdomen
Rectus abdominis
External oblique
Internal oblique
Transversus abdominis

Muscles of the Back and Shoulder
Trapezius
Deltoid
Supraspinatus
Infraspinatus
Teres minor
Teres major
Levator scapulae
Rhomboideus major
Rhomboideus minor
Latissimus dorsi
Subscapularis

Muscles of the Upper Arm
Biceps brachii
Brachioradialis
Brachialis
Triceps brachii

Muscles of the Forearm
Pronator teres
Flexor group
    Flexor carpi ulnaris
    Palmaris longus
    Flexor carpi radialis
Extensor group
    Extensor carpi radialis longus
    Extensor carpi radialis brevis
    Extensor digitorum communis
    Anconeus
    Extensor carpi ulnaris

Muscles of the Hip and Thigh
Psoas minor
Psoas major
Sartorius
Gracilis
Adductor magnus
Adductor longus
Pectineus
Gluteus maximus
Gluteus medius
Gluteus minimus
Tensor fasciae latae
Quadriceps femoris group
    Rectus femoris
    Vastus lateralis
    Vastus medialis
    Vastus intermedius
Hamstring group
    Biceps femoris
    Semitendinosus
    Semimembranosus

Muscles of the Lower Leg
Tibialis anterior
Peroneus longus
Peroneus brevis
Extensor digitorum longus
Gastrocnemius
Soleus

Figure 6.2

Human cadaver, selected muscles

1. epicranius, frontalis portion
2. temporalis
3. orbicularis oculi
4. nasalis
5. zygomaticus major
6. orbicularis oris
7. sternocleidomastoid
8. pectoralis major
9. deltoid
10. skin, reflected
11. latissimus dorsi
12. linea alba (connective tissue)
13. rectus abdominis
14. internal oblique
15. external oblique

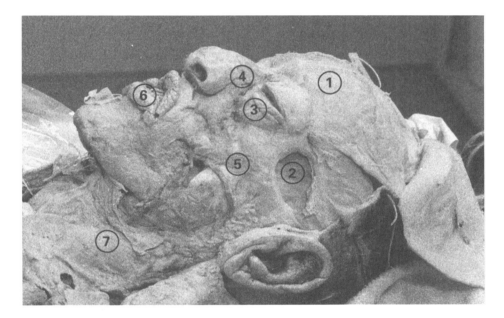

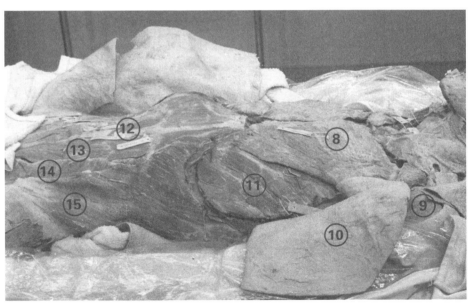

Muscles of the Head   (Figures 6.2–6.4)

Name and Description	Origin	Insertion	Action
Epicranius (ep′-i-KRĀ-nē-us)			
A muscle consisting of two parts, frontalis and occipitalis, covering the superior aspect of the scalp	Cranial aponeurosis (frontalis); occipital bone (occipitalis)	Skin of eyebrows and crest of nose (frontalis); cranial aponeurosis (occipitalis)	Raises eyebrows and wrinkles forehead (frontalis); pulls scalp posteriorly (occipitalis)
Corrugator (KOR-oo-gā-ter)			
A superficial facial muscle; fibers run from the midline in a transverse direction to superciliary arch	Medial portion of superciliary arch	Skin superior to the arch	Draws eyebrows downward and inward
Orbicularis oculi (or-bik′-yoo-LA-ris OK-yoo-li)			
Sphincter muscle of eyelids	Frontal and maxillary bones and ligaments surrounding orbit	Tissue of eyelids	Closes eyes, causes blinking, squinting, and depression of eyebrows
Temporalis (tem′-po-RA-lis)			
A fan-shaped muscle covering temporal bone	Temporal fossa	Coronoid process of mandible	Closes jaw and elevates mandible
Nasalis (nā-ZA-lis)			
Superficial muscle covering bridge of nose	Maxilla superior to incisor and canine teeth	Skin of anterior nares and nasolabial indentation	Constricts nostrils
Zygomaticus (zī′-gō-MAT-i-kis) **major**			
A slender band of muscle descending diagonally from zygomatic bone to angle (corner) of mouth	Zygomatic bone	Superior lateral portion of orbicularis oris muscle at angle of mouth	Draws corners of mouth superiorly and posteriorly
Zygomaticus minor			
A thin band of muscle superior and medial to zygomaticus major; descends diagonally from zygomatic bone to upper lip	Zygomatic bone	Superior lateral portion of orbicularis oris muscle, medial to angle of mouth	Draws upper lip superiorly and laterally
Orbicularis oris (OR-is)			
Sphincter muscle of lips; fibers run in many directions	Maxilla and mandible	Muscle and skin at angles of mouth	Closes mouth, purses and protrudes lips
Buccinator (BUK-si-nā′-ter)			
Horizontal cheek muscles; found deep to masseter	Molar area of maxilla and mandible	Orbicularis oris	Draws corners of mouth laterally
Masseter (MASS-e-ter)			
A muscle extending across the mandible	Zygomatic process (arch)	Angle and ramus of mandible	Closes jaw and elevates mandible
Mentalis (men-TA-lis)			
Oblique muscle bands running from medial chin toward angle of mandible	Mandible, inferior to lateral incisor	Skin of chin	Moves skin of chin superiorly

Which muscle encircles
your eye?

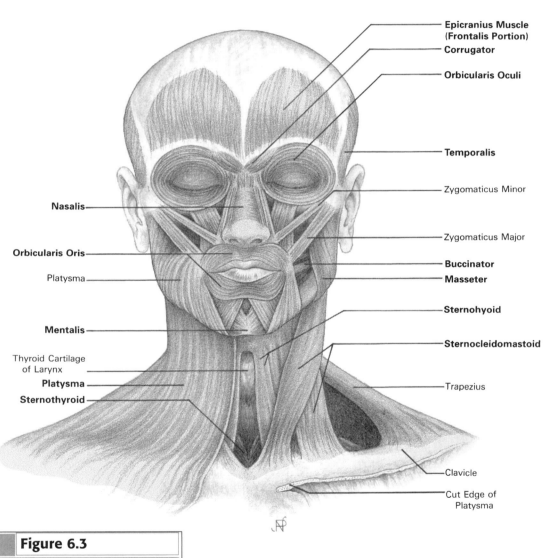

Epicranius Muscle
(Frontalis Portion)

Corrugator

Orbicularis Oculi

Temporalis

Zygomaticus Minor

Nasalis

Zygomaticus Major

Orbicularis Oris

Buccinator

Platysma

Masseter

Sternohyoid

Mentalis

Sternocleidomastoid

Thyroid Cartilage
of Larynx

Platysma

Trapezius

Sternothyroid

Clavicle

Cut Edge of
Platysma

Figure 6.3

Head and neck muscles, anterior
view

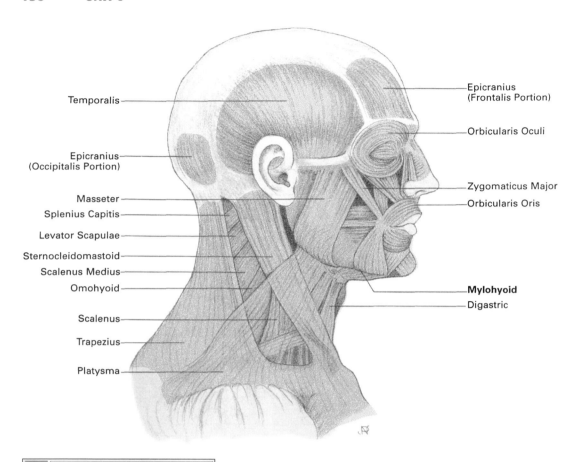

Temporalis

Epicranius
(Occipitalis Portion)

Masseter

Splenius Capitis

Levator Scapulae

Sternocleidomastoid

Scalenus Medius

Omohyoid

Scalenus

Trapezius

Platysma

Epicranius
(Frontalis Portion)

Orbicularis Oculi

Zygomaticus Major

Orbicularis Oris

Mylohyoid

Digastric

Figure 6.4

Head and neck muscles, lateral view

Muscles of the Neck   (Figures 6.3 and 6.4)

Name and Description	Origin	Insertion	Action
Platysma (pla-TIZ-ma) A thin, broad, sheetlike superficial muscle of neck	Fascia of deltoid and pectoralis major muscles	Lower jaw	Depresses lower lip
Sternocleidomastoid (ster′-nō-klī-dō-MAS-toyd) A two-headed muscle deep to platysma on anterolateral surfaces of neck	Body of sternum and clavicle	Mastoid process of temporal bone	Rotates head and flexes neck
Sternohyoid (ster′-nō-HĪ-oyd) A straplike muscle extending medially along neck	Inferior surface of manubrium	Body of hyoid bone	Depresses hyoid bone
Sternothyroid (ster′-nō-THĪ-royd) A straplike muscle lateral to sternohyoid	Manubrium and medial end of clavicle	Thyroid cartilage of larynx	Depresses larynx
Mylohyoid (mī′-lō-HĪ-oyd) A muscle forming the floor of the mouth; located deep to digastric	Medial surface of mandible	Hyoid bone	Elevates hyoid bone and tongue
Scalenus (skā-LĒ-nus) group A group of three muscles: anterior, medius, and posterior; deep to platysma along lateral aspect of neck	Transverse process of cervical vertebrae	First two ribs	Rotates head and elevates upper ribs

Question 6A.2

Which muscle depresses your lower lip?

Question 6A.3

Which muscle attached to the first two ribs assists you in breathing?

Muscles of the Eyeball   (Figure 8.1)

Name and Description	Origin	Insertion	Action
The following narrow bands of muscle connect different aspects of the eyeball to the bony orbit.			
Superior rectus A band of muscle on superior aspect of eyeball with fibers running in a straight pattern anterior to posterior	Posterior optic orbit; superior rim of optic foramen	Anterior superior region of eyeball	Rotates eyeball superiorly and medially
Inferior rectus A band of muscle on inferior aspect of eyeball with fibers running in a straight pattern anterior to posterior	Posterior optic orbit; inferior rim of optic foramen	Anterior inferior region of eyeball	Rotates eyeball inferiorly and medially
Medial rectus A band of muscle on medial aspect of eyeball with fibers running in a straight pattern anterior to posterior	Posterior optic orbit; medial rim of optic foramen	Anterior medial region of eyeball	Adduction of eyeball; rotates eyeball medially
Lateral rectus A band of muscle on medial aspect of eyeball with fibers running in a straight pattern anterior to posterior	Posterior optic orbit; lateral rim of optic foramen	Anterior lateral region of eyeball	Abduction of eyeball; rotates eyeball laterally
Superior oblique (ō-BLĒK) A band of muscle overlying the superior rectus muscle	Posterior optic orbit; superior rim of optic foramen and lesser wing of sphenoid bone	Anterior lateral region of eyeball; muscle lies medially and tendon passes through pulley and attaches to lateral eyeball between superior and lateral rectus muscles	Rotates eyeball inferiorly and laterally
Inferior oblique A band of muscle overlying the inferior rectus muscle	Anterior floor of optic orbit	Anterior lateral region of eyeball between inferior and lateral rectus muscles	Rotates eyeball superiorly and laterally

Muscles of the Chest   (Figure 6.5)

Name and Description	Origin	Insertion	Action
Pectoralis (pek´-tō-RA-lis) major			
A large fan-shaped muscle covering the upper portion of the chest	Anterior surface of clavicle, sternum, and ribs 1–6	Greater tubercle and crest of humerus	Flexes, adducts, and rotates arm medially
Pectoralis minor			
A flat, thin muscle underlying the pectoralis major	Ribs 3–5	Coracoid process of scapula	Depresses shoulder and draws scapula forward and down
External intercostals (in´-ter-KOS-tulz)			
Eleven paired muscles located between the ribs; fibers are directed obliquely toward the sternum	Inferior margin of ribs	Superior margin of rib below	Elevate ribs during inspiration
Internal intercostals			
Eleven paired muscles located between the ribs; fibers are directed obliquely away from the sternum	Inner surface of ribs	Superior margin of rib below	Draw ribs together during expiration
Serratus (se-RĀ-tus) anterior			
A muscle located beneath the pectorals; arises anteriorly as ribbons of muscle; extends posteriorly toward scapula	Superior margin of rib 1 to rib 8 or 9	Ventral surface of vertebral border of scapula	Draws scapula forward

Question 6A.4

Do the external intercostals elevate the ribs in inspiration or expiration?

Question 6A.5

Do the internal intercostals draw the ribs together during expiration or inspiration?

Muscles of the Abdomen   (Figure 6.5)

Name and Description	Origin	Insertion	Action
Rectus abdominis (ab-DOM-i-nus)			
A paired, medial, superficial, elongated muscle; appears segmented, extends from pubis to ribs	Pubic crest and pubic symphysis	Cartilages of ribs 5, 6, and 7	Flexes vertebral column and assists in compressing abdomen
External oblique			
A paired, superficial, lateral muscle; fibers run inferiorly and medially	Inferior margin of lower eight ribs	Iliac crest and pubis	Compresses abdomen and assists in rotating the vertebral column
Internal oblique			
A paired muscle underlying the external oblique; fibers run inferiorly and laterally	Iliac crest and lumbodorsal fascia	Xiphoid process and costal cartilages of ribs 3 and 4	Compresses abdomen
Transversus abdominis			
The most internal flat muscle of the abdominal wall; lateral portions underlie internal oblique	Iliac crest; lumbodorsal fascia and inner surface of last six ribs	Xiphoid process and linea alba	Compresses abdomen

Question 6A.6

In which direction do the fibers of the rectus abdominis muscle run?

Question 6A.7

In which direction do the fibers of the transversus abdominis run?

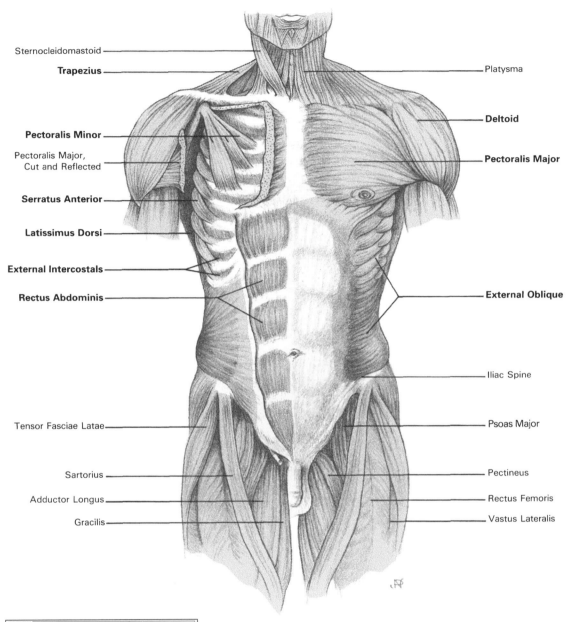

Sternocleidomastoid

Trapezius

Pectoralis Minor

Pectoralis Major,
Cut and Reflected

Serratus Anterior

Latissimus Dorsi

External Intercostals

Rectus Abdominis

Tensor Fasciae Latae

Sartorius

Adductor Longus

Gracilis

Platysma

Deltoid

Pectoralis Major

External Oblique

Iliac Spine

Psoas Major

Pectineus

Rectus Femoris

Vastus Lateralis

Figure 6.5

Superficial and deep muscles of the
trunk, anterior view

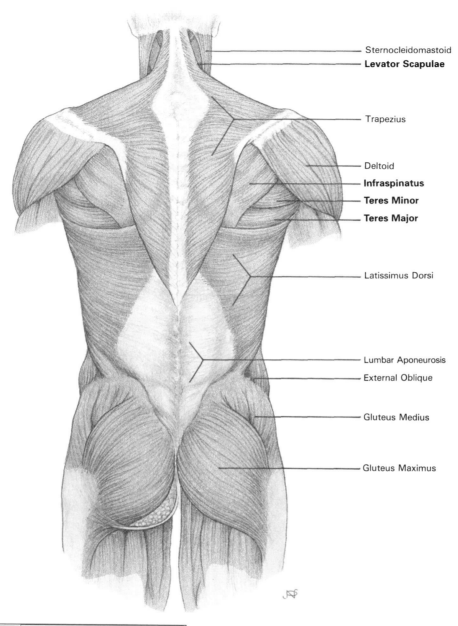

Sternocleidomastoid
Levator Scapulae

Trapezius

Deltoid
Infraspinatus
Teres Minor
Teres Major

Latissimus Dorsi

Lumbar Aponeurosis
External Oblique

Gluteus Medius

Gluteus Maximus

Figure 6.6

Superficial muscles of the trunk,
posterior view

Question 6A.8

With reference to the
scapular spine, where is
the supraspinatus located?

Question 6A.9

With reference to the
scapular spine, where is
the infraspinatus located?

Muscles of the Back and Shoulder   (Figures 6.6 and 6.7)

Name and Description	Origin	Insertion	Action
Trapezius (tra-PĒ-zē-us)			
A broad superficial muscle of the posterior neck and trunk	Occipital bone; spines of C7–T12 vertebrae	Scapular spine and posterior lateral portion of clavicle	Extends head, draws scapula toward midline; upper portion elevates scapula, lower portion depresses scapula
Deltoid (DEL-toyd)			
A thick triangular muscle at the lateral end of shoulder	Lateral portion of clavicle, spine, and acromion process of scapula	Deltoid tubercle of humerus	Abducts arm, aids in rotation, flexion and extension of humerus
Supraspinatus (soo′-pra-spi-NA-tus)			
A muscle occupying the supraspinous fossa of the scapula; deep to trapezius	Medial two thirds of scapular spine	Superior portion of greater tubercle of humerus	Abducts arm
Infraspinatus (in′-fra-spi-NA-tus)			
A thick, triangular muscle occupying the infraspinous fossa of the scapula; deep to deltoid and trapezius	Medial portion of infraspinous fossa	Greater tubercle of humerus	Rotates arm; superior fibers abduct arm; inferior fibers adduct arm
Teres (TER-ez) **minor**			
A narrow, elongated muscle extending from the surface of the axillary border of the scapula to the proximal humerus	Dorsal surface of axillary border of scapula	Greater tubercle of humerus	Adducts arm; rotates arm laterally
Teres major			
A thick, flattened muscle extending from the vertebral border of the scapula to the proximal humerus	Dorsal surface of inferior angle of scapula	Superior medial aspect of humerus at lesser tubercle	Adducts, extends, and rotates arm medially
Levator scapulae (la-VA-ter SKAP-yoo-le)			
An elongated muscle at the posterior lateral portion of the neck; lies beneath the trapezius	Posterior tubercles and transverse processes of C1–C4	Superior vertebral (medial) border of scapula	Elevates and rotates scapula; bends neck laterally
Rhomboideus (rom-BOY-dē-us) **major**			
Bands of muscle located beneath the trapezius and posterior to levator scapulae	Spinous processes of T2–T5	Vertebral border of scapula, beneath scapular spine	Pulls scapula medially
Rhomboideus minor			
Bands of muscle located beneath trapezius and posterior to levator scapulae; superior to rhomboideus major	Spinous processes of C7–T1	Vertebral border of scapula, at root of scapular spine	Pulls scapula medially
Latissimus dorsi (la-TIS-i-mus DOR-sē)			
A flat, broad muscle of the lower back	Spines of T6–T12 vertebrae by means of lumbar aponeurosis; spines of lumbar and sacral vertebrae and iliac crest	Inferior portion of intertubercular groove	Extends, adducts, and rotates arm medially; draws shoulder backward and downward
Subscapularis (sub-skap-yoo-LA-ris)			
Overlies anterior surface of scapula; fills subscapular fossa	Lateral border of scapula	Lesser tubercle of humerus and anterior portion of capsular ligament of shoulder	Rotates humerus medially

Figure 6.7

Muscles of trunk and upper arm.
A. Superficial muscles of upper
trunk and right arm. B. Superficial
muscles of upper dorsal trunk: left
side. C. Muscles of upper dorsal
trunk: cut to show deep muscles

1. levator scapulae
2. trapezius
3. deltoid
4. infraspinatus
5. teres major
6. latissimus dorsi

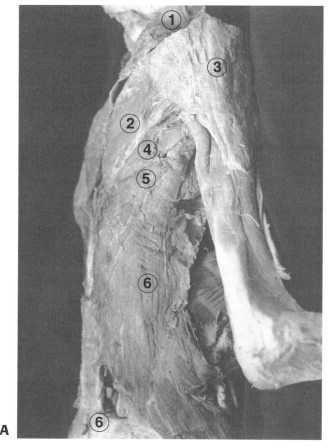

A

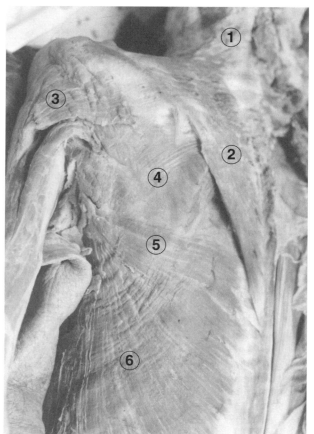

B

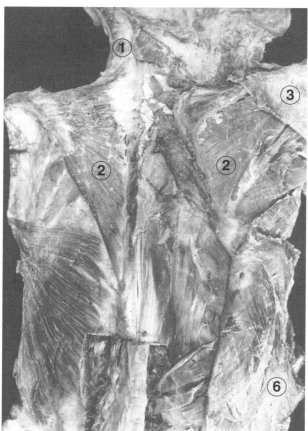

C

Muscles of the Upper Arm   (Figures 6.7–6.10)

Name and Description	Origin	Insertion	Action
Biceps brachii (BĪ-seps BRA-kē-ī) A prominent muscle on anterior surface of humerus; comprised of short and long heads	Coracoid process of scapula (short head); intertubercular groove of humerus (long head)	Radial tuberosity	Flexes elbow and shoulder; supinates hand
Brachioradialis (brā′-kē-ō-rā′dē-Ā-lis) A superficial muscle of lateral aspect of forearm	Lateral supracondylar ridge of humerus	Styloid process of radius	Flexes forearm
Brachialis (brā-kē-Ā-lis) A muscle lying deep to biceps brachii	Distal anterior aspect of humerus	Coronoid process of ulna	Flexes forearm
Triceps (TRĪ-seps) **brachii** A muscle overlying posterior surface of humerus; comprised of long, lateral, and medial heads	Inferior aspect of glenoid fossa (long head); posterior aspect of humerus (lateral head); radial groove (medial head)	Olecranon process of ulna	Extends forearm

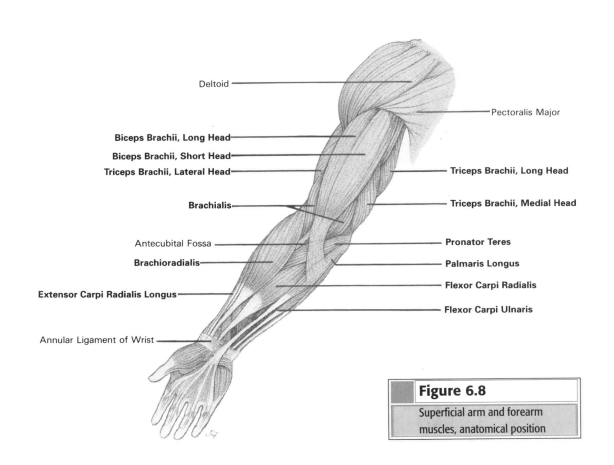

Deltoid

Pectoralis Major

Biceps Brachii, Long Head

Biceps Brachii, Short Head

Triceps Brachii, Lateral Head

Triceps Brachii, Long Head

Triceps Brachii, Medial Head

Brachialis

Antecubital Fossa

Pronator Teres

Brachioradialis

Palmaris Longus

Flexor Carpi Radialis

Extensor Carpi Radialis Longus

Flexor Carpi Ulnaris

Annular Ligament of Wrist

Figure 6.8
Superficial arm and forearm muscles, anatomical position

Figure 6.9

Superficial arm and forearm muscles, posterior view

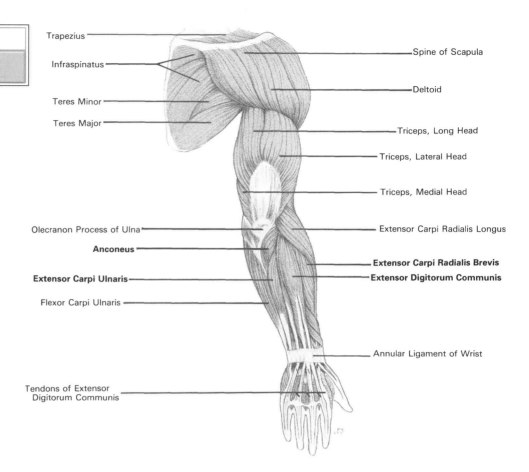

Trapezius

Infraspinatus

Teres Minor

Teres Major

Spine of Scapula

Deltoid

Triceps, Long Head

Triceps, Lateral Head

Triceps, Medial Head

Olecranon Process of Ulna

Anconeus

Extensor Carpi Ulnaris

Flexor Carpi Ulnaris

Extensor Carpi Radialis Longus

Extensor Carpi Radialis Brevis

Extensor Digitorum Communis

Annular Ligament of Wrist

Tendons of Extensor Digitorum Communis

Question 6A.10

Identify numbered superficial muscles.

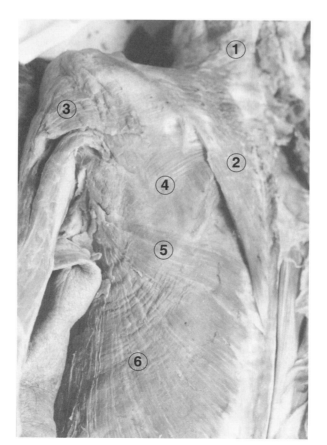

Figure 6.10

Muscles of right arm: Left. Superficial right arm muscles, dorsal view;
Right. Superficial right arm muscles: rotated medially

1. deltoid
2. biceps brachii
3. triceps
4. brachioradialis

5. flexor carpi ulnaris
6. extensor carpi ulnaris
7. flexor carpi radialis

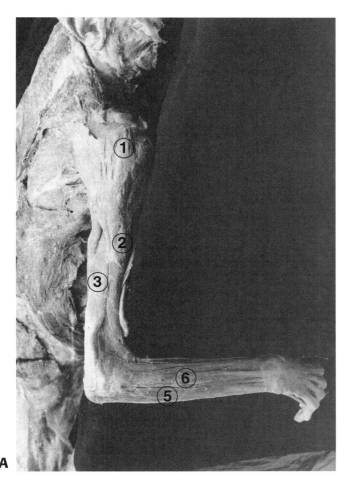

A

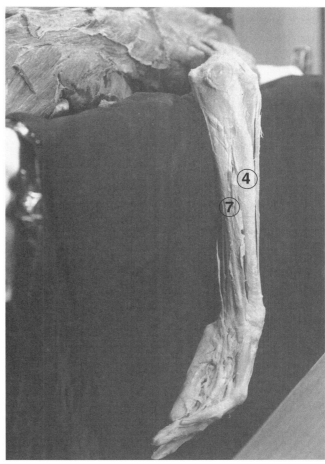

B

Question 6A.11

Which muscle flexes the elbow and shoulder?

Question 6A.12

Which muscle extends the forearm?

Question 6A.13

Identify the muscle of the forearm that: adducts the hand, extends the wrist, and inserts at the base of the fifth metacarpal.

Muscles of the Forearm   (Figures 6.8–6.10)

Name and Description	Origin	Insertion	Action
Pronator (pro-NĀ-ter) **teres**			
A superficial muscle passing obliquely across the forearm, medially to laterally	Medial epicondyle of humerus and coronoid process of ulna	Midlateral surface of radius	Pronates hand and flexes forearm
The following three muscles are in the **flexor group:**			
Flexor carpi ulnaris (FLEK-ser KAR-pi UL-na-ris)			
A muscle located medially to flexor carpi radialis	Olecranon process of ulna; medial epicondyle of humerus	Base of fifth metacarpal, pisiform, and hamate bones of wrist	Flexes wrist
Palmaris (pal-MA-ris) **longus**			
A slender muscle found medial to the flexor carpi radialis; terminates in a long, flattened tendon	Medial epicondyle of humerus	Palmar aponeurosis, transcarpal ligament	Flexes wrist
Flexor carpi radialis (rā-dē-A-lis)			
A superficial muscle of the anterior forearm; fibers run diagonally and longitudinally down forearm	Medial epicondyle of humerus	Bases of second and third metacarpals	Flexes forearm and hand; abducts hand
The following five muscles are in the **extensor group:**			
Extensor carpi radialis longus			
A superficial muscle running along the posterior lateral surface of the forearm	Lateral supracondylar ridge of humerus	Dorsal surface of base of second metacarpal	Extends and abducts wrist
Extensor carpi radialis brevis (BREV-is)			
A superficial muscle posterior to the extensor carpi radialis longus	Lateral epicondyle of humerus	Dorsal surface of base of the third metacarpal	Extends and abducts wrist
Extensor digitorum communis			
A muscle found between the extensor carpi radialis brevis and extensor carpi ulnaris	Lateral epicondyle of humerus	Second and third phalanges of the four fingers	Extends wrist and fingers
Anconeus (an-KŌ-neus)			
A superficial triangular muscle on superior forearm; fibers run obliquely from medial to lateral	Posterior surface of lateral epicondyle	Olecranon process of ulna	Extends forearm
Extensor carpi ulnaris			
A superficial muscle located on the medial aspect of posterior forearm	Lateral epicondyle of humerus; dorsal surface of ulna	Base of fifth metacarpal	Extends and adducts wrist

Muscles of the Hip and Thigh   (Figures 6.11–6.13)

Name and Description	Origin	Insertion	Action
Psoas (SŌ-us) minor A deep, straplike muscle lateral to the lumbar region of the vertebral column	Bodies of T12 and L1 vertebrae	Iliac fascia and superior lateral portion of pubis	Flexes and rotates thigh laterally; flexes vertebral column; is often absent
Psoas major A long, deep spindle-shaped muscle inferior to the psoas minor	Transverse processes of L1–L5; bodies and fibrocartilages of T12–L5 vertebrae	Lesser trochanter of femur	Flexes and rotates thigh laterally; flexes vertebral column and bends it laterally
Sartorius (sar-TŌR-ē-us) An elongated superficial band of muscle running obliquely and inferiorly across the anterior thigh	Anterior superior iliac spine	Superior medial tibia	Flexes and rotates thigh laterally; flexes and rotates leg medially
Gracilis (GRAS-i-lis) A thin, elongated, superficial band of muscle running along the medial aspect of thigh	Inferior ramus of pubis and pubic symphysis	Superior medial tibia	Flexes and rotates leg medially; adducts thigh
Adductor magnus The largest of the adductor muscles; a large triangular muscle underlying the adductor longus and adductor brevis	Pubic ramus and inferior surface of ischium	Linea aspera and medial epicondyle of femur	Adducts, flexes, and medially rotates thigh
Adductor longus A flat, roughly triangular muscle extending inferiorly and laterally from the pubis toward the femur	Anterior aspect of pubis	Medial portion of linea aspera of femur	Adducts, flexes, and medially rotates thigh
Pectineus (pek-TIN-ē-us) A deep, short, flat band of muscle in the superior thigh, lateral to the adductor longus	Pubic spine	Pectineal line of femur	Adducts leg; flexes thigh
Gluteus maximus (GLOO-tē-us MAK-si-mus) The largest and most superficial gluteus muscle; comprises the major portion of buttock mass	Ilium, sacrum, and coccyx	Gluteal tuberosity of femur; fascia lata of lateral thigh muscles	Extends hip; rotates thigh laterally
Gluteus medius (MĒ-dē-us) A thick, broad muscle at the lateral surface of the pelvis; posterior one-third is covered by the gluteus maximus; common site of intramuscular injections	Superior lateral surface of ilium	Greater trochanter of femur	Abducts and rotates thigh medially

Muscles of the Hip and Thigh (Continued)

Name and Description	Origin	Insertion	Action
Gluteus minimus (MIN-i-mus)			
The smallest gluteus muscle; underlies the gluteus maximus and gluteus medius	Inferior surface of ilium	Greater trochanter of femur	Abducts and rotates thigh medially
Tensor fasciae latae (FASH-ē-e LA-te)			
A muscle of the superior lateral thigh; enclosed between layers of thick fascia	Anterior iliac crest	Iliotibial band of fascia lata	Flexes, abducts, and rotates thigh medially

The following four muscles comprise the four heads of the **quadriceps** (KWOD-ri-seps) **femoris group:**

Name and Description	Origin	Insertion	Action
Rectus femoris			
A superficial muscle of the anterior thigh; extends straight down thigh	Anterior inferior iliac spine; superior rim of acetabulum	Base of patella	Extends leg and flexes thigh
Vastus lateralis			
A large muscle forming the lateral portion of the thigh	Greater trochanter and linea aspera of femur	Lateral aspect of patella	Extends leg
Vastus medialis			
A large muscle forming the medial portion of the thigh	Linea aspera of femur	Medial aspect of patella	Extends leg
Vastus intermedius			
An elongated muscle lying beneath the rectus femoris	Anterolateral surface of femur	Base of patella	Extends leg

The following three muscles comprise the **hamstring group:**

Name and Description	Origin	Insertion	Action
Biceps femoris			
A large, long muscle located on the posterolateral aspect of the thigh; consists of two heads (long and short)	Ischial tuberosity (long head) and linea aspera of femur (short head)	Head of fibula and lateral condyle of tibia; by means of a tendon	Extends and rotates thigh laterally; flexes knee
Semitendinosus (sem-ē-ten-din-ō-sus)			
A superficial muscle located medial to the biceps femoris	Ischial tuberosity	Medial surface of superior tibial shaft	Extends thigh and flexes knee
Semimembranosus (sem-ē-mem-bran-ō-sus)			
An elongated muscle underlying the semitendinosus	Ischial tuberosity by means of a large tendon	Posterior medial condyle of tibia	Extends thigh and flexes knee

Question 6A.14

Which muscle do you sit on?

....

Figure 6.11

Superficial muscles of hip and leg, anterior view

1. sartorius
2. vastus lateralis
3. rectus femoris
4. tibialis anterior
5. gastrocnemius
6. soleus

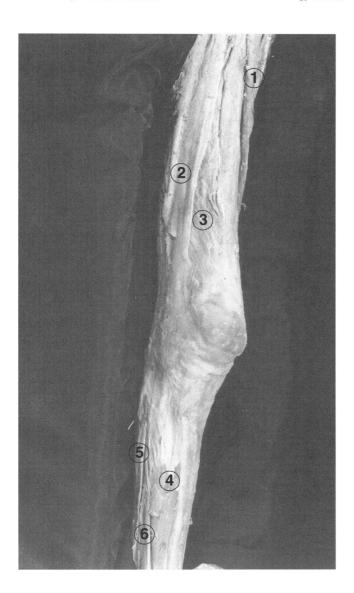

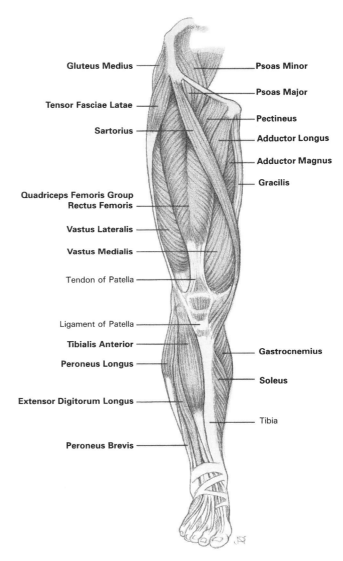

Gluteus Medius

Tensor Fasciae Latae

Sartorius

Quadriceps Femoris Group
Rectus Femoris

Vastus Lateralis

Vastus Medialis

Tendon of Patella

Ligament of Patella

Tibialis Anterior

Peroneus Longus

Extensor Digitorum Longus

Peroneus Brevis

Psoas Minor

Psoas Major

Pectineus

Adductor Longus

Adductor Magnus

Gracilis

Gastrocnemius

Soleus

Tibia

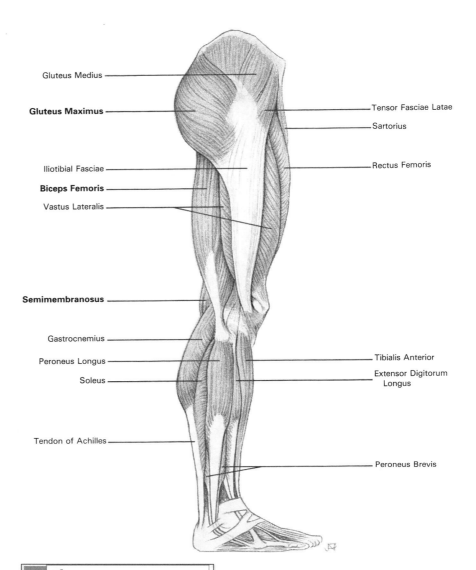

Gluteus Medius

Gluteus Maximus

Iliotibial Fasciae

Biceps Femoris

Vastus Lateralis

Semimembranosus

Gastrocnemius

Peroneus Longus

Soleus

Tendon of Achilles

Tensor Fasciae Latae

Sartorius

Rectus Femoris

Tibialis Anterior

Extensor Digitorum
Longus

Peroneus Brevis

Figure 6.12

Superficial muscles of hip and leg,
lateral view

Figure 6.13

Superficial muscles of hip and leg, posterior view

1. adductor magnus
2. semitendinosus
3. biceps femoris
4. semimembranosus
5. gastrocnemius (cut)
6. soleus

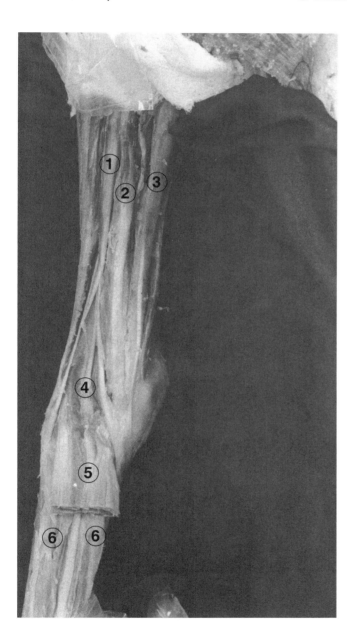
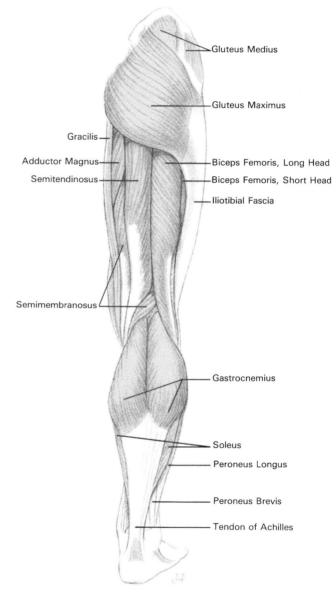

Muscles of the Lower Leg   (Figures 6.11–6.13)

Name and Description	Origin	Insertion	Action
Tibialis (tib-ē-A-lis) anterior			
Elongated superficial muscle of anterior leg	Tibial condyle and interosseous membrane	First cuneiform and first metatarsal	Dorsiflexes and inverts foot
Peroneus (per-ō-NĒ-us) longus			
Superficial muscle running along lateral aspect of lower leg	Tibial condyle and head and middle of fibula	First cuneiform and first metatarsal	Plantar flexes and everts foot
Peroneus brevis			
A small muscle lying beneath peroneus longus	Lateral surface of fibular shaft	Fifth metatarsal and fibular tuberosity	Plantar flexes and everts foot
Extensor digitorum longus			
An elongated muscle running along lateral, anterior surface of lower leg; partially covered by tibialis anterior	Tibial condyle and posterior crest of fibula	Bases of phalanges of toes	Dorsiflexes and everts foot, extends toes
Gastrocnemius (gas-tro-NĒ-mē-us or gas-trok-NĒ-mē-us)			
Large, thick, paired, superficial muscles of posterior lower leg	Medial and lateral condyles of femur	Through Achilles tendon to calcaneus	Flexes leg and foot
Soleus (SŌ-lē-us)			
A flattened muscle lying deep to the gastrocnemius	Proximal end of fibula	Through Achilles tendon to calcaneus	Flexes foot

Question 6A.15

Name the large paired muscles of the calf of the leg.

Human Muscles: Origins/Insertions/Actions

Complete this chart:

Muscle	Origin	Insertion	Action
Epicranius (frontalis)	Cranial aponeurosis	1	Raises eyebrows, wrinkles forehead
Epicranius (occipitalis)	2	Cranial aponeurosis	3
Orbicularis oris	Maxilla and mandible	4	5
6	Frontal and maxillary bones	Tissue of eyelids	7
Temporalis	8	9	Closes jaw, elevates mandible
Sternocleidomastoid	10	Mastoid process of temporal bone	11
12	Anterior surface of clavicle and sternum, ribs 1–6	13	Flexes, adducts, and rotates arm medially
External oblique	14	Iliac crest and pubis	15
16	17	Cartilages of ribs 5, 6, and 7	Flexes vertebral column and compresses abdomen
Latissimus dorsi	Spines of T6–T12 vertebrae per lumbar aponeurosis, spines of lumbar and sacral vertebrae and iliac crest	18	19
Deltoid	20	Deltoid tuberosity of humerus	21
22	Occipital bone, spines of C7–T12 vertebrae	23	Extends head, draws scapula to midline, elevates and depresses scapula
Biceps brachii	24	Radial tuberosity	25

Muscle	Origin	Insertion	Action
26	Inferior aspect, glenoid fossa (long) posterior humerus (lateral), radial groove (medial)	27	Extends forearm
Gluteus maximus	28	Gluteal tuberosity of femur, fasciae latae of lateral thigh muscles	29
Gracilis	Inferior ramus of pubis and pubic symphysis	30	31
32	Anterior superior iliac spine	33	Flexes and rotates thigh laterally, flexes and rotates leg medially
Gluteus medius	34	Greater trochanter of femur	35
36	Greater trochanter and linea aspera of femur	37	Extends leg
Vastus medialis	38	Medial aspect of patella	39
40	Anterior inferior iliac spine, superior rim or acetabulum	Base of patella	41
Semitendinosus	42	Medial surface of superior tibial shaft	43
Biceps femoris	Ischial tuberosity (long head), linea aspera of femur (short head)	44	45
46	47	First cuneiform and first metatarsal	Dorsiflexes and inverts foot
48	Medial and lateral condyles of femur	49	Flexes leg and foot
Soleus	Proximal end of fibula	50	51

After completing this chart, you may apply your knowledge by finding these muscles on a cadaver, labeling models on an articulated skeleton, or by affixing labels on yourself.

B. Cat Musculature

PURPOSE

Unit 6B will give you an understanding of the names and gross anatomical relationships of cat muscles.

OBJECTIVES

After completing Unit 6B, you will be able to

- remove the skin and fascia from the cat.
- identify the major muscles of the head, neck, trunk, and limbs.
- recognize similarities and differences in muscles found in the cat and human being.

MATERIALS

preserved cats	disposable gloves
dissection instruments	lab coat or smock
scalpel with extra blades	straight pins
scissors	muscle labels
blunt metal probe	identification tags
forceps	oil of wintergreen (methyl salicylate)
dissection trays	(optional)
newspapers	safety glasses
Scotch brand Magic Tape	twist ties
cotton-tipped applicators	disinfectant solution
Glycerol-Lysol preservative	plastic storage bags
solution	

PROCEDURE

EXERCISE 1   Skinning the Cat

Clear your work area of everything except your lab manual, newspapers placed under a dissection tray, embalmed cat, and dissecting kit. Be sure to wear gloves. Before skinning your cat, you may wish to rinse the fur under a slow stream of cold running water. Do not saturate the fur. This will reduce some of the odor of the preservative that was used to embalm the cat. Do not allow the preservative to get into your eyes.

The first step in skinning your cat is to grasp the fur in the posterior inferior neck region, along with the underlying skin, and make a superficial 1/2″ horizontal incision at the nape of the neck with your scissors or scalpel. (*Remember:* You can always cut more, but if you cut too much, you cannot put it back together.) If your cat's blood vessels were injected through vessels in the neck, you may be able to extend that same ventral incision around the neck, provided that you do not cut too deeply, damaging underlying muscles.

CAUTION
BLOODBORNE PATHOGENS

Be careful when you are making the incision that you do not cut yourself. In the event that you do cut yourself, exercise proper cleanup and disposal of body fluids.

Make the incisions described in the following list and shown in **Figure 6.14A**, **B**, and **C** (note variations in skin thickness as you cut):

A: After making the initial incision, continue cutting around the neck (1). Then make a series of collarlike incisions in the following order: around the tail (2), around the wrists (3), and around the ankles (4).

B: If your cat is a female, make a midventral incision from the incision around the neck to the anus (1). In the case of a male cat, make a shorter midventral incision: then cut around the external genitalia (2).

C: Make four incisions extending from the midventral incision, one to each of the wrist collars (1,2) and ankle collars (3,4).

Now you need to remove the skin from the underlying muscle. (As you do this, notice the tissues that are written in boldface in the following descriptions.)

Grasp the skin with your fingers or forceps, and with your blunt metal probe or the blade of the scalpel facing the skin (so that you do not accidentally cut muscle) gently cut or probe the thin **fascia**, which is part of the subcutaneous layer of the skin that binds the skin to underlying muscle. In the subcutaneous layer, you may also find **adipose tissue**; remove this. Continue to grasp and pull the skin firmly with your opposite thumb and index finger as you gently cut or probe. You may see a thin sheet of **platysma muscle**, which is brown in color, usually in the neck region, immediately under the skin; gently remove this, along with the skin. If you are working with a partner, one partner should dissect the anterior end of the cat, the other the posterior end, so that you don't get in each other's way or accidentally nick each other with your scalpels.

Continue to pull the skin with one hand and cut or probe the underlying fascia with the other, and remove the skin in one piece. If your cat is a female, you may at this point see dense gray masses of tissue overlying the chest and

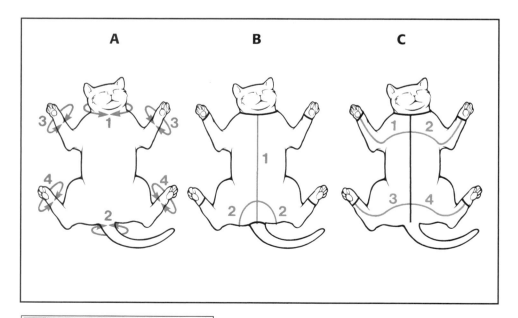

Figure 6.14

Steps in skinning the cat

abdominal muscles on each side of the midline; these are the mammary glands, which you can remove with forceps.

There are two methods for maintaining your specimen in good condition between laboratory sessions. Some instructors suggest using the skin to wrap the cat; if you do this, secure the skin in position with pins. Others do not recommend saving the skin but instead spray a preservative solution on the body of the cat and then wrap it in cloths or paper towels that have been saturated with water. Whichever method you use, the last step is to place your specimen in a plastic bag, tag it for identification, and store it.

Be sure you roll up the newspapers, rinse your dissecting pan under running water, and wipe your lab bench with a disinfectant solution *after each session of working with the cat or other specimens.*

Question   **6B.1**

Why is it important to avoid allowing the cat to dry out?

EXERCISE **2**   Identification of Muscles

Before attempting to identify muscles on your cat, you may find it helpful to prepare a label for each muscle listed in the tables of cat muscles that follow. Unless you use labels that have been made water resistant in some manner, the ink will blur and become illegible. You can use an imprinter that makes plastic labels, or use Scotch brand Magic Tape to cover labels typed or written in ink on file card quality paper.

Because muscles are paired, you can identify them on either the left or the right side of your cat. Knowledge of the general description, **origin** (usually the proximal end of the muscle), **insertion** (usually the distal end), and **action** of each muscle, as listed in the various tables in this unit will aid in muscle identification. **Figure 6.15** illustrates the bones of the cat skeleton that serve as origins and insertions of muscles. The tables that follow will assist you in finding muscles of major body regions. The tables are sequenced to allow you to find muscles that are more easily dissected first.

In order to see the muscles clearly, you will need to remove additional fascia that covers each muscle itself and also binds it to adjacent muscles. This is a tedious process that is best done by gently picking away the fascia covering the muscle with a forceps. You may find it helpful to swab the surface of the fascia-covered muscles in the area you are working with oil of wintergreen, which helps to loosen the fascia, before removing it. After the fascia has been removed, look for separations between muscles. These may appear as places where muscle fibers run in different directions, or exhibit slight color or textural differences. These separations will help determine where one muscle ends and another begins. After you have determined a line of demarcation between two adjacent muscles, insert your blunt probe between the muscles to separate them. You may also need to gently cut the fascia between muscles with a scalpel to separate them. Certain muscles, such as those of the chest wall, are thinly layered in a manner similar to shingles of a roof, and need to be handled gently. Remember that you are to separate muscles, not cut into them or remove them.

After you have identified each muscle, pin a label on it. Many students find it easiest to locate muscles of the thigh and lower leg first, because they are the most resilient. You should also observe other students' dissections and compare them with your specimen.

Muscles are named in various ways, and often the name gives you valuable information about it. A muscle may be named for

- its shape
- its action
- its location
- its origin and insertion
- the direction of its fibers
- the number of divisions it has
- any combination of these

Figure 6.15

Cat skeleton, lateral view

1. skull
2. cervical vertebrae
3. scapula
4. thoracic vertebrae
5. lumbar vertebrae
6. sacrum
7. caudal vertebrae
8. ilium
9. ischium
10. pubis

11. femur
12. fibula
13. tibia
14. calcaneus
15. metatarsals
16. metacarpals
17. radius
18. ulna
19. olecranon process
20. humerus

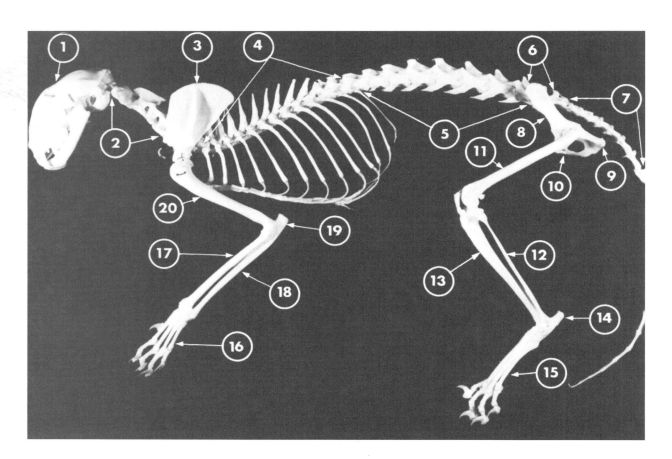

Muscles of the Neck and Throat   (Figures 6.16–6.18)

Name and Description	Origin	Insertion	Action
Sternomastoid			
A superficial band of muscle extending cranially and laterally on the ventral surface of the neck	Manubrium of sternum and midventral raphe at base of neck	Mastoid process and lambdoidal ridge of skull	Turns head when single muscle contracts; lowers head when both muscles contract
Cleidomastoid (klī′-dō-MAS-toyd)			
A flat thin muscle; fibers run beneath and lateral to the sternomastoid	Clavicle	Mastoid process (reversible origin and insertion)	Together the muscles lower the head toward the neck; singly each turns head
Sternohyoid			
A thin, superficial longitudinal band of muscle covered posteriorly by the sternomastoid; lies over the trachea	First costal cartilage and manubrium of sternum	Hyoid bone	Draws hyoid posteriorly
Masseter			
Cheek muscle	Zygomatic arch	Coronoid fossa of mandible	Elevates mandible
Digastric			
A band of muscle under chin; medial to the masseter	Mastoid process of skull	Mandible	Depresses mandible
Mylohyoid			
A thin, broad superficial sheet running transversely across the neck	Cricoid cartilage of larynx	Median raphe	Raises the floor of the mouth and draws hyoid bone forward
Stylohyoid (stī-lō-HĪ-oyd)			
A thin, ribbonlike superficial muscle running transversely over the posterior portion of the mylohyoid	Styloid process of hyoid bone	Hyoid bone	Elevates hyoid
Sternothyroid			
A thin flat band of muscle lying under the sternohyoid	Costal cartilage of sternum	Thyroid cartilage of larynx	Draws larynx posteriorly
Thyrohyoid (thī′-rō-HĪ-oyd)			
A narrow short muscle lying anterior to the sternohyoid	Thyroid cartilage of larynx	Greater cornu of hyoid bone	Elevates larynx
Cricothyroid (krī′-kō-THĪ-royd)			
A thin narrow muscle that covers the ventral surface of the larynx	Cricoid cartilage of larynx	Thyroid cartilage of larynx	Causes vocal cords to become taut

Question 6B.2

Which two muscles are used in chewing?

Question 6B.3

Which three muscles insert on the hyoid bone?

Figure 6.16

Neck region, lateral view

1. external jugular vein
2. transverse jugular vein
3. posterior facial vein
4. anterior facial vein
5. lymph node
6. submaxillary gland
7. parotid gland
8. parotid duct
9. masseter
10. clavodeltoid

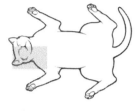

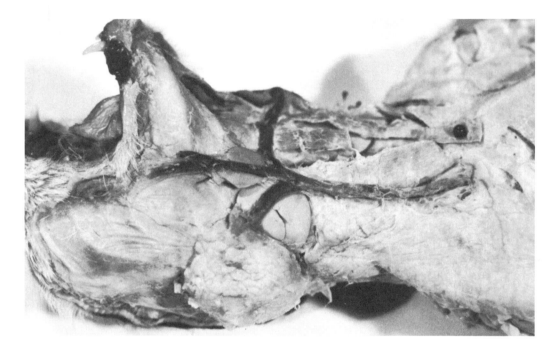

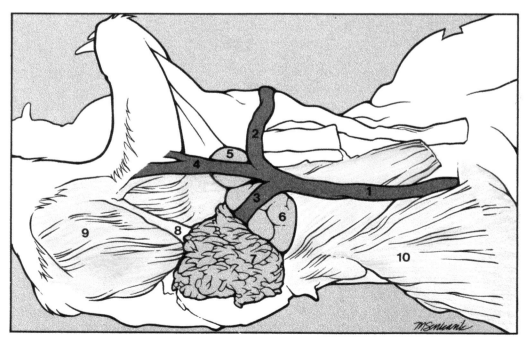

Figure 6.17

Superficial muscles of the neck and throat, ventral aspect

1. sternomastoid
2. sternohyoid
3. masseter
4. mandible bone
5. digastric
6. mylohyoid
7. clavotrapezius
8. clavodeltoid

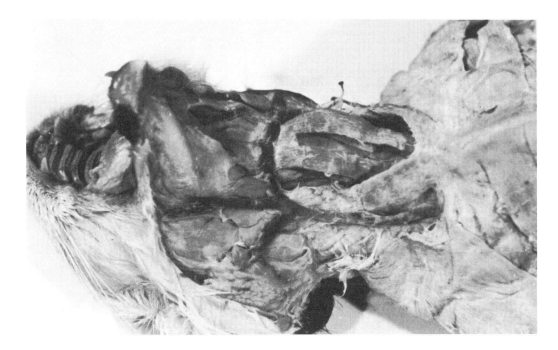
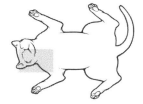
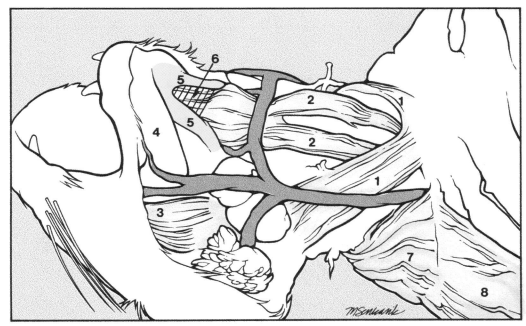

Figure 6.18

Deep muscles of the neck, ventral view

1. thyroid cartilage of larynx
2. cricoid cartilage of larynx
3. trachea
4. sternohyoid muscles, cut and reflected
5. sternothyroid
6. thyrohyoid
7. common carotid artery
8. vagus nerve
9. external jugular vein

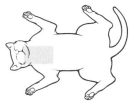

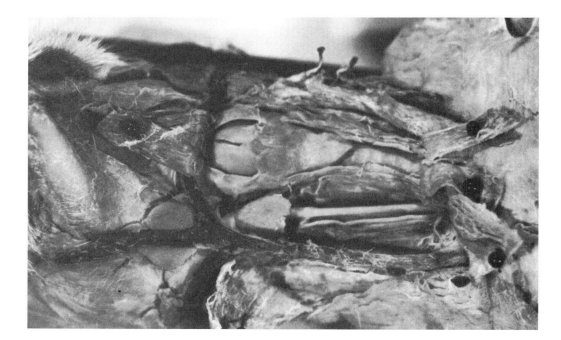

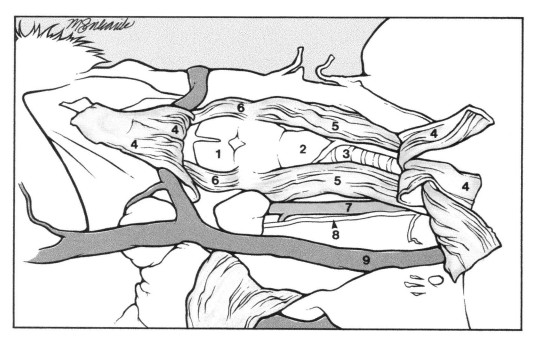

Muscles of the Chest (Pectoralis Group)   (Figures 6.19 and 6.20)

Name and Description	Origin	Insertion	Action
Pectoantebrachialis (pek′-tō-an′-tē-brā′-kē-A-lis)			
A superficial band of muscle about 1 cm wide that extends laterally from the midline and across the chest	Manubrium of sternum	Proximal end of forearm	Adducts arm
Pectoralis major			
A superficial triangular muscle caudal to the pectoantebrachialis; a portion of the pectoralis major lies under the pectoantebrachialis	Sternum and median ventral raphe	Shaft of humerus	Adducts humerus
Pectoralis minor			
A superficial muscle caudal to the pectoralis major; a portion lying deep to the pectoralis major; pectoralis minor is larger than the pectoralis major in the cat	Sternum	Bicipital groove of humerus	Adducts humerus
Xiphihumeralis (zif-ē-hyoo′-mer-AL-is)			
A band of muscle caudal to the pectoralis minor; fibers extend laterally and anteriorly and then pass deep to the pectoralis minor	Xiphoid process of sternum	Bicipital groove of humerus	Adducts humerus
Serratus (se-RĀ-tus) **ventralis**			
Fan-shaped slips of muscle deep to the pectoralis group covering the ventrolateral surface of the ribs	First nine or ten ribs and transverse processes of last five cervical vertebrae	Scapula near vertebral border	Draws scapula cranially, ventrally, and against thoracic wall

Question 6B.4

Which of the pectoralis muscles of the cat is the larger?

Figure 6.19

Superficial muscles of the chest

1. clavodeltoid (clavobrachialis)
2. pectoantebrachialis
3. pectoralis major
4. pectoralis minor

5. xiphihumeralis
6. serratus ventralis
7. latissimus dorsi

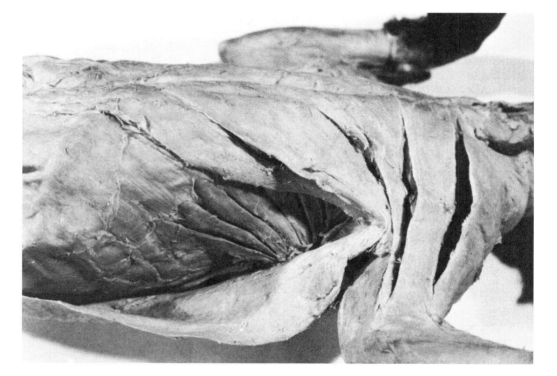

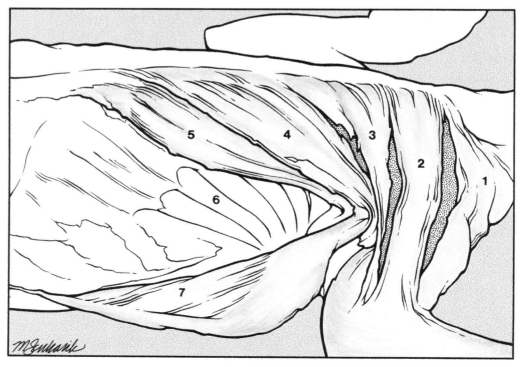

Figure 6.20

Superficial muscles of the chest, lateral view

1. xiphihumeralis
2. serratus ventralis
3. latissimus dorsi

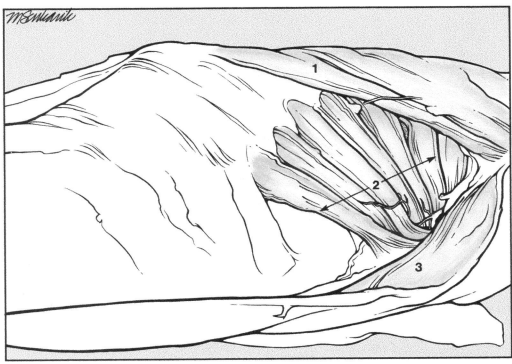

Muscles of the Abdominal Wall   (Figure 6.21)

Name and Description	Origin	Insertion	Action
External oblique A superficial sheet of muscle forming the outermost muscle of the abdominal wall	Lumbodorsal fascia and posterior ribs	Linea alba by way of aponeurosis	Constricts abdomen
Internal oblique A sheetlike muscle immediately beneath the external oblique; its muscle fibers run in a different direction from those of external oblique	Lumbodorsal fascia and border of pelvic girdle	Aponeurosis to linea alba	Compresses abdomen
Transversus abdominis Sheet of muscle beneath the fleshy portion of the internal oblique; fibers run in a transverse direction	Similar to internal oblique	Similar to internal oblique	Compresses abdomen
Rectus abdominis A superficial longitudinal band of muscle lying immediately lateral to the midventral line of abdomen	Anterior portion of pubic symphysis	Sternum and costal cartilages	Retracts ribs and sternum and compresses abdomen

Question 6B.5

Which of the above four muscles is the deepest?

Figure 6.21

Abdominal muscle detail

1. external oblique
2. internal oblique
3. transversus abdominis
4. rectus abdominis

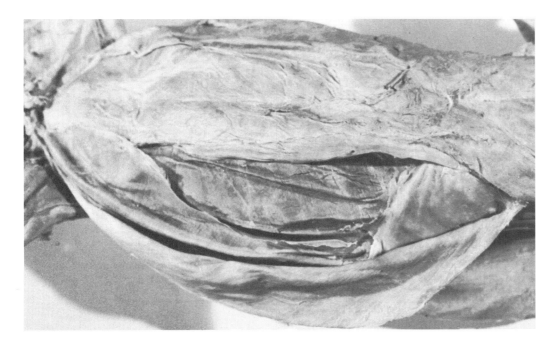

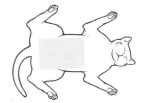

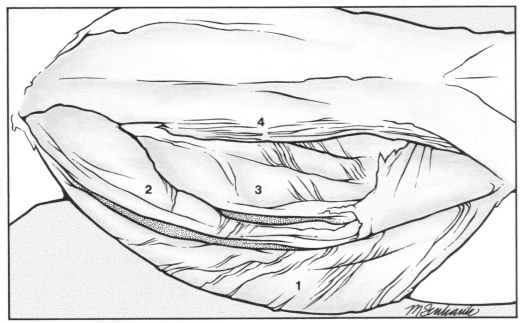

Muscles of the Upper Back, Shoulder, and Back of Neck   (Figures 6.22–6.26)

Name and Description	Origin	Insertion	Action
Latissimus dorsi A superficial, flat, triangular muscle extending from the middorsal line to the medial surface of the humerus; a portion of this muscle is covered dorsally by a smaller triangular muscle, the spinotrapezius	Neural spines of fourth thoracic to sixth lumbar vertebrae and lumbodorsal fascia	Medial surface of humerus	Retracts humerus dorsally and posteriorly
Spinotrapezius (spī-nō-tra-PĒ-zē-us) A triangular superficial muscle overlying the latissimus dorsi in the middorsal region	Spinous processes of thoracic vertebrae and spines of cervical vertebrae (1–4)	Fascia of scapula	Draws scapula medially and posteriorly toward the tail
Acromiotrapezius (a-krō-mē-ō-tra-PĒ-zē-us) A flat superficial muscle covering the scapula	Spinous process of axis to third thoracic vertebra	Metacromion process and spine of scapula	Draws scapula medially
Clavotrapezius (klav'-ō-tra-PĒ-zē-us) A flat superficial muscle on the dorsal surface of the neck	Superior nuchal line and median dorsal line of neck	Clavicle	Draws clavicle dorsally and anteriorly
Levator scapulae ventralis (ven-TRAL-is) A longitudinal band of muscle on the side of the neck; between the clavotrapezius and acromiotrapezius	Transverse occipital bone and process of atlas	Metacromion process of scapular spine	Draws scapula cranially
Supraspinatus A thick muscle occupying the supraspinous fossa of the scapula; lies deep to the acromiotrapezius	Supraspinous fossa of scapula	Greater tubercle of humerus	Draws humerus cranially
Infraspinatus A muscle occupying the infraspinous fossa of the scapula; lies deep to acromiotrapezius	Infraspinous fossa of scapula	Greater tubercle of humerus	Rotates humerus laterally
Clavodeltoid (klav'-ō-DEL-toyd) *or* **clavobrachialis** (klav'-ō-brā-kē-Ā-lis) A superficial muscle of the shoulder; continuous with the clavotrapezius at its proximal end; extends into the arm	Clavotrapezius and clavicle	Proximal end of ulna	Flexes forearm
Acromiodeltoid (a-krō'-mē-ō-DEL-toyd) A muscle posterior to the clavodeltoid; its distal portion lies under the clavodeltoid	Acromion process of scapula	Deltoid ridge of humerus	Raises and rotates humerus with spinodeltoid

Muscles of the Upper Back, Shoulder, and Back of Neck (Continued)

Name and Description	Origin	Insertion	Action
Spinodeltoid (spī′-nō-DEL-toyd)			
A band of muscle posterior to the acromiodeltoid and levator scapulae ventralis	Scapular spine	Deltoid ridge of humerus	Raises and rotates humerus
Rhomboideus			
A narrow, elongated muscle located beneath the clavo- and acromiotrapezius; separated into two parts, capitis (cranial portion) and caudalis (caudal portion)	Cervical spines to midthoracic spines	Medial border of scapula	Draws the scapula medially toward the vertebral column

Muscles of the Upper Arm   (Figures 6.22–6.27)

Name and Description	Origin	Insertion	Action
Epitrochlearis (ep′-i-trok′-lē-A-ris)			
A flat superficial muscle on the medial side of humerus	Lateral surface of latissimus dorsi	Olecranon process of ulna	Extends forearm
Biceps brachii			
A muscle lying on the ventromedial surface of the humerus; it can be more easily seen by transecting and reflecting the pectoralis muscles near the humerus; has medial and lateral heads	Glenoid fossa of scapula	Radius	Flexes forearm
Triceps brachii longus			
A muscle covering the posterior surface of the humerus	Axillary border of scapula	Olecranon process of ulna	Extends forearm
Triceps brachii lateralis			
A band of muscle covering the lateral surface of the humerus	Greater tuberosity of humerus	Olecranon process of ulna	Extends forearm
Triceps brachii medialis			
A muscle lying deep to the triceps brachii lateralis (transect the triceps lateralis to see the triceps medialis)	Shaft of humerus	Olecranon process of ulna	Extends forearm
Brachialis			
A muscle lying along the outer lateral surface of the humerus; it can be exposed by transecting and reflecting the lateral head of the triceps	Lateral surface of humerus	Ulna	Flexes forearm

Question 6B.6

How many trapezius muscles are in the cat? the human?

Figure 6.22

Detail of superficial muscles of left shoulder

1. acromiotrapezius
2. levator scapulae ventralis
3. spinodeltoid
4. acromiodeltoid
5. brachialis
6. clavodeltoid
7. clavotrapezius
8. triceps, long head
9. triceps, lateral head

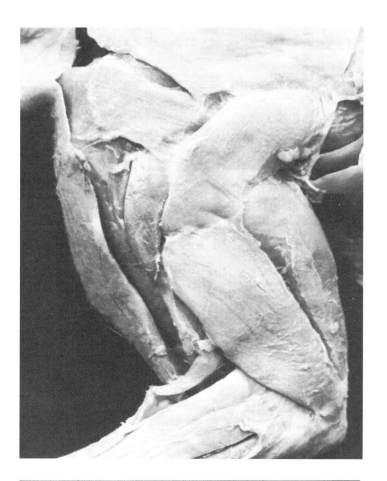

Question 6B.7

Which of the muscles listed in Figure 6.22 is absent in the human?

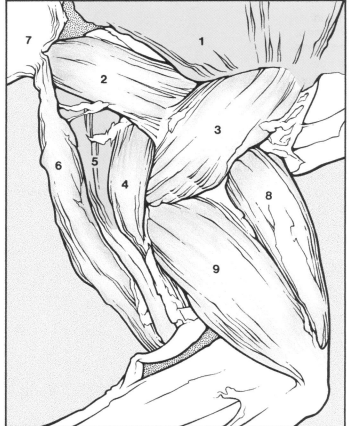

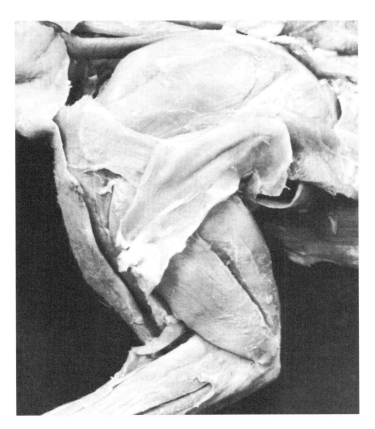

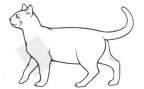

Figure 6.23

Muscles of scapular region after transection of the acromiotrapezius and spinotrapezius

1. acromiotrapezius, cut and reflected
2. spinotrapezius, cut and reflected
3. supraspinatus
4. infraspinatus
5. rhomboideus
6. clavotrapezius

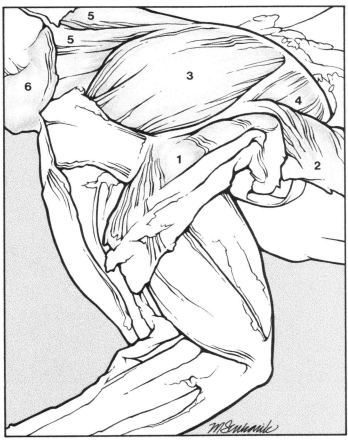

Figure 6.24

Superficial muscles of the back and upper arm, dorsal aspect

1. clavotrapezius
2. acromiotrapezius
3. spinotrapezius
4. latissimus dorsi
5. triceps brachii longus
6. supraspinatus
7. infraspinatus
8. clavodeltoid
9. acromiodeltoid
10. triceps brachii lateralis
11. levator scapulae ventralis
12. spinodeltoid

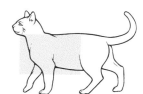

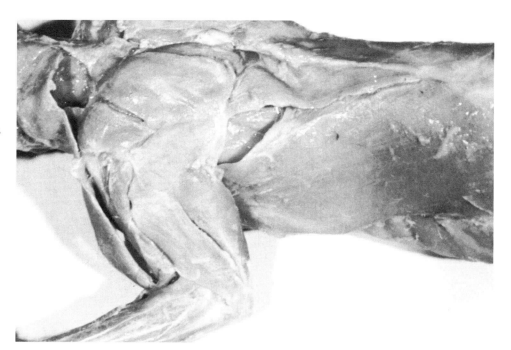

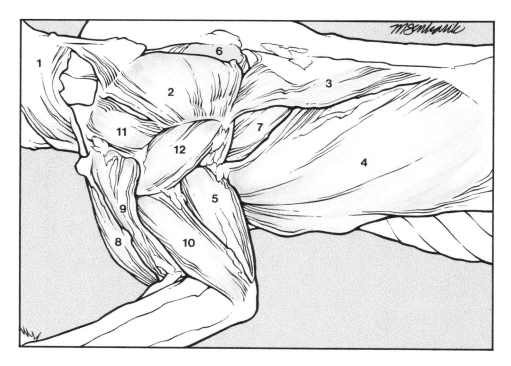

Figure 6.25

Superficial muscles of shoulder and forelimb, dorsal view

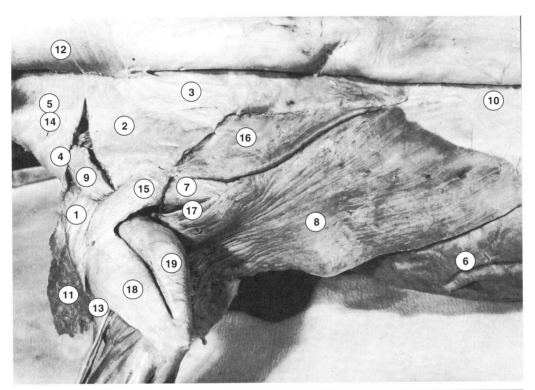

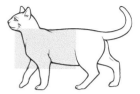

1. acromiodeltoid
2. acromiotrapezius
3. aponeurosis between right and left acromiotrapezius
4. clavodeltoid
5. clavotrapezius
6. external oblique
7. infraspinatus
8. latissimus dorsi
9. levator scapulae ventralis
10. lumbar aponeurosis
11. pectoralis major, cut
12. platysma
13. radial nerve, superficial branch
14. raphe between clavodeltoid and clavotrapezius
15. spinodeltoid
16. spinotrapezius
17. teres major
18. triceps brachii, lateral head
19. triceps brachii, long head

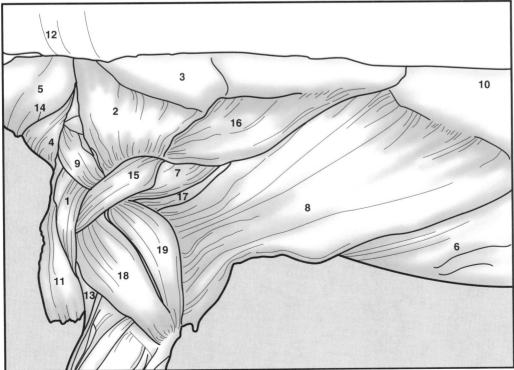

Figure 6.26

Deep muscles of the shoulder, dorsal aspect

1. acromiotrapezius, cut and reflected
2. spinotrapezius, cut and reflected
3. clavotrapezius
4. rhomboideus caudalis
5. rhomboideus capitis
6. supraspinatus
7. infraspinatus
8. teres major
9. triceps brachii, long head

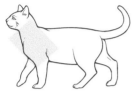

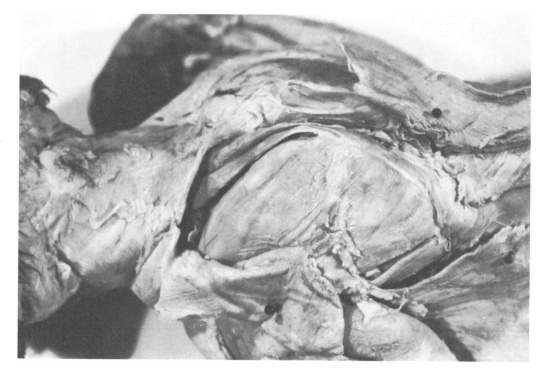

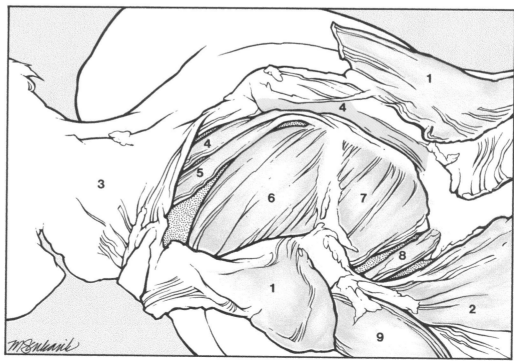

Figure 6.27

Muscles of upper arm, ventrolateral view

1. epitrochlearis
2. biceps brachii

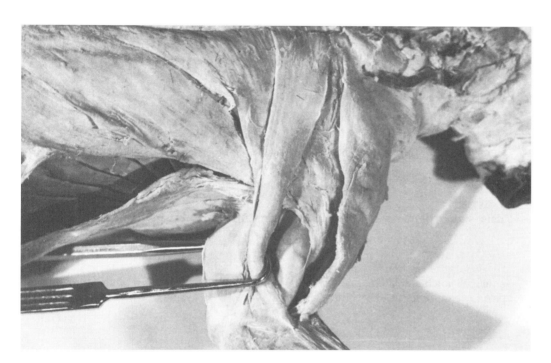

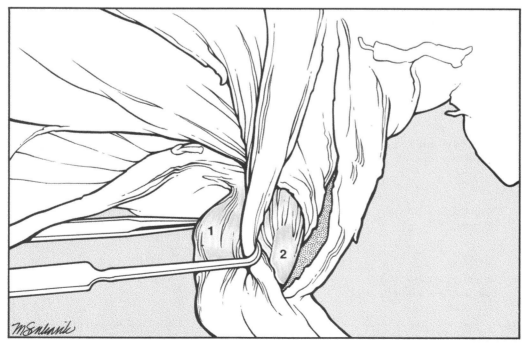

Muscles of the Forearm, Lateral Surface   (Figure 6.28)

Name and Description	Origin	Insertion	Action
Brachioradialis The most superior ribbonlike muscle of the forearm group	Middle of humerus on the dorsal surface	Distal end of radius	Supinator of hand
Extensor carpi radialis Narrow band of muscle that lies superior to the extensor digitorum communis; has two heads—short and long	Middle of humerus	Attachment by means of tendon to second and third metacarpals	Extends hand
Extensor digitorum communis (kom-YOO-nis) A thin muscle that lies superior to the extensor digitorum lateralis	Lateral surface of humerus	Attachment by means of tendons to third and fourth digits	Extends third and fourth digits
Extensor digitorum lateralis A long wedge-shaped muscle that lies superior to the extensor carpi ulnaris	Lateral surface of humerus	Attachment by means of tendons to third and fourth digits	Extends third and fourth digits
Extensor carpi ulnaris A long muscle that forms the most inferior portion of the lateral forearm	Semilunar notch and lateral epicondyle of humerus	Proximal portion of fifth metacarpal	Extends fifth digit and ulnar side of wrist

Muscles of the Forearm, Medial Surface   (Figure 6.29)

Name and Description	Origin	Insertion	Action
Pronator teres A short wedge-shaped muscle that lies above the flexor carpi radialis	Medial epicondyle of humerus	Medial surface of radius	Pronates hand (rotates radius)
Flexor carpi radialis A large narrow muscle that lies above the palmaris longus	Medial epicondyle of humerus	Second and third metacarpals	Flexes second and third metacarpals
Palmaris longus Largest of the medial forearm group; lies above the flexor carpi ulnaris	Medial epicondyle of humerus	Attachment by means of tendons to all digits	Flexes digits
Flexor carpi ulnaris A large band of muscle that covers the medial surface of the ulna	Two heads—medial epicondyle of humerus and olecranon process	Pisiform bone of wrist	Flexes digits

Question 6B.8

Which muscle opposes (has the opposite action) the brachioradialis?

Figure 6.28

Muscles of left forearm, lateral (dorsal) surface

1. brachioradialis
2. extensor carpi radialis, short head
3. extensor carpi radialis, long head
4. extensor digitorum communis
5. extensor digitorum lateralis
6. extensor carpi ulnaris

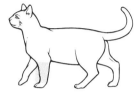

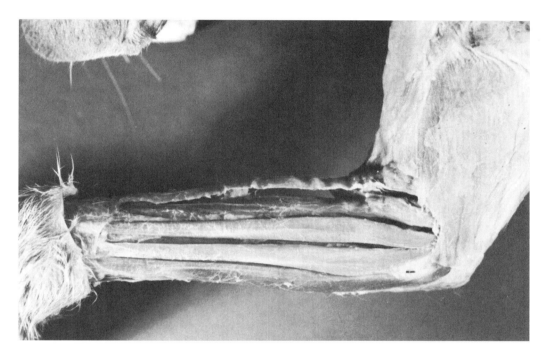

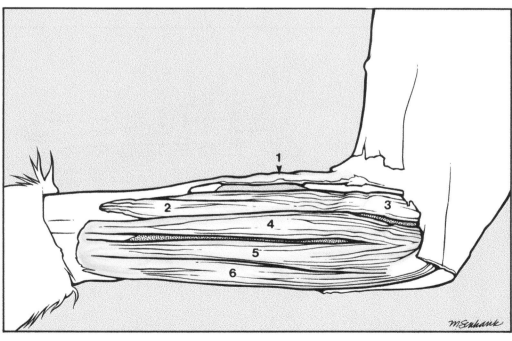

Figure 6.29

Muscles of left forearm, medial surface

1. brachioradialis
2. extensor carpi radialis, long head
3. extensor carpi radialis, short head
4. pronator teres
5. flexor carpi radialis
6. palmaris longus
7. flexor carpi ulnaris, second head
8. flexor carpi ulnaris, first head

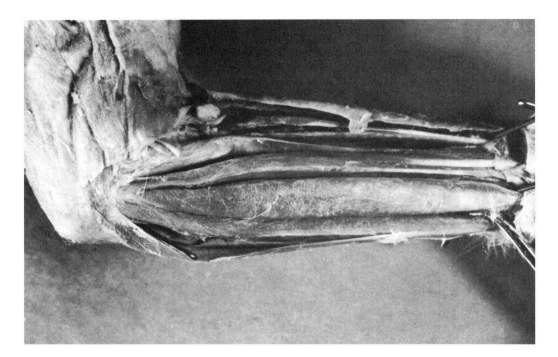

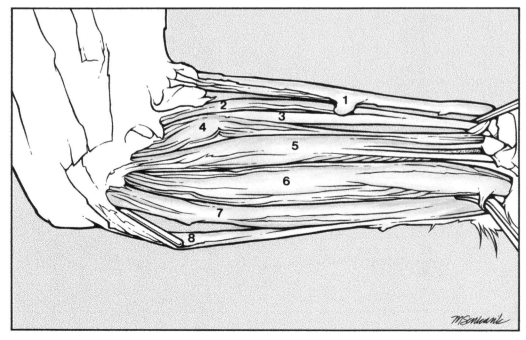

Muscles of the Thigh, Dorsal Aspect   (Figures 6.30–32)

Name and Description	Origin	Insertion	Action
Tensor fasciae latae A triangular muscle on the lateral side of the hip medial to the tough white fascia (fascia lata) covering the anterior portion of the thigh	Ilium and fascia	Fascia lata	Tightens fascia lata, extends thigh
Biceps femoris A large thick muscle covering the lateral surface of the thigh, caudal to the tensor fasciae latae; one of the hamstring muscles	Tuberosity of ischium	Patella to shaft of tibia	Abducts thigh and flexes shank
Caudofemoralis (kaw′-dō-fem′-or-AL-is) A small muscle cranial to the dorsal portion of the biceps femoris	Transverse processes of second and third caudal vertebrae	Patella	Abducts thigh, extends shank
Gluteus maximus A small muscle cranial to the caudofemoralis	Fascia and transverse processes of last sacral and first caudal vertebrae	Fascia lata and slightly on greater trochanter	Abducts thigh
Gluteus medius A thick muscle dorsal to the tensor fasciae latae	Crest and lateral surface of ilium	Greater trochanter	Abducts thigh

Question 6B.9

Which of the above muscles is most posterior?

Figure 6.30

Superficial thigh muscles, dorsal aspect of left leg

1. sartorius, cut
2. gluteus medius
3. gluteus maximus
4. caudofemoralis
5. tensor fasciae latae
6. vastus lateralis
7. biceps femoris, cut

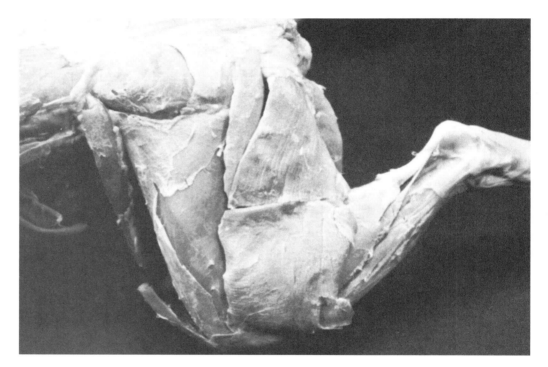

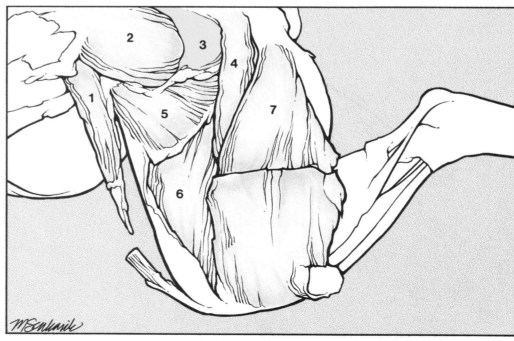

Figure 6.31

Superficial muscles of thigh, lateral view

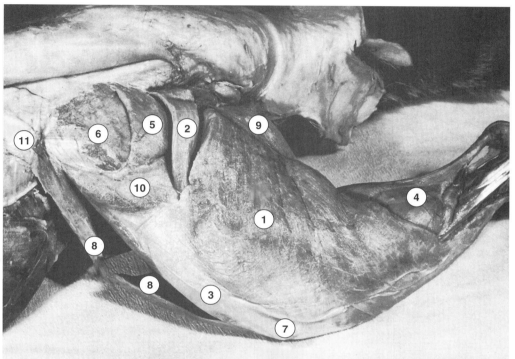

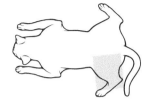

1. biceps femoris
2. caudofemoralis
3. fascia lata
4. gastrocnemius
5. gluteus maximus
6. gluteus medius
7. patella
8. sartorius, cut
9. semitendinosus
10. tensor fasciae latae
11. thoracolumbar fascia

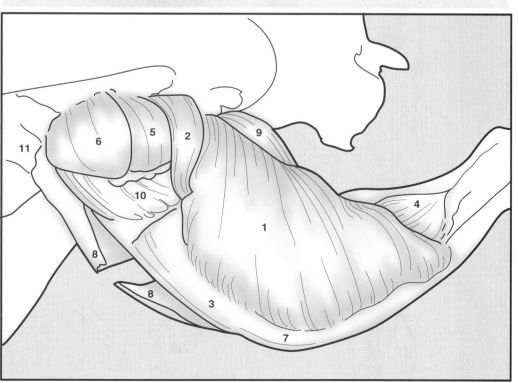

Figure 6.32

Deep muscles of left thigh, dorsal aspect

1. sartorius
2. gluteus medius
3. gluteus maximus
4. caudofemoralis
5. tensor fasciae latae
6. vastus lateralis, exposed
7. biceps femoris, cut
8. semitendinosus
9. sciatic nerve
10. gastrocnemius
11. tendon of Achilles

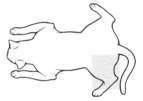

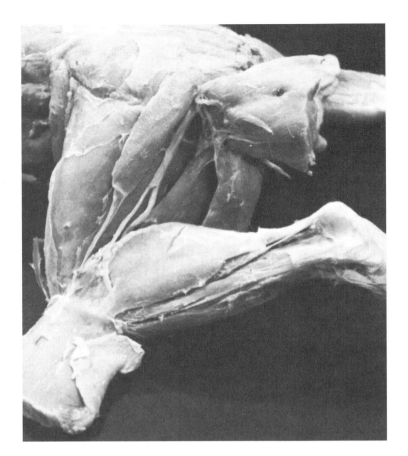

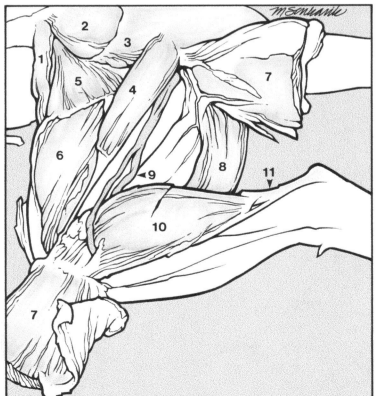

Muscles of the Thigh, Ventral Aspect   (Figures 6.33–6.35)

Name and Description	Origin	Insertion	Action
Sartorius			
A flat superficial band of muscle covering the lateral ventral portion of the thigh	Crest and ventral border of ilium	Proximal end of tibia	Adducts and rotates thigh, extends shank
Gracilis			
A flat superficial band of muscle covering the medial ventral portion of the thigh	Pubic symphysis and ischium	Aponeurosis to tibia	Adducts and retracts leg
The following four muscles comprise the four heads of the **quadriceps femoris group:**			
Vastus lateralis			
This large muscle can be seen better from the dorsal aspect; it is covered by the fascia lata of the tensor fasciae latae muscle	Greater trochanter and shaft of femur	Patella and patellar ligament	Extends shank
Rectus femoris			
An elongated muscle bordered laterally by the vastus lateralis and medially by the vastus medialis	Ilium anterior to acetabulum	Patella and patellar ligament	Extends shank
Vastus intermedius			
A flat muscle deep to the rectus femoris	Shaft of femur	Patella and patellar ligament	Extends shank
Vastus medialis			
A band of muscle on the medial portion of the thigh, deep to the sartorius	Femur	Patella	Extends shank
Adductor longus			
A narrow muscle band lying lateral to the adductor femoris	Pubis	Femur	Adducts thigh
Adductor femoris			
A muscle deep to the gracilis, proximal to the trunk of the cat	Inferior rami of pubis and ischium	Shaft of femur	Adducts thigh
Semimembranosus			
A large thick muscle deep to the gracilis, caudal to the adductor femoris; one of the hamstring muscles	Ischium	Medial epicondyle of femur	Extends thigh
Semitendinosus			
A stout band of muscle caudal to the semimembranosus; more easily seen from the ventral surface; one of the hamstring muscles	Ischial tuberosity	Proximal end of tibia	Flexes shank

Figure 6.33

Superficial muscles of right thigh, ventral aspect

1. sartorius
2. gracilis
3. gastrocnemius
4. tibialis anterior
5. flexor digitorum longus
6. tibia

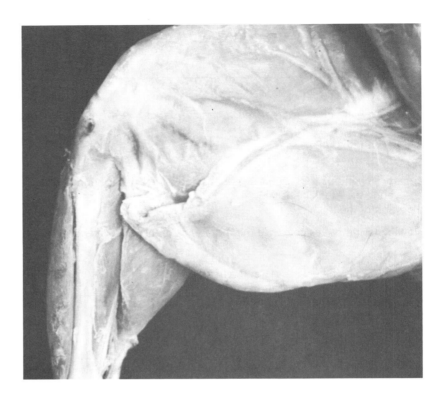

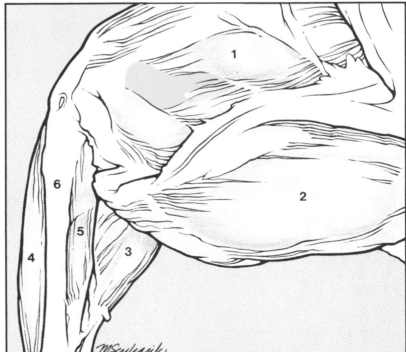

Question 6B.10

What is the action of the quadriceps femoris group?

Figure 6.34

Superficial muscles of thigh, medial view

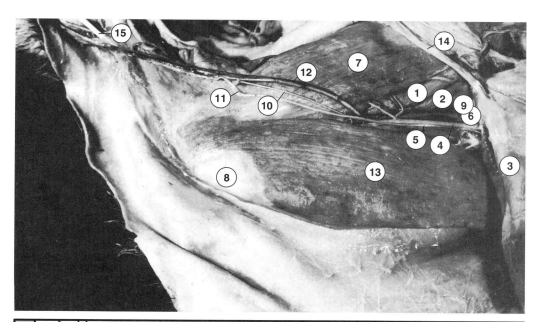

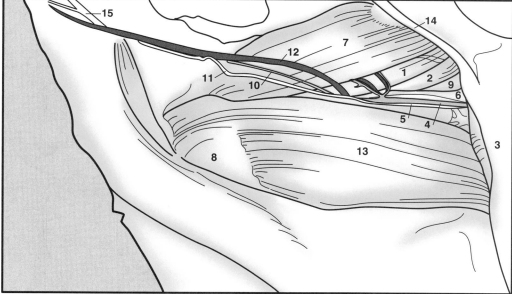

1. adductor femoris
2. adductor longus
3. aponeurosis of external oblique
4. femoral artery
5. femoral nerve
6. femoral vein
7. gracilis
8. patella
9. pectineus
10. saphenous artery
11. saphenous nerve
12. saphenous vein
13. sartorius
14. spermatic cord
15. tendon of tibialis anterior

Figure 6.35

Deep muscles of left thigh, ventral aspect

1. sartorius, cut and transected
2. gracilis, cut and transected
3. tensor fasciae latae
4. vastus lateralis
5. rectus femoris
6. vastus medialis
7. adductor longus
8. adductor femoris
9. semimembranosus
10. semitendinosus

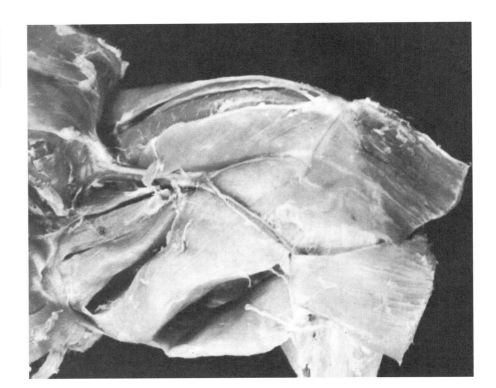

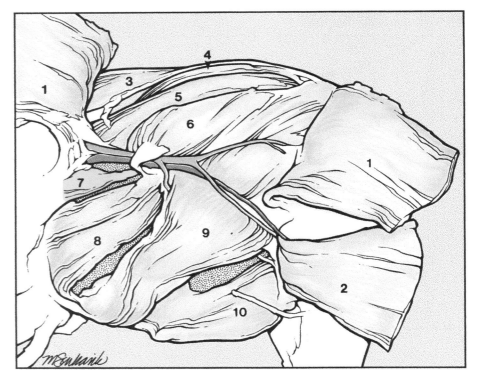

Muscles of the Lower Leg   (Figure 6.36)

Name and Description	Origin	Insertion	Action
Tibialis anterior A thin muscle lying along the lateral border of the tibia	Proximal portion of tibia	First metatarsal	Flexes foot
Flexor digitorum longus An elongated muscle running medial to the tibialis anterior	Posterior surface of tibial shaft below popliteal line	Distal phalanx of toes	Flexes and inverts foot
Extensor digitorum longus A flat spindle shaped muscle lying ventrolateral to the tibialis anterior muscle	Tibia and fibula	All four digits	Extends phalanges
Peroneus group (longus, brevis, tertius) A slender cylinder-shaped muscle located on the lateral surface of the leg	Lateral portion of fibula	Metatarsals and digits	Extends foot and flexes phalanges
Gastrocnemius A large muscle mass on the dorsal surface of the foreleg; has medial and lateral heads	Lateral and medial condyles of femur	Tendon of Achilles on calcaneus	Extends foot
Soleus A flattened muscle lying deep to the gastrocnemius	Lateral portion of fibula	Tendon of Achilles on calcaneus	Extends foot

Question 6B.11

Which of the above muscles originates on a bone in the thigh?

Figure 6.36

Muscles of right lower leg, lateral aspect

1. gastrocnemius
2. soleus
3. peroneus longus
4. peroneus brevis
5. extensor digitorum longus
6. tibialis anterior
7. tendon of Achilles

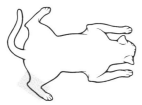

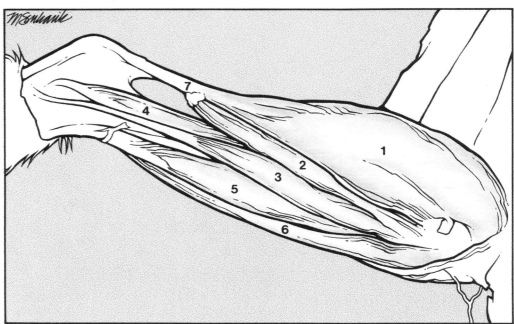

C. The Physiology of Muscle Contraction

PURPOSE

The purpose of Unit 6C is to enable you to understand the physiology of muscle contraction.

OBJECTIVES

After completing Unit 6C, you will be able to

- name the ions that must be present for normal muscle contraction to occur.
- understand the importance of ATP in muscle contraction.
- calculate the degree of contraction in a psoas muscle preparation.
- induce and interpret various responses to muscle stimulation by electrical and chemical methods.

MATERIALS

microscope slides   electronic recording device
glass probes or sharp probes   5% acetic acid
psoas muscle-preparation kit*   millimeter rulers
ATP, KCl, MgCl$_2$ solutions from kit

PROCEDURE

CHARACTERISTICS OF MUSCLE CONTRACTION

EXERCISE 1   **Chemistry of Muscle Contraction**

In order for a muscle to contract, **ATP (adenosine triphosphate)** must be broken down into ADP and phosphate with the release of energy. Sources of ATP in muscle include glycogen and creatine phosphate. Certain ions also must be present.

1. Follow your instructor's directions for the preparation of needed solutions or use the solutions in your prepared kit.
2. Place a glycerinated psoas muscle fiber in a small amount of glycerol on a microscope slide.
3. Measure the lengths of the muscle fibers in millimeters.
4. Flood the fibers with several drops of the solutions containing ATP, potassium chloride, and magnesium chloride.
5. After 30 seconds, measure the length of the fibers again. If there is a difference, calculate the percentage contraction.
6. Repeat steps 2–5, using (1) ATP alone and (2) potassium and magnesium salts alone.

*This kit may be obtained from Carolina Biological Supply Co., Burlington, N.C.

EXERCISE 2   **Muscle Tone and Graded Muscle Contraction**

A healthy muscle always exhibits **muscle tone**, which is a reflex reaction resulting in a slight tautness. A series of involuntary, continuous stimuli to the muscle prevents it from becoming totally relaxed.

Graded muscle contractions are responses of whole muscles and are caused by the contraction of varying numbers of motor units, the frequency of stimuli and the condition of the muscle fibers at the time of the contraction.

Have your lab partner perform the following activities. Record his or her responses in the following table.

Muscle Tone and Graded Muscle Contraction	
Activities	**Responses** **(or Degree of Contraction: none, +, ++, +++)**
1. Have your lab partner rest his/her flexed arm on the lab table. Palpate his/her biceps, then record the degree of contraction.	
2. Keeping the arm flexed, have your partner hold one lab manual. Palpate his/her biceps, then record the degree of contraction.	
3. Keeping the arm flexed, have your partner hold three lab manuals. Palpate his/her biceps, then record the degree of contraction.	
4. In which activity did your partner exhibit muscle tone?	
5. In which activity was the degree of contraction the greatest? Why?	
6. Explain the difference between muscle tone and a stronger contraction.	

EXERCISE 3 Muscle Fatigue

Muscle fatigue results when a muscle can no longer respond to stimuli and contract. Some causes of fatigue are depletion of ATP and/or metabolic energy sources and buildup of metabolic waste products such as lactic acid.

Have your lab partner perform the following activities. Record each response in the table below.

Muscle Fatigue

Activities	Response
1. Record the time you start.	
2. Count the number of times your partner, holding his/her lab book, flexes and extends his/her arm with resting the arm on the table. Continue until he/she can no longer do so.	
3. Record the time at which he/she can no longer flex and extend his/her arm.	
4. How long did it take for your lab partner's muscles to fatigue?	

5. Have your lab partner rest for 2 minutes. Repeat steps 1 through 4 again; record your data.
  — start time:
 	
  — number of flexions and extensions:
 	
  — length of time to fatigue:	

6. What muscles fatigued?	
7. Was there a difference between the length of time for the muscles to fatigue the first time? The second time?	
8. Explain the difference in time between the first and second fatigue.	

EXERCISE 4   Phases of Skeletal Muscle Contraction

A single muscle contraction is known as a **muscle twitch**. A normal muscle twitch has three phases, as shown in the graph following. Point 1 is the point of stimulation of the muscle by electrical, mechanical, or chemical means. Distance 1–2 is the **latent period**, during which electrical and chemical changes take place within the muscle prior to actual contraction. Distance 2–3 represents the **contraction period**, when the muscle contracts, and distance 3–4 is the **relaxation period**, when the muscle returns to its resting length. Notice that the muscle twitch lasts approximately 0.1 second. The amount of stimulus necessary for a muscle to contract is referred to as the **threshold stimulus** or **liminal stimulus**. A stimulus of less than threshold strength is known as **subthreshold** or **subliminal**. According to the graph, which phase is the longest, and why?

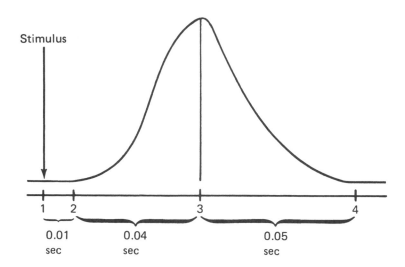

The Muscular System

MATCHING

Name _____

Choose the correct response for questions 1–5 from the following muscles:

Section _____

Date _____

    a. sternocleidomastoid   **c.** masseter
    b. gastrocnemius   **d.** deltoid

_____   **1.** Has clavicle and scapula as its origin and humerus as its insertion

_____   **2.** Major muscle of chewing

_____   **3.** Extends foot

_____   **4.** Muscle used to turn the head

_____   **5.** Calf muscle

Choose the correct response for questions 6–8 from the following terms:

    a. muscle twitch   **c.** ATP
    b. tetanus   **d.** lactic acid

_____   **6.** Sustained contraction of a muscle

_____   **7.** Single muscle contraction produced by a single stimulus

_____   **8.** Muscle fatigue is caused by an accumulation of this substance

MULTIPLE CHOICE

_____   **9.** Which of these muscles is present in the cat and absent in the human?
    a. pectoralis major   **c.** epitrochlearis
    b. gluteus maximus   **d.** sartorius

_____   **10.** Which of these muscles is named according to its function?
    a. sternomastoid   **c.** triceps brachii
    b. semimembranosus   **d.** adductor femoris

_____   **11.** The origin of a flexor muscle is located
    a. proximal to the insertion.   **c.** lateral to the insertion.
    b. distal to the insertion.   **d.** at its insertion.

_____   **12.** Which muscle type can be microscopically identified by the presence of prominent A bands and peripherally located nuclei?
    a. cardiac   **c.** skeletal
    b. smooth   **d.** all of the above

_____ 13. Pointing the toes (plantar flexion) involves contraction of which of these muscles?
a. tibialis anterior and extensor digitorum longus
b. gastrocnemius and soleus
c. peroneus longus and semitendinosus
d. calcaneus and peroneus tertius

_____ 14. The longest phase of a skeletal muscle twitch is the
a. latent period.
b. contraction period.
c. relaxation period.
d. they are all the same length.

_____ 15. Which of these represents largest to smallest in diameter?
a. fascicle, fiber, myofibril, myofilament
b. fiber, fascicle, fibril, actin
c. actin, myofilament, fiber, fascicle
d. actin, myosin, myofilament, fibril

_____ 16. Which of these is *not* a muscle of the quadriceps femoris group?
a. rectus femoris
b. vastus lateralis
c. vastus intermedius
d. rectus abdominis

_____ 17. In an adult, intramuscular injections are often given into which of these muscles?
a. gracilis
b. gluteus medius
c. latissimus dorsi
d. biceps femoris

_____ 18. In the chemistry of muscle contraction exercise, the muscle fibers that contracted to the greatest extent were those treated with
a. ATP alone
b. ATP and magnesium
c. magnesium and potassium
d. ATP, magnesium, and potassium

_____ 19. Energy sources for skeletal muscular contraction include
a. creatine phosphate.
b. ATP.
c. glycogen.
d. all of the above.

_____ 20. The actual contractile elements of a skeletal muscle cell are the
a. tonofibrils.
b. sarcoplasm.
c. desmosomes.
d. myofibrils.

_____ 21. Which of the following is associated only with cardiac muscle?
a. intercalated discs
b. the lack of distinct myofibrils
c. reduced sarcoplasmic reticulum
d. sarcomeres

_____ 22. The point of anatomical proximity between a nerve and muscle is the
a. sarcoplasm.
b. motor end plate.
c. epineurium.
d. epimysium.

_____ 23. Which of the following is *not* a phase of a simple muscle twitch?
a. period of relaxation
b. period of contraction
c. latent period
d. summation period

_____ 24. Which muscle is *not* located on the posterior surface of the arm?
   a. extensor carpi ulnaris   c. flexor carpi ulnaris
   b. triceps longus   d. extensor carpi radialis

_____ 25. The sheet of dense fibrous connective tissue that laterally braces the thigh to maintain balance is the
   a. vastus lateralis   c. semimembranosus
   b. iliotibial fasciae   d. tensor fasciae latae

DISCUSSION QUESTIONS

26. Draw a sarcomere. Label the following: A band, I band, and Z lines.

27. Name the muscles in the following groups:
   a. hamstring _____

   _____

   b. quadriceps femoris _____

   _____

28. Listed below are several cat muscles. Compare them to human muscles with respect to presence or absence, location, number of heads, divisions, or size:
   a. pectoralis major _____

   _____

   b. pectoralis minor _____

   _____

   c. pectoantebrachialis _____

   _____

   d. xiphihumeralis _____

   _____

   e. sternomastoid _____

   _____

   f. clavotrapezius, acromiotrapezius, spinotrapezius _____

   _____

   g. levator scapulae ventralis _____

   _____

   h. clavodeltoid, acromiodeltoid, spinodeltoid _____

   _____

   i. biceps brachii _____

   _____

   j. caudofemoralis _____

   _____

   k. gluteus maximus _____

   _____

   l. sartorius _____

   _____

   m. adductor femoris _____

   _____

   n. epitrochlearis _____

   _____

CASE STUDY

MUSCULAR SYSTEM

Michael Johnson is a 25-year-old weight lifter. He pulled his rotator cuff during an exercise workout.

1. **What muscles were injured?**

2. **What tissues were involved?**

3. **Describe the problems involved in normal healing in this kind of injury.**

The Nervous System

A. Nervous System Anatomy

PURPOSE

Unit 7A will familiarize you with the histological and gross anatomical features of the nervous system.

OBJECTIVES

After completing Unit 7A, you will be able to
- identify the characteristics of a neuron.
- identify the features of the spinal cord.
- differentiate between cerebral and cerebellar tissue.
- identify the meninges.
- identify the major gross anatomical characteristics of the brain.
- identify selected cranial and spinal nerves.

MATERIALS

prepared slides of motor neurons (nerve smear), and myelinated nerve fibers
slides of cross sections of spinal cord, cerebrum, and cerebellum
preserved sheep or beef brains with meninges attached (if available)
ox spinal cord (preserved)
preserved cats
dissecting microscope
compound microscopes
disposable gloves
human cadaver (if available)
model of human vertebral column with spinal nerves
models of human motor neuron, spinal cord, and brain
dissecting kits

PROCEDURE

Anatomically, the nervous system is divided into the central nervous system (CNS) and the peripheral nervous system (PNS). The CNS is composed of the brain and the spinal cord, which is a continuation of the brain. The PNS is composed of 12 pairs of cranial nerves, which extend outward from the brain, and 31 pairs of spinal nerves, which are extensions from the spinal cord. The autonomic nervous system (ANS) is part of the peripheral nervous system.

The basic structural and functional cell of all nervous tissue is the **neuron**, which is irritable and capable of conducting electrochemical impulses. Neurons are characterized in a number of ways: by the direction of impulse flow, by the number of processes or extensions, and by the presence or absence of myelin sheaths.

Another group of cells is composed of specialized connective tissue called glia or neuroglia. Glia support neurons by performing various functions. For example, **astrocytes** help form the blood–brain barrier that regulates the movement of materials into the brain. **Ependymal cells** produce protective myelin sheaths around axons in the central nervous system. **Microglia** are responsible for phagocytosis.

In each of these exercises, find the structures printed in boldface.

Figure 7.1

Motor nerve cells of the spinal cord (430×) (Color Plate 31)

1. dendrites
2. cytoplasm of cell body
3. nucleolus
4. nucleus
5. Nissl substance
6. axon
7. nuclei of neuroglial cells

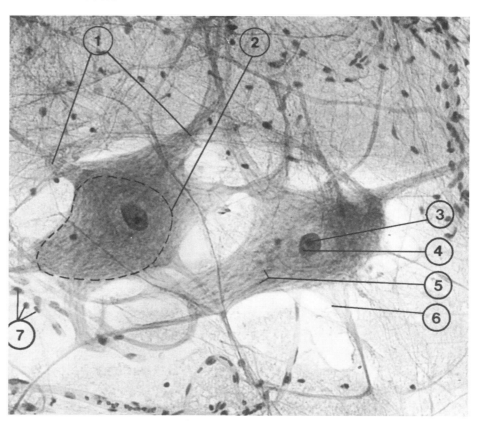

EXERCISE **1**   **Cytological Characteristics of Neurons**

Examine a prepared slide of neurons. The neurons on your slide are probably **motor neurons**, which carry impulses *away* from the spinal cord. Using high power carefully focus your slide. Using oil immersion or high power locate the large, stellate (starlike) **cell body**. Extending from the cell body are **processes**. Some neurons in the embryo have one process, an axon, and are therefore unipolar. The neurons of the retina of the eye are bipolar, in that they have two processes, an axon and a dendrite. Most neurons are **multipolar**, in that they have many processes. **Dendrites** transmit impulses to the cell body; **axons** transmit impulses away from the cell body. In multipolar neurons there is only one axon and many dendrites.

The large **nucleus** within the cell body contains a **nucleolus**. In the **cytoplasm**, **Nissl substance**, which is composed of granules of RNA, can be seen (**Figure 7.1** and **7.3**). On the slide of myelinated nerve fibers locate the **Schwann cells**, which appear as larger oval-shaped areas, with strands of **myelin sheath** running parallel to them (**Figure 7.2**).

On the motor neuron model, find the boldfaced structures listed above. On the model, the **myelin sheath** may appear as a segmented, yellowish fat-containing protective wrapping around the **axon**. Observe the **neurilemma**, which plays a role in axon regeneration of peripheral nerves, that surrounds the myelin sheath. Large **Schwann cells**, from which both myelin and neurilemma originate, are peripherally located along the axon. The constrictions between myelin sheaths are **nodes of Ranvier**. Axons may branch into **collaterals**. **Synaptic end bulbs** terminate the axons and collaterals. These end bulbs function in the transmission of impulses across the synapse, which is the space between neurons. Find the **axon hillock**, which connects the cell body with the axon, the **axis cylinder**, and **axon terminals** (**Figure 7.3**).

Question **7A.1**

Draw and label a neuron and its main components.

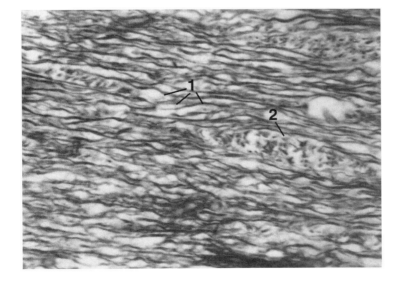

Figure 7.2

Longitudinal section of myelinated nerves (1000×)

1. myelin sheath
2. Schwann cell

Figure 7.3

Motor neuron

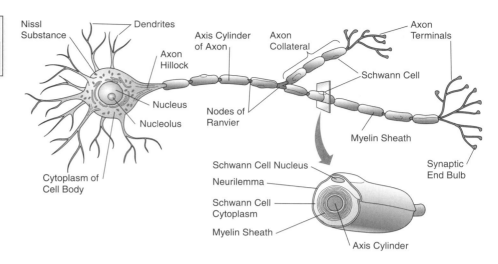

EXERCISE **2**

Microscopic Examination of Cerebrum and Cerebellum (Color Plates 32 and 33)

Observe a slide of cerebrum. The **cerebral cortex** forms the surface of the cerebrum of the brain. It is arranged in layers from the outer **gray matter** under the **pia mater,** which is the deepest of the three meningeal membranes covering the brain, to the inner **white matter** layer of the **corpus callosum.** In the middle layers, **pyramidal cells** predominate (**Figure 7.4**). Note that the **dendrites** of the pyramidal cells tend to project peripherally from the **perikaryon,** and the **axon** projects interiorly (**Figures 7.4** and **7.5**).

Both gray and white matter are found in the central nervous system. Gray matter is composed of cell bodies of neurons and nonmyelinated fibers. White matter is arranged in bundles of myelinated nerve fibers known as **tracts.**

Observe a slide of cerebellar tissue. The cortical region is composed of gray matter, which has two layers: the outer molecular layer and the deeper granular layer. **Purkinje cells** are found in a narrow layer at the interface between these two layers. White matter of the **cerebellum** underlies the granular layer of the cortex. (**Figures 7.6–7.8**). The axons of the Purkinje cells project interiorly into the white matter. The dendrites of these cells tend to branch repeatedly and project peripherally.

Question 7A.2

Where are pyramidal cells located?

Question 7A.3

Where are Purkinje cells located?

Figure 7.4

Human cerebral cortex (100×)
(Color Plate 32)

1. pyramidal cells of cortex
2. white matter

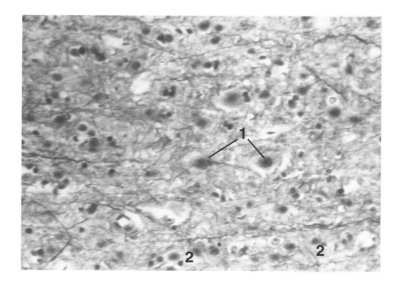

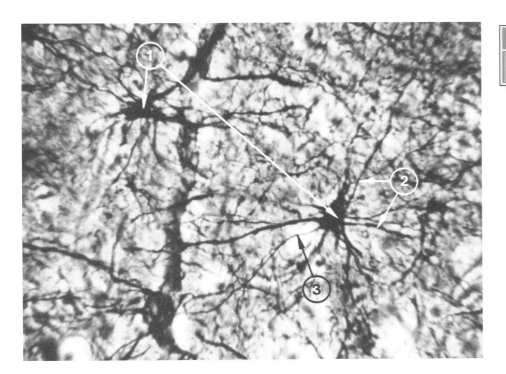

Figure 7.5

Detail of pyramidal cells in cerebrum (1000×)

1. pyramidal cells
2. dendrites of pyramidal cell
3. axon of pyramidal cell

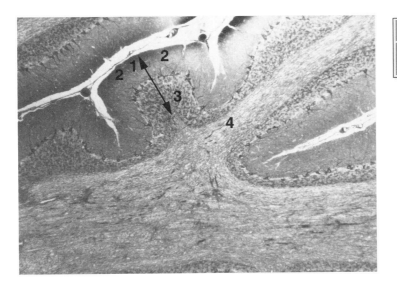

Figure 7.6

Gray and white matter of cerebellum (40×)

1. gray matter of cortex
2. cortex, molecular layer
3. cortex, granular layer
4. white matter

Figure 7.7

Purkinje cells of cerebellum (100×)

1. Purkinje cells
2. molecular layer
3. granular layer
4. white matter

Figure 7.8

Detail of Purkinje cells of cerebellum (400×)

1. nucleus of Purkinje cell
2. dendrite of Purkinje cell
3. molecular layer
4. granular layer

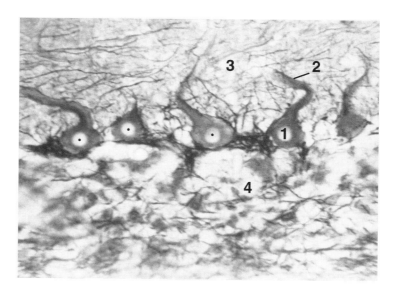

Question 7A.4

Draw and label a cross section of a spinal cord.

EXERCISE 3   **The Spinal Cord**

   Observe a cross section of **spinal cord** under a dissecting microscope. Note the more pronounced anterior indentation, so that you can position the slide with this feature toward you. This indentation is the **anterior median sulcus.** The **posterior median sulcus** is not as deep. Notice the central gray matter, which is shaped like the letter H. The **central canal** is in the center of the horizontal band of gray matter. The central canal is part of the circuit through which cerebrospinal fluid circulates in the central nervous system. It is surrounded by white matter that is divided into **tracts.** The **dorsal gray columns,** or **horns,** are elongated, dorsal extensions of the "H," and the **ventral horns** are rounded, ventral extensions. **Lateral horns** are triangular lateral extensions between the dorsal and ventral horns. Fibers of **afferent,** or **sensory, nerves** enter the spinal cord at the dorsal horns. Fibers of the **efferent,** or **motor nerves** exit the spinal cord from the ventral horns (**Figures 7.9** and **7.10**).

Figure 7.9

Slide of human spinal cord, transverse section at the cervical region (100×)

1. posterior median sulcus
2. posterior median septum
3. white matter tract
4. posterior gray horn
5. motor neurons

6. anterior gray horn
7. central canal
8. gray matter
9. anterior median sulcus
10. lateral horn

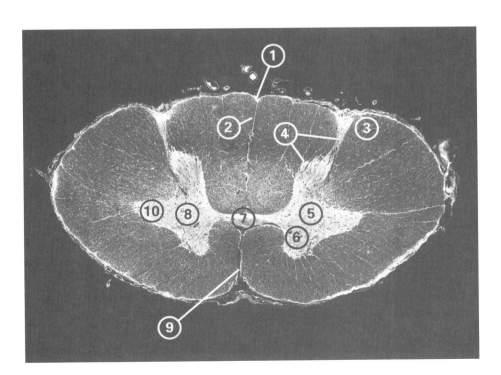

Figure 7.10

Spinal cord model showing:

1. dorsal (posterior) median sulcus
2. dorsal root ganglion (sensory)
3. ventral root (motor)
4. white matter
5. ventral median fissure
6. ventral horn
7. dorsal horn
8. central canal
9. lateral horn

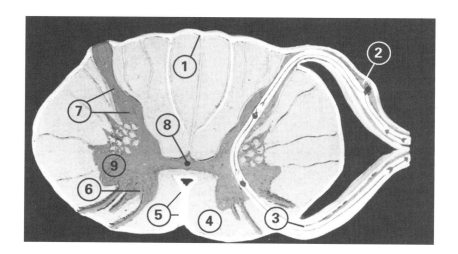

EXERCISE 4   Examination of Meninges

The **meninges** are protective coverings around the brain and spinal cord. Examine a preserved sheep or beef brain. The thick, whitish, outer layer is the **dura mater**. Gently make first a horizontal incision and then a vertical incision through this layer and reflect it back. Beneath the dura mater, you will see a white-to-transparent covering with a spider-web pattern. This is the **arachnoid layer**. Gently reflect this layer laterally and observe the innermost layer, which is adherent to the sulcus and the fissures of the brain itself. This third innermost layer is the **pia mater**. Observe the spinal cord specimen and locate the meninges if they are present.

Between the meninges are spaces called the **subdural** and **subarachnoid spaces**. The subarachnoid space contains **cerebrospinal fluid**. Blood sinuses of the skull are in the dura mater, and dried blood may be obvious on your specimen. You may observe fine vascularization in the pia mater.

EXERCISE 5 Examination of a Brain (Color Plate 46)

Identify the structures printed in boldface on a preserved human brain or model (**Figures 7.11–7.13**) or a preserved sheep brain (**Figures 7.14–7.16**). The brain is composed of six divisions: the cerebrum, cerebellum, diencephalon, midbrain, pons, and medulla oblongata. The diencephalon is divided into the thalamus and hypothalamus. The midbrain, pons, and medulla are often referred to as the brainstem. The brainstem continues inferiorly to form the spinal cord.

The **cerebrum** is the largest brain division. It is positioned superiorly and anteriorly in the cranial cavity. The **longitudinal fissure** divides it into two cerebral hemispheres. The outer region, the **cortex**, is composed of **gray matter**. The **corpus callosum** and **fornix** of the cerebrum are white matter and connect the two hemispheres. On the cerebral surface, notice the grooves or **sulci** and the rolling ridges, which are also known as **gyri** or **convolutions**. Each cerebral hemisphere is divided into lobes: **frontal, parietal, temporal,** and **occipital**. A fifth lobe, the Isle of Rheil, or insula, is deep within the cerebrum and therefore is not visible. Locate the **central sulcus** separating the **frontal** and **parietal lobes**, the **precentral gyrus** of the frontal lobe, and the **postcentral gyrus** of the parietal lobe. Immediately posterior to the frontal lobe on the lateral surface is the **lateral sulcus**.

Though the functions of the cerebral cortex are varied in the human being, three main areas have been identified: sensory, motor, and association. **Sensory areas** receive input from various body regions and perceive different sen-

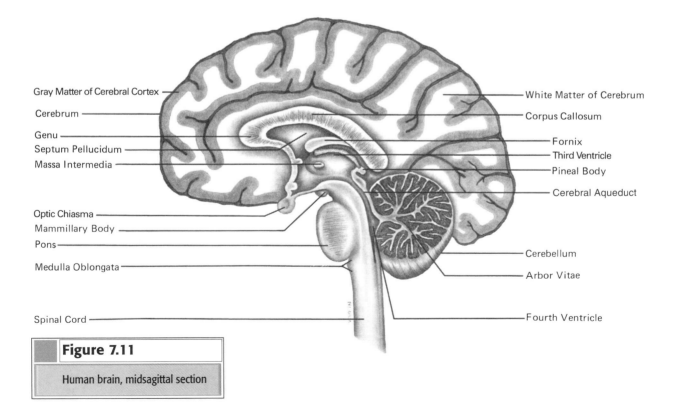

Gray Matter of Cerebral Cortex
Cerebrum
Genu
Septum Pellucidum
Massa Intermedia
Optic Chiasma
Mammillary Body
Pons
Medulla Oblongata
Spinal Cord

White Matter of Cerebrum
Corpus Callosum
Fornix
Third Ventricle
Pineal Body
Cerebral Aqueduct
Cerebellum
Arbor Vitae
Fourth Ventricle

Figure 7.11

Human brain, midsagittal section

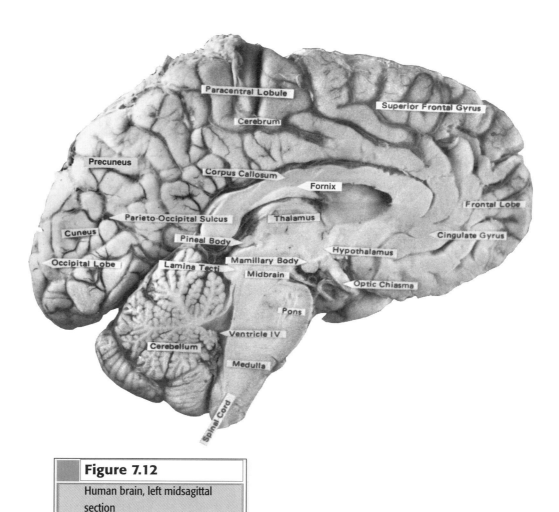

Paracentral Lobule

Superior Frontal Gyrus

Cerebrum

Precuneus

Corpus Callosum

Fornix

Frontal Lobe

Parieto-Occipital Sulcus

Thalamus

Cuneus

Cingulate Gyrus

Pineal Body

Hypothalamus

Occipital Lobe

Lamina Tecti

Mamillary Body

Midbrain

Optic Chiasma

Pons

Ventricle IV

Cerebellum

Medulla

Spinal Cord

Figure 7.12

Human brain, left midsagittal section

sations. Vision and hearing, as well as taste and smell, are localized here. **Motor areas** control various skeletal muscle activities. Higher functions, such as personality and reasoning, are centered in the **association areas**. The precentral gyrus contains a primary motor area and the postcentral gyrus contains a primary sensory area. The right cerebral hemisphere controls left body functions and the left cerebral hemisphere controls right body functions.

On the ventral surface of the brain, identify the **olfactory nerves** and **optic nerves**. Note the **optic chiasma**, the X-shaped structure formed by the crossover of right and left optic nerves.

The **cerebellum** is a bilobed mass inferior to the occipital lobes of the cerebrum. It is smaller than the cerebrum but is also divided into two hemispheres. Coordination of muscular activities is controlled by the cerebellum. Like all six brain divisions, the cerebellum contains both gray and white matter. Gray matter is peripheral to the **arbor vitae**, a medial, treelike configuration, which is composed of white matter.

Question **7A.5**

What is the functional difference between the cerebral cortical, sensory, and motor regions?

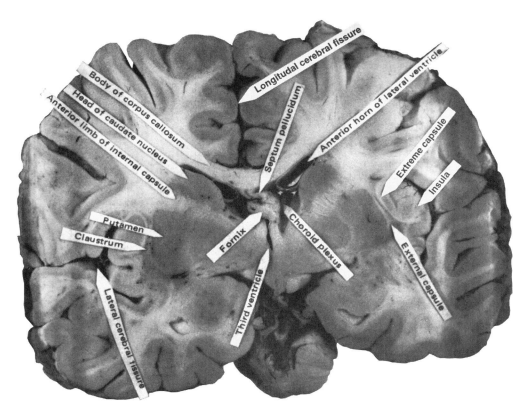

Figure 7.13

Human brain, coronal section

The **diencephalon** is formed by the union of the **thalamus** and the **hypo-thalamus**. It lies between the cerebrum and midbrain. The thalamus, the larger region above the hypothalamus, controls the relaying of sensory impulses to higher brain regions and motor impulses from the cerebrum to the rest of the body. It also transmits impulses to the limbic system, which is responsible for emotions. The hypothalamus is responsible for the regulation and maintenance of internal homeostasis by controlling body temperature, appetite, fluid balance, and sexual drive (**Figure 7.12**). A stalk known as the **infundibulum** extends inferiorly from the hypothalamus. The **pituitary gland** or **hypophysis** is at the distal end of the infundibulum.

Figure 7.14

Sheep brain, dorsal aspect

1. postcentral gyrus of cerebrum
2. cerebellar hemisphere
3. spinal cord
4. longitudinal cerebral fissure
5. sulcus

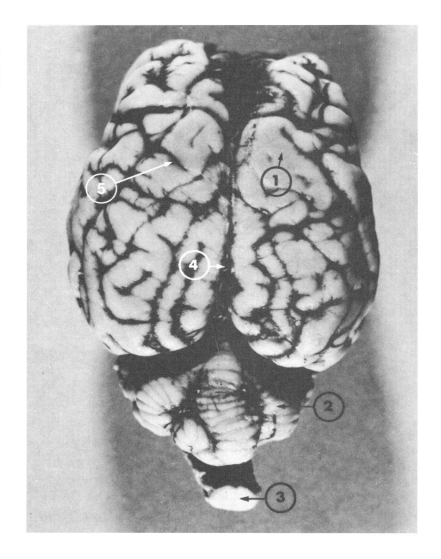

The **midbrain**, a region of the brainstem, extends inferiorly between the diencephalon and the pons. Both sensory and motor nerve fibers, which carry impulses between the spinal cord and the brain, are found here. The **cerebral aqueduct**, the connection between the **third** and **fourth ventricles**, separates the midbrain into ventral and dorsal parts.

There are four ventricles in the brain in which cerebrospinal fluid is produced. The **septum pellucidum** is a thin membrane that separates the **right** and **left lateral ventricles** in the right and left cerebral hemispheres, respectively. These ventricles are also known as the **first** and **second ventricles**. The **third ventricle** is in the diencephalon and the **fourth** is between the pons and the cerebellum (**Figures 7.11–7.13**).

Figure 7.15

Sheep brain, ventral aspect, showing gross detail

1. frontal lobes
2. olfactory bulb
3. left optic nerve
4. optic chiasma
5. pons
6. medulla oblongata
7. spinal cord
8. cerebellum
9. temporal lobe
10. optic tract
11. right optic nerve

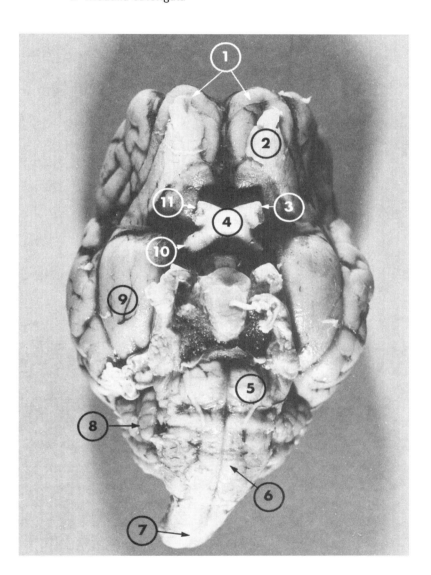

Question 7A.6

Identify the brain region responsible for each of the following functions:

motor coordination

primary sensations

internal homeostasis

State the functions of the following:

pons

medulla

hypothalamus

The **pons**, or bridge, is the middle section of the brainstem above the medulla. It bulges slightly on the ventral surface. The apneustic and pneumotaxic respiratory centers are located here. In addition to controlling breathing, the pons is the center that arouses the cerebrum to the conscious state.

The most inferior portion of the brainstem, the **medulla oblongata**, connects the brain to the spinal cord as it exits the cranium through the foramen magnum of the occipital bone. These structures were identified when you studied the skeletal system in Unit 5. Note the **pyramids**, which are rounded masses on the ventral surface of the medulla. **Pyramidal tracts** are nerve fibers to the skeletal muscles. Respiratory, cardiac, and vasomotor control centers, as well as reflex centers controlling functions such as sneezing and hiccuping, are located in the medulla.

Figure 7.16

Sheep brain, midsagittal section, showing internal structures

1. lateral ventricle
2. fornix
3. corpus callosum
4. olfactory bulb
5. optic nerve
6. optic chiasma
7. pituitary gland (hypophysis)
8. midbrain
9. pons
10. medulla oblongata
11. brainstem
12. central canal of spinal cord
13. fourth ventricle
14. medullary body (arbor vitae) of cerebellum
15. gray matter of cerebellum
16. third ventricle
17. corpora quadrigemina
18. intermediate mass of thalamus
19. infundibulum

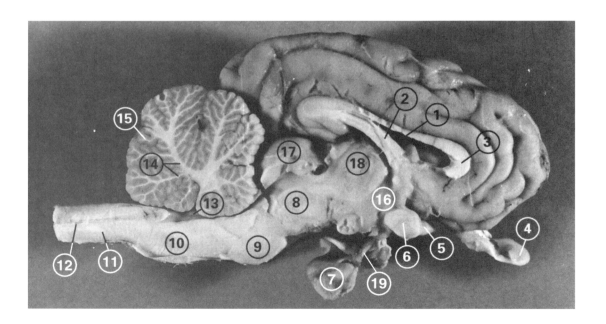

On a preserved sheep brain, make a midsaggital incision through the longitudinal fissure of the cerebrum and continue the incision posteriorly through the cerebellum and spinal cord. Locate the ventricles, septum pellucidum, corpus callosum, fornix, and the six major brain divisions (**Figure 7.16**).

Make a coronal incision through the cerebrum. Observe the white and gray matter. Identify the lateral ventricles, corpus callosum, and septum pellucidum (**Figures 7.11–7.13, 7.16**).

EXERCISE 6   Examination of Cranial Nerves

On a preserved brain with intact cranial nerves or on a brain model, locate the cranial nerves. Some contain only sensory fibers and are called **sensory nerves** (pairs I, II, and VIII). **Mixed nerves** (pairs V, VII, IX, and X) contain both sensory and motor fibers. Others contain motor fibers and are called **motor nerves** (pairs III, IV, VI, XI, and XII). Note that, of the 12 pairs of cranial nerves, the distal 10 pairs emerge from the brainstem (**Figure 7.17**).

I. Olfactory
II. Optic
III. Oculomotor
IV. Trochlear
V. Trigeminal
VI. Abducens
VII. Facial
VIII. Vestibulocochlear (Acoustic)
IX. Glossopharyngeal
X. Vagus
XI. Accessory (Spinal Accessory)
XII. Hypoglossal

Question **7A.8**

Which cranial nerves do not emerge from the brainstem?

EXERCISE 7   Examination of Spinal Nerves

On a preserved spinal cord or a model, note the emerging nerves. In the human being, there are 31 pairs: 8 cervical, 12 thoracic, 5 lumbar, 5 sacral, and 1 coccygeal.

To expose the **brachial plexus** in your cat or a human cadaver, carefully separate the muscles, blood vessels, and connective tissue from the axillary region and chest on one side of the body. The brachial plexus (which consists of interconnected white, tough nerves emerging from the last three cervical and first thoracic vertebrae) should now be visible. After removing the muscles, blood vessels, and connective tissue from the lumbosacral region, you should be able to see the **lumbosacral plexus**. It is composed of branches from the last three lumbar and first sacral nerves. **Figure 7.18** depicts the major spinal nerves of the cat. Observe their exit from the spinal cord.

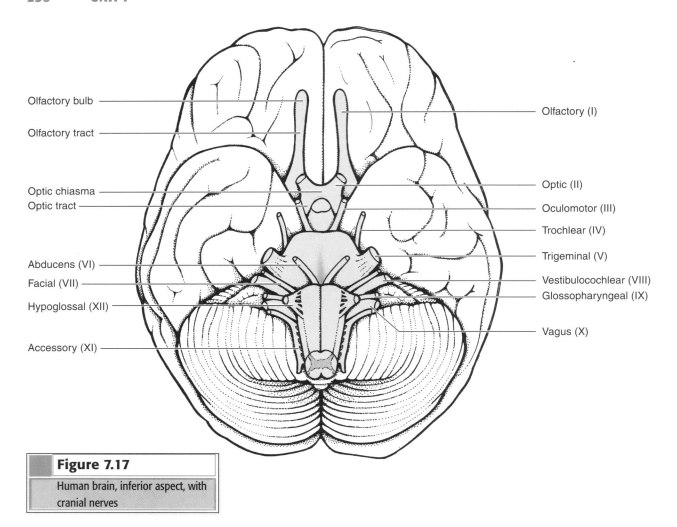

Figure 7.17

Human brain, inferior aspect, with cranial nerves

Olfactory bulb
Olfactory tract
Optic chiasma
Optic tract
Abducens (VI)
Facial (VII)
Hypoglossal (XII)
Accessory (XI)

Olfactory (I)
Optic (II)
Oculomotor (III)
Trochlear (IV)
Trigeminal (V)
Vestibulocochlear (VIII)
Glossopharyngeal (IX)
Vagus (X)

Identification 7.1

 7.1

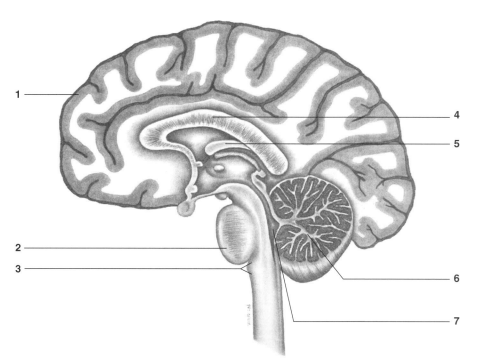

Figure 7.18

Major spinal nerves, cat

1. olfactory bulb
2. eyeball
3. cerebrum
4. cerebellum
5. brachial plexus
6. spinal cord
7. nerves to arm: ulnar, medial, radial
8. lumbar nerves
9. lumbosacral plexus
10. sciatic nerves
11. caudal nerves
12. cervical plexus
13. vagus nerve

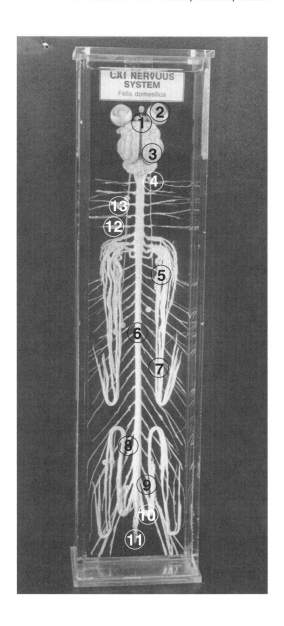

B. Nervous System Physiology

PURPOSE

Unit 7B will familiarize you with some of the physiological aspects of the nervous system.

OBJECTIVES

After completing Unit 7B, you will be able to
- identify and demonstrate selected reflexes in the human being.
- perform tests for cranial nerve and cerebellar function.
- identify selected characteristics of nerve impulse transmission.

MATERIALS

1% HCl solution	penlight
Ringer's solution	coffee, tobacco, or spices
rubber reflex hammer	cotton
10% NaCl solution	cotton-tipped applicator
10% sugar solution	stopwatch
tongue depressors	tuning fork (512 cycles per second)
disposable gloves	swivel chair
facial tissues	

PROCEDURE

Nerves conduct impulses upon stimulation. For a nerve to be excitable and to change from a resting state to a conducting state, a difference in ionic concentration must exist between the external and internal surfaces of the cell membrane. Conductivity results from a change in ionic concentration from the point of stimulation and extends along the nerve.

EXERCISE 1   Human Reflexes

Question 7B.1

Try to stop yourself from swallowing. Explain your response.

A reflex is a rapid, involuntary motor response to a specific stimulus. A reflex arc consists of five components: a receptor, sensory neuron, interneuron, motor neuron, and effector (muscle or gland) (**Figure 7.19**).

THE SWALLOWING REFLEX
Swallow the saliva in your mouth and immediately swallow again. Explain your result and compare it to the rapid succession of swallowing demonstrated by rapidly drinking a glass of water.

THE PATELLAR REFLEX
Sit so that your legs hang down freely from the knees. Have another student strike the patellar ligament (just below the knee) with the flat posterior portion of a rubber reflex hammer. This experiment may require a little patience, and it is best to divert the subject's attention. Notice that the leg is extended by the contraction of the quadriceps muscle group. Repeat on another student. Record your data (**Figure 7.19**).

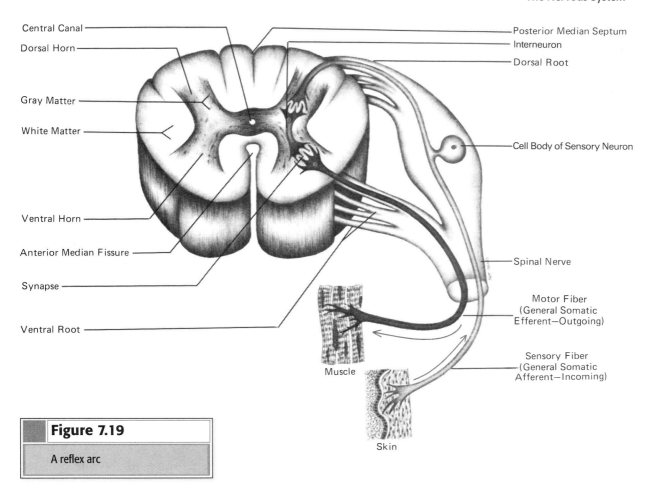

Central Canal

Dorsal Horn

Gray Matter

White Matter

Ventral Horn

Anterior Median Fissure

Synapse

Ventral Root

Posterior Median Septum

Interneuron

Dorsal Root

Cell Body of Sensory Neuron

Spinal Nerve

Motor Fiber
(General Somatic
Efferent—Outgoing)

Sensory Fiber
(General Somatic
Afferent—Incoming)

Muscle

Skin

Figure 7.19

A reflex arc

PHOTO-PUPIL REFLEX

Close your eyes for two minutes. While facing a bright light, open them and let another student examine your pupils immediately. Describe the response.

THE ACCOMMODATION REFLEX

In a moderate light, look at a distant object (20 ft or more removed) and have another student examine your pupils. Now look at a pencil held about 10″ from your face (without changing the illumination) and have your partner note your pupils. Result?

CONVERGENCE REFLEX

Look at a pencil held 36″ away. Have another student note the position of your eyeballs. Have your partner slowly bring the pencil closer until it nearly touches your nose. What change is observed in your eyeballs? This effect is called convergence.

Question 7B.2

What is the purpose of photo-pupil reflex action?

Question 7B.3

What is the purpose of the accommodation reflex?

Question 7B.4

What reflex results?

CAUTION

Use care when touching your eye.

Question 7B.5

What is the purpose of the corneal reflex?

Question 7B.6

What is the result?

Question 7B.7

What is the result?

Question 7B.8

What is the result?

THE ACHILLES OR ANKLE JERK

Kneel on a chair; let your feet hang freely over the edge of the chair. Bend your foot to increase the tension of the gastrocnemius muscle. Have your partner tap the tendon of Achilles with a rubber reflex hammer.

CORNEAL REFLEX

Gently touch the cornea of your eye with a piece of facial tissue.

EXERCISE 2   Tests for Human Cranial Nerve Function

A superficial assessment of cranial nerve function can be made by performing relatively simple procedures. Working in pairs, test each other's cranial-nerve function in the following manner. Record your results in the space provided as you perform each test and then fill in the table entitled Cranial Nerves and Their Functions.

OLFACTORY NERVE (CRANIAL NERVE I)

Ask the subject to identify the odor of coffee, tobacco, or spices. The test may not be valid if the subject has a cold. (Additional tests of olfactory discrimination will be performed in Unit 8C, Exercise 1.)

OPTIC NERVE (CRANIAL NERVE II)

Ask the subject to read a portion of a printed page with each eye, wearing glasses if necessary. (Additional vision tests will be performed in Unit 8A, Exercises 3–5.)

OCULOMOTOR NERVE (CRANIAL NERVE III)

Ask your partner to follow your finger, a pencil, or a penlight with both eyes, keeping his or her head still as you slowly move the object up, then down. In addition to innervating eye movements through the extrinsic rectus ocular muscles, cranial nerve III also innervates the upper eyelid and provides parasympathetic stimulation to the pupils. Therefore, you should observe your subject for signs of **ptosis** (drooping of one or both eyelids) and, with the room lights darkened, for pupillary reaction to light, if you did not already do so in Exercise 1. To test for reaction to light, better results will be obtained if you bring the penlight in from the side rather than from the front of the subject. Observe the pupil.

TROCHLEAR NERVE (CRANIAL NERVE IV) AND ABDUCENS NERVE (CRANIAL NERVE VI)

Have your partner follow your finger, a pencil, or a penlight with both eyes, keeping his or her head still, as you slowly move the object laterally in each direction. Both eyes should follow the object as it is moved from side to side.

TRIGEMINAL NERVE (CRANIAL NERVE V)

To test the motor responses of this nerve, ask the subject to clench his teeth. As you provide resistance by holding your hand under the subject's chin, ask the subject to open his mouth. He should be able to clench his teeth and open his mouth against resistance.

To test the sensory responses of the trigeminal nerve, have the subject close his eyes. Test for a response to light touch by whisking a piece of dry cotton over the mandibular, maxillary, and ophthalmic areas of the face. Wet the cotton with cold water and determine whether the subject is able to discriminate temperature in these same areas. Now gently touch the subject's cornea with a piece of dry cotton or facial tissue, if you did not already do so in Exercise 1.

FACIAL NERVE (CRANIAL NERVE VII)

To test the motor response of this nerve, ask the subject to wrinkle his forehead, raise his eyebrows, puff his cheeks, and smile showing his teeth. Look for any asymmetry.

To test the sensory response of this nerve, touch a cotton-tipped applicator stick that has been dipped into a 10% NaCl solution to the tip of the subject's tongue. Repeat with 10% sugar solution on the anterior surface of the tongue. (Additional exercises involving taste will be done in Unit 8C, Exercise 2.)

ACOUSTIC OR VESTIBULOCOCHLEAR NERVE (CRANIAL NERVE VIII)

To test the cochlear portion of this nerve, determine the subject's ability to hear a ticking stopwatch and repeat a whispered sentence. The Weber and Rinne tests (Unit 8B, Exercise 3), which use a tuning fork, may also be done at this time.

In general, if the subject is able to keep his balance while walking, the vestibular branch of cranial nerve VIII is functioning.

GLOSSOPHARYNGEAL NERVE (IX) AND VAGUS NERVE (X)

These nerves may be tested together. If you wish (and your subject is willing), test the gag reflex by touching the subject's uvula with a cotton-tipped applicator.

The motor portion of these nerves may be tested by: (1) asking the subject to swallow some water, (2) depressing his tongue with a tongue depressor and asking him to say "ah" (the uvula should move), and (3) noticing any hoarseness when he speaks.

ACCESSORY NERVE (CRANIAL NERVE XI)

This procedure tests the strength of the trapezius and sternocleidomastoid muscles, which are innervated by this nerve. To check the strength of the trapezius, place your hands on the subject's shoulders and determine whether he can raise them against resistance. To test the strength of the sternocleidomastoid muscle, place your hands on each side of the subject's head and ask him to turn his head to each side against resistance.

HYPOGLOSSAL NERVE (CRANIAL NERVE XII)

Ask the subject to stick out his tongue. The tongue should protrude straight, with no deviation.

Question 7B.9

What is the result?

Use care when touching your eye.

Question 7B.10

What is the result?

Question 7B.11

What is the result of testing the cochlear branch of cranial nerve VIII?

Question 7B.12

What is the result of testing cranial nerves IX and X?

Question 7B.13

What is the result of testing cranial nerve XI?

Complete the following chart for cranial nerves:

Cranial Nerves and Their Functions

Number	Name and Branches	Sensory *or* Motor *or* Both	Function
1	Optic Nerve	2	3
4	5	6	Raise Eyebrows Taste
XI	7	8	9
V	10	Both	11
12	13	14	Smell
15	Trochlear Nerve	16	17
VIII	18	19	20

EXERCISE 3   Tests for Human Cerebellar Functions

The cerebellum functions to maintain coordination, posture, and gait. After asking the subject to perform each of the following, check off in this table whether he or she was able to perform the task (+) or was not able to perform it (−).

Cerebellar Functions

Test	+	−
With eyes closed, touch index finger of each hand to nose		
With eyes open, touch examiner's fingers		
Move hands and fingers fast		
Looking straight ahead, move the heel of one foot down the front of the other leg		
Looking straight ahead, touch outstretched hand with toes of corresponding foot		
Stand with feet together and eyes closed without losing balance		
While walking, swing arms slightly		
While looking straight ahead, walk in tandem (heel to toe) without losing balance		

The Nervous System

MULTIPLE CHOICE

Name

Section

Date

_____ 1. The structural and functional unit of the nervous system is the
   a. Schwann cell
   b. neuroglia
   c. neuron
   d. telodendria

_____ 2. The ventral root of a spinal nerve
   a. conducts motor impulses to effectors.
   b. is synonymous with the anterior root.
   c. conducts impulses that are voluntarily controlled.
   d. does all of these.

_____ 3. The trochlear nerve (IV) functions in
   a. facial movements.
   b. eye movements.
   c. tongue movements.
   d. hearing.

_____ 4. Which lobe of the brain contains the primary motor area?
   a. occipital
   b. parietal
   c. frontal
   d. temporal

_____ 5. The posterior lobe of the brain and the bone protecting it are named
   a. occipital.
   b. parietal.
   c. frontal.
   d. temporal.

_____ 6. Which of these may be used to test the sensory function of the trigeminal nerve?
   a. vestibular reflex
   b. corneal reflex
   c. accommodation reflex
   d. swallowing reflex

_____ 7. The cell body of a neuron is also known as the
   a. axon.
   b. perikaryon.
   c. neurofibril.
   d. dendrite.

_____ 8. Identify the correct superior to inferior sequence of spinal nerves:
   a. cervical, thoracic, lumbar, sacral, coccygeal
   b. thoracic, lumbar, cervical, sacral, coccygeal
   c. coccygeal, sacral, lumbar, thoracic, cervical
   d. lumbar, sacral, coccygeal, cervical, thoracic

_____ 9. In the spinal cord, the tracts are
   a. divisions of the gray matter.
   b. divisions of the white matter.
   c. same as ventral horns.
   d. same as dorsal horns.

_____ 10. The hypoglossal nerve functions in
   a. hearing
   b. tongue movements
   c. facial movements
   d. eye movements

_____ 11. Which of these is a mixed cranial nerve (that is, both sensory and motor)?
   a. trigeminal
   b. acoustic
   c. olfactory
   d. hypoglossal

_____ 12. Which meninge is the most external?
   a. pia mater
   b. dura mater
   c. arachnoid
   d. subarachnoid

_____ 13. A term that is synonymous with pituitary gland is
   a. infundibulum.
   b. optic chiasma
   c. hypophysis.
   d. pons.

_____ 14. The arbor vitae is located in the
   a. cerebellum.
   b. cerebrum.
   c. diencephalon.
   d. medulla oblongata.

_____ 15. The brachial plexus consists of which nerves?
   a. cervical only
   b. cervical and thoracic
   c. cervical and lumbar
   d. thoracic

_____ 16. Which of the following cranial nerves influences maintenance of equilibrium?
   a. glossopharyngeal
   b. acoustic
   c. olfactory
   d. vagus

_____ 17. Which reflex protects eye tissues by causing blinking?
   a. photo-pupil
   b. convergence
   c. accommodation
   d. corneal

_____ 18. Which of these divides the cerebrum into left and right hemispheres?
   a. longitudinal fissure
   b. corpus callosum
   c. central sulcus
   d. transverse fissure

_____ 19. If the anterior (ventral) root of a spinal nerve were cut, how would the regions served by the nerve be affected?
   a. They would be painful.
   b. There would be loss of sensation.
   c. There would be loss of movement.
   d. There would be loss of sensation and movement.

_____ 20. Which of the following would be capable of regenerating?
   a. facial nerve
   b. cerebral tracts
   c. spinal tracts
   d. none could regenerate

_____ 21. Which cranial nerves innervate the extrinsic eye muscles?
   a. I, II, and III
   b. III, IV, and V
   c. III, IV, and VI
   d. II, III, and VI

_____ 22. Which of these is (are) not necessary for a reflex to occur?
   a. afferent spinal nerves   c. cerebral cortex
   b. efferent spinal nerves   d. effectors

_____ 23. Inability to touch the index finger to the nose with the eyes closed may indicate a disorder of the
   a. cerebrum.   c. eighth cranial nerve.
   b. basal ganglia.   d. cerebellum.

_____ 24. Where are cell bodies of afferent neurons located?
   a. dorsal root ganglia   c. ventral horns of the spinal cord
   b. pyramidal cells of the cerebrum   d. at the effector

_____ 25. Which of these terms is synonymous with _efferent?_
   a. sensory   c. autonomic
   b. motor   d. neuron

DISCUSSION QUESTIONS

26. Give reasons why mature neurons cannot undergo mitosis following injury.

_____

_____

_____

27. Give the function of the following:

    a. olfactory bulbs

    _____

    b. pyramidal tracts

    _____

    c. optic chiasma

    _____

    d. thalamus

    _____

    e. hypothalamus

    _____

    f. cerebellum

    _____

28. Which is a spinal reflex? Give an example.

    _____

    _____

    _____

29. Which ions are involved in impulse transmission?

    _____

    _____

CASE STUDY

NERVOUS SYSTEM

While cycling, Mark, a freshman pre-med major, hit a crack in the road, catapulted off his bicycle, and landed on his back after bumping his head. He tried to stand and could not. His vision was blurred and he had a difficult time righting his body position.

Explain his symptoms relative to the accident.

What brain parts that could have been impacted are at the posterior aspect of the skull?

The Special Senses

A. The Eye and Vision

PURPOSE

Unit 8A will familiarize you with the basic anatomy and physiology of the eye.

OBJECTIVES

After completing Unit 8A, you will be able to
- identify the structural components of the eye.
- perform selected vision tests.

EXERCISE 1 Anatomy of the Eyeball

MATERIALS

dissecting kits
dissecting trays
preserved sheep or cow eyes

eye models
disposable gloves

PROCEDURE

First study an eye model and, using **Figure 8.1** as a guide, locate the following structures printed in boldface. Note the **extrinsic eye muscles**, which, as you recall from Unit 6, are skeletal muscles that move the eyeball itself. Locate the **superior rectus** muscle, which inserts at the anterior superior aspect of the eyeball. The **inferior rectus** inserts at the anterior inferior aspect of the eyeball. The insertions of the **medial rectus** and **lateral rectus** muscles are at the anterior medial and anterior lateral regions of the eye, respectively. The **superior oblique** muscle inserts at the anterior lateral region of the eyeball but first passes through a cartilaginous pulleylike loop, superior to the superior rectus muscle. Likewise, the **inferior oblique** muscle inserts at the anterior lateral region but first passes through a similar loop of tissue on the inferior surface of the eye.

Observe the lateral aspect of the model. The outer layer is the **sclera**, commonly known as the white of the eye. Beneath the sclera, observe the middle **choroid** layer, which is highly vascular and darkly pigmented. The inner **retina** is a thin, pigmented layer of tissue that is composed primarily of neurons, rods, and cones that are stimulated by light waves impinging upon them. The neurons then transmit impulses to the occipital lobe of the brain for interpre-

Question 8A.1

What is the action of the inferior oblique muscle?

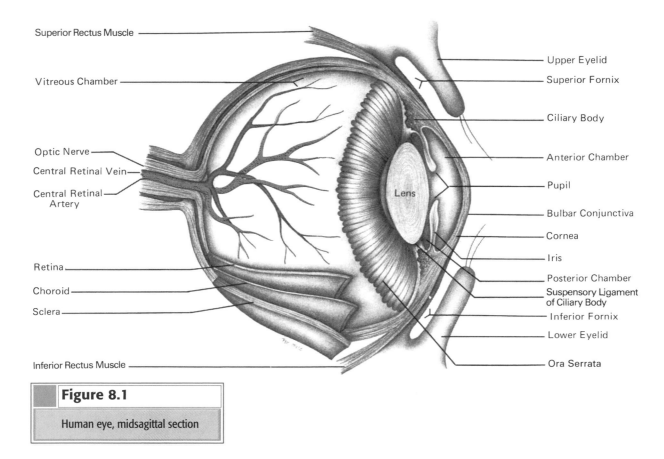

Superior Rectus Muscle

Vitreous Chamber

Optic Nerve
Central Retinal Vein
Central Retinal Artery

Retina
Choroid
Sclera

Inferior Rectus Muscle

Upper Eyelid
Superior Fornix
Ciliary Body
Anterior Chamber
Pupil
Bulbar Conjunctiva
Cornea
Iris
Posterior Chamber
Suspensory Ligament of Ciliary Body
Inferior Fornix
Lower Eyelid
Ora Serrata

Lens

Figure 8.1

Human eye, midsagittal section

Question 8A.2

Differentiate between rods and cones with respect to function.

tation of visual images. Observe the anterior aspect of the eyeball. The transparent **cornea** covers the pigmented ring of tissue known as the **iris**. The anterior interior portion of the choroid layer is continuous with the **ciliary body**, which is located at the **ora serrata**, which encircles the anterior margin of the retina. Attached to the ciliary body is the **suspensory ligament**, which is composed of transparent threadlike structures that hold the transparent biconvex **lens** in place. As you look directly into the anterior aspect of the eye, the blackened area surrounded by the iris is the **pupil**. The iris contains circular and radial muscle fibers that contract, causing the pupil of the eye to constrict and dilate, respectively.

The lens and suspensory ligament divide the eye into an **anterior cavity** and a **posterior cavity**. The anterior cavity in turn, is divided into an **anterior chamber** located between the iris and the cornea, and a **posterior chamber**, between the iris and the lens. Both chambers are filled with a fluid known as **aqueous humor.** The larger, posterior cavity of the eye is behind the lens and extends to the retina. This cavity is filled with a transparent gel known as **vitreous,** or **vitreous humor,** which prevents the eyeball from collapsing. The vitreous also refracts light.

Observe the posterior aspect of the eye model and locate the outwardly projecting cylinder of tissue, the **optic nerve,** which carries impulses to the optic chiasma. The optic tracts within the brain then carry the impulses to the visual centers located in the occipital lobe.

Question **8A.3**

Draw a midsagittal section of an eyeball and label these structures that refract light: cornea, lens, aqueous humor, and vitreous.

Obtain a preserved cow or sheep eye and place it on a dissecting tray. Use **Figures 8.2** through **8.5** as references to find the structures printed in bold-face. Tease away the adipose and excess connective tissue on its posterior and lateral surfaces. Look for the **optic nerve**, a tough cylindrical stub encased in connective tissue, protruding from the posterior aspect of the eye. The outer layer of whitish-gray connective tissue surrounding the eyeball on all except the anterior aspect is the **sclera**. The transparent layer of tissue covering the anterior aspect of the eye is the **cornea**. Using the pointed end of your scissors or the sharp tip of the blade of your scalpel, pierce, then cut, the sclera around the periphery of the cornea. Do not cut too deeply. Handle the eyeball gently and try to remain within two millimeters of the rim of the cornea when doing this. Remove the cornea and set it aside. Watery fluid may leak out of the anterior aspect of the eyeball as you do this. The fluid is **aqueous humor**. Look into the anterior eye itself and find the ring-shaped **iris** surrounding the opening you made after removing the cornea. Cut around the periphery of the iris to remove it. Then remove the hardened, amber-colored, kernel-like structure, which is the **lens**. Using your forceps, carefully extract the gelatinous **vitreous** from the posterior cavity of the eyeball. Be careful not to injure either the innermost, transparent-to-beige layer, the **retina**, or its underlying bluish-to-black **choroid** layer while working.

Figure 8.2

Lateral view of cow eye

1. optic nerve
2. sclera
3. cornea

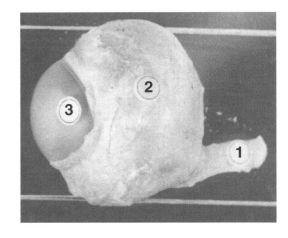

Figure 8.3

Frontal view of cow eye

2. sclera
3. cornea
4. choroid
6. lens

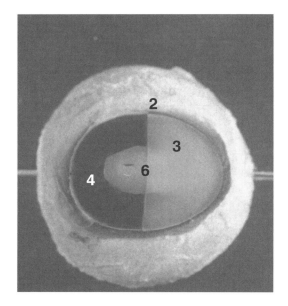

Figure 8.4

Sagittal section of cow eye

4. choroid
5. iris
6. lens

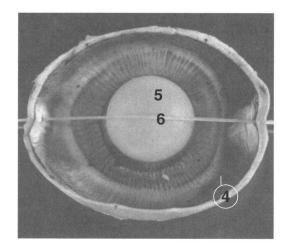

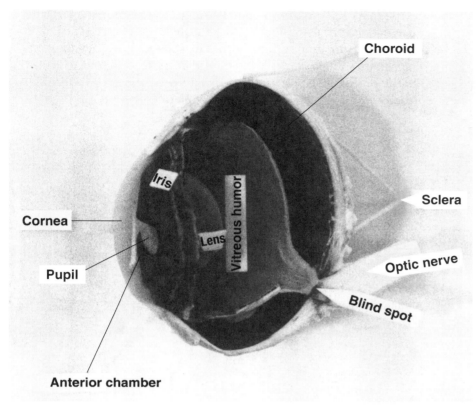

Choroid

Iris

Vitreous humor

Sclera

Cornea

Lens

Pupil

Optic nerve

Blind spot

Anterior chamber

Figure 8.5

Dissected cow eye

EXERCISE 2   Microscopic Examination of the Eye

MATERIALS

microscopic slides of the eye (sagittal section)
compound microscope
dissecting microscope

PROCEDURE

The eyes function to form visual images and to stimulate nerve impulses that are conducted to the visual area of the cerebral cortex for interpretation. The image upon the retina is formed by the refraction of light rays as they move through the eyeball from the cornea to the retina. The region of the retina in which vision is most acute is toward the center, in an area known as the **fovea centralis.** It is at the fovea that cones, which function in color vision, are most concentrated. Rods are photoreceptors that function in black-and-white vision. Rods become more diffuse toward the periphery of the retina, resulting in decreased visual acuity. At the point at which the optic nerve enters the retina, there is a complete absence of rods and cones. This region is referred to as the blind spot.

Question **8A.4**

Which structure separates the anterior cavity from the posterior cavity?

Examine a prepared microscope slide of a sagittal section of an eye under a dissecting microscope and, using **Figures 8.1** through **8.5** for reference, observe the **cornea**, **aqueous humor**, **lens**, **vitreous**, **sclera**, **choroid**, and **retina**.

Using a compound microscope, focus on the posterior wall of the same slide and find the **retina**, **fovea centralis**, **rods** and **cones** layer of the retina, **pigment epithelium** of the retina, **choroid**, and **sclera** (**Figure 8.6**).

Figure 8.6

Fovea centralis area of retina (430×) (Color Plate 36)

1. fovea centralis
2. retina
   3. inner ganglion, plexiform, and nuclear layer
   4. outer plexiform and nuclear layer
5. rods and cones
6. pigment epithelium of retina
7. choroid
8. sclera

Question **8A.5**

Where are rods and cones located in relation to the choroid?

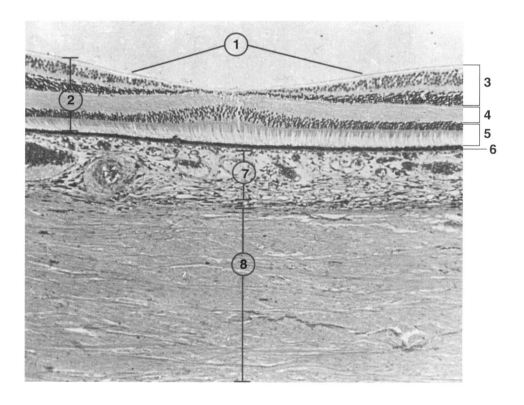

EXERCISE 3   **Visual Acuity**

PROCEDURE

Have another student hold your lab manual 20 ft away from you (or position the manual against a wall 20 ft away). Read the letters on the Snellen chart (**Figure** 8.7). By covering each eye in turn, you can determine the approximate acuity of vision in your right and left eyes. Record your results. Normal vision is 20/20, which means that an individual being tested can read letters at a distance of 20 ft that a person with normal vision can see at 20 ft. If vision were 20/50, the person being tested would see at 20 ft what an individual with normal vision could see at 50 ft. These measurements are specific to each eye.

your result
_____
(normal) 20

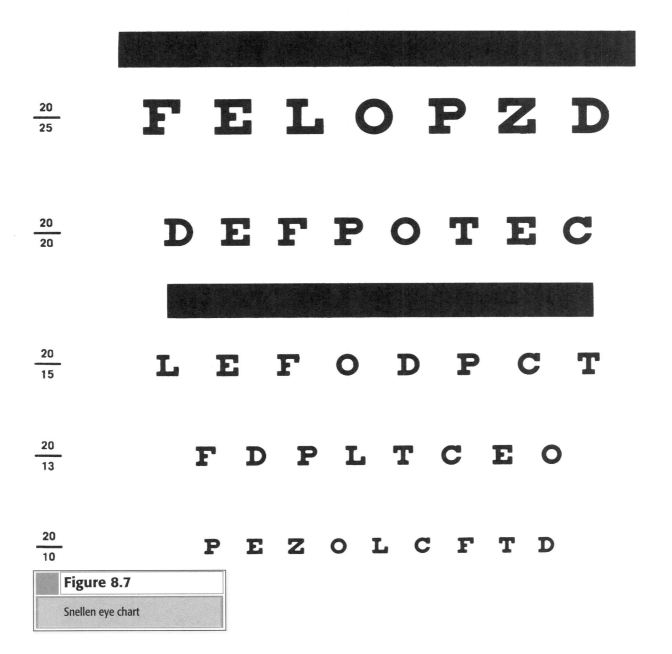

$\dfrac{20}{25}$ F E L O P Z D

$\dfrac{20}{20}$ D E F P O T E C

$\dfrac{20}{15}$ L E F O D P C T

$\dfrac{20}{13}$ F D P L T C E O

$\dfrac{20}{10}$ P E Z O L C F T D

Figure 8.7

Snellen eye chart

EXERCISE 4 Location of the Blind Spot

MATERIAL

Tape measure

PROCEDURE

This exercise will enable you to determine the approximate distance from your eye of an object whose image falls onto the blind spot of each eye.

To determine the blind spot in your left eye, hold this page, containing the cross and dot, 18 inches in front of your face. Close your right eye, and focus on the circle. Slowly bring the page closer, still focusing on the circle. The point at which the cross disappears indicates the blind spot, where the optic nerve enters your left eye. When the cross disappears, measure and record the distance from your eye to the page.

To determine the blind spot in your right eye, hold this page 18 inches in front of your face. Close your left eye, and focus on the cross. Slowly bring the page closer, still focusing on the cross. The point at which the circle disappears indicates the blind spot, where the optic nerve enters your right eye. When the circle disappears, measure and record the distance from your eye to the page.

EXERCISE 5 Tests for Color Blindness

MATERIALS

Ishihara color-blindness plates

PROCDEDURE

Examine various Ishihara plates that test for color blindness. Follow the directions with reference to each plate.

EXERCISE 6   Afterimages

MATERIALS

small (approximately $1'' \times 1.5''$) rectangles of brightly colored green, yellow, orange, and blue cards or pads of paper

PROCEDURE

When sensory receptors are continuously stimulated, they undergo adaptation and become fatigued. This may be demonstrated with vision by stimulating photoreceptors in the retina of the eye.

Stare fixedly at one of the colored cards or papers for 15 seconds, then immediately look at a blank, light-colored wall or a sheet of white paper for a few seconds. Is the color you see on the wall or white paper the same as the original color? Is the shape the same? Repeat with the remaining cards or rectangles of paper, then record your results:

Original Color	Complementary Color	Shape
Green		
Yellow		
Orange		
Blue		

The complementary color you saw on the wall or white paper is an **afterimage** and may be explained by adaptation and fatigue. Cones of the retina that are sensitive and have adapted to one color (e.g., green) have become fatigued, but the red-sensitive (complementary color to green) cones have not. Therefore, the unadapted complementary cones become stimulated to a greater extent when looking at the wall.

B. The Ear, Hearing, and Equilibrium

PURPOSE

Unit 8B will familiarize you with the basic anatomy and physiology of the ear.

OBJECTIVES

After completing Unit 8B, you will be able to
- identify the anatomical features of the ear.
- perform physiological exercises related to hearing and equilibrium.

 EXERCISE 1   Anatomy of the Ear

MATERIALS

ear models

PROCEDURE

Auditory structures include the ears, auditory nerves, and auditory areas of the temporal lobes of the cerebrum. Observe an ear model and find the structures printed in boldface. Also refer to **Figures 8.8, 8.9,** and **8.10** to help you find these structures.

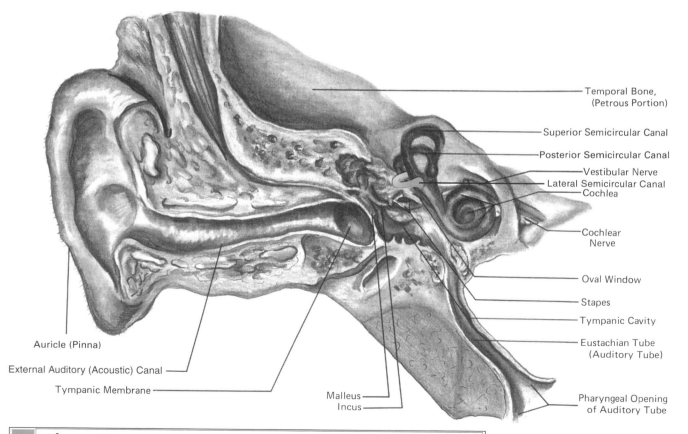

Figure 8.8

Frontal section of right ear showing external, middle, and internal structures

Each ear consists of an external, a middle, and an internal ear. The **external ear** has two divisions: the **auricle** (or **pinna**), which is a flap of tissue on the outside of the ear that directs sound waves into the ear, and the **external auditory** (or **acoustic**) **meatus**, which is a canal in the temporal bone that terminates at the **tympanic membrane**, or **eardrum**. The tympanic membrane is a tense membrane of circular and radial fibers that separates the external ear from the middle ear. This membrane transforms sound waves into mechanical vibrations that are transmitted into first the middle and then the internal ear. The tympanic membrane also serves as a barrier to prevent the entry of foreign bodies, water, and microorganisms into deeper structures.

The **middle ear** is medial to the tympanic membrane and is a **tympanic cavity** within the temporal bone. The middle ear contains three tiny bones, called **auditory ossicles** that vibrate when sound waves impinge upon the tympanic membrane.

The ossicles are the **malleus**, or **hammer**, which is attached to the tympanic membrane; the middle **incus**, or **anvil**; and the **stapes**, or **stirrup**, which spans the space between the incus and the **oval window**, the entry to the inner ear (**Figures 8.8** and **8.9**). The flattened footplate of the stapes covers the oval window and acts as a piston to set fluid in the inner ear into motion. The **auditory** (or **Eustachian**) **tube** runs obliquely and inferiorly from the middle ear, linking it with the nasopharynx. The auditory tube functions to equalize pressure on the two sides of the tympanic membrane.

The **inner ear** functions in hearing and in maintaining equilibrium. It contains the **bony labyrinth**, a series of channels coursing through the medial aspect of the temporal bone. The **cochlea, vestibule,** and **semicircular canals** are located within the bony labyrinth. The medium through which sound waves travel in the inner ear is fluid, rather than air.

The **cochlea** has a spiral shape that resembles a snail. Depending on the detail of your particular model, you may or may not see the following parts of this structure. The apex of the spiral of the cochlea contains a channel, the **helicotrema**. If the cochlear portion of your model opens, you will see the three chambers contained in each spiral of the cochlea: the **scala vestibuli**, the **scala media** (or **cochlear duct**), and the **scala tympani**. The scala vestibuli begins at the oval window and extends to the helicotrema. The scala media contains the **organ of Corti**, the receptor for hearing. The organ of Corti contains hair cells, which, when set into motion, initiate nerve impulses to the brain and are interpreted as various sounds. The **tectorial membrane**, located superior to the hair cells, may be visible on the model. The scala tympani begins at the helicotrema and extends to the **round window**, an opening that is smaller and inferior to the oval window. Figure 8.10 shows a sagittal section of the cochlea, illustrating the relationship between the helicotrema, scala vestibuli, scala media, and scala tympani.

The **vestibule** is in the center of the bony labyrinth, posterior to the cochlea and anterior to the larger looplike, semicircular canals. The vestibule contains two sacs, the **utricle**, which is proximal to the semicircular canals, and the **saccule**, which is proximal to the cochlea. These structures contain receptors for equilibrium and changes in head position.

There are three **semicircular canals** in each ear. These also function to maintain equilibrium by responding to movements of the head. The **superior**,

Question **8B.1**

In which structure does wax, or cerumen, accumulate and possibly impair hearing?

Question **8B.2**

Which ossicle is the smallest?

Question **8B.3**

In which scala are hair cells located?

Question **8B.4**

Name two inner ear structures that respond to head movements.

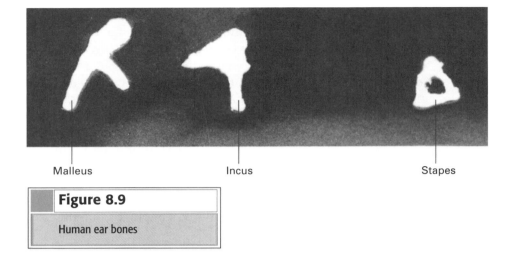

Malleus   Incus   Stapes

Figure 8.9

Human ear bones

or **anterior, semicircular canal** extends the farthest superiorly. The **inferior, posterior, semicircular canal** also extends vertically, and is at right angles to the superior semicircular canal. The **lateral semicircular canal** is inferior to the other two and is positioned horizontally in the temporal bone.

Observe **cranial nerve VIII** (**vestibulocochlear** or **acoustic**), which is formed by the **cochlear nerve** from the cochlea and by branches of the **vestibular nerve** from the vestibule and semicircular canals.

Figure 8.10

Section of cochlea (100×)

1. helicotrema
2. scala media
3. organ of Corti
4. scala tympani
5. scala vestibuli
6. spiral ganglion
7. cochlear nerve

See Figure 8.11

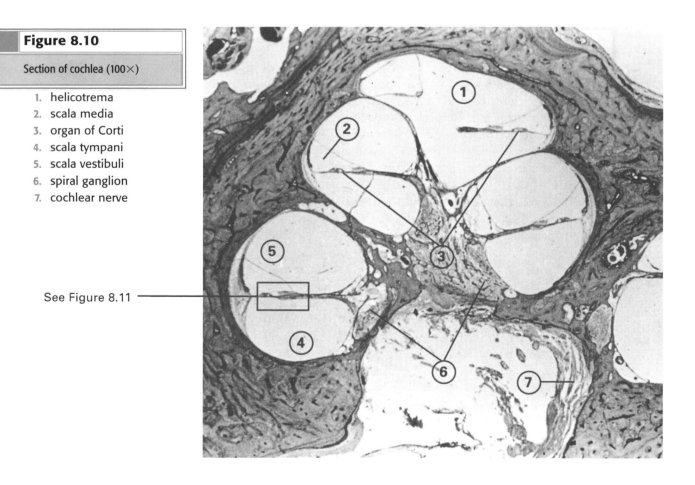

EXERCISE 2 **Microscopic Examination of the Ear**

MATERIALS

microscope slides of middle and inner ear
compound microscopes

PROCEDURE

Examine a prepared slide of the inner ear. Under low power use **Figure 8.10** as a guide to identify the **helicotrema, scala media, organ of Corti, scala tympani, scala vestibuli, spiral ganglion** of the cochlear nerve, and the **cochlear nerve**. Under high power and using **Figure 8.11** for reference, find the **tectorial membrane** and **hair cells** of the organ of Corti. Also locate the **basilar membrane**, which supports this structure.

Figure 8.11	
Organ of Corti (430×)	

1. scala media	4. scala tympani
2. tectorial membrane	5. basilar membrane
3. spiral ganglion cells of cochlear nerve	6. hair cells

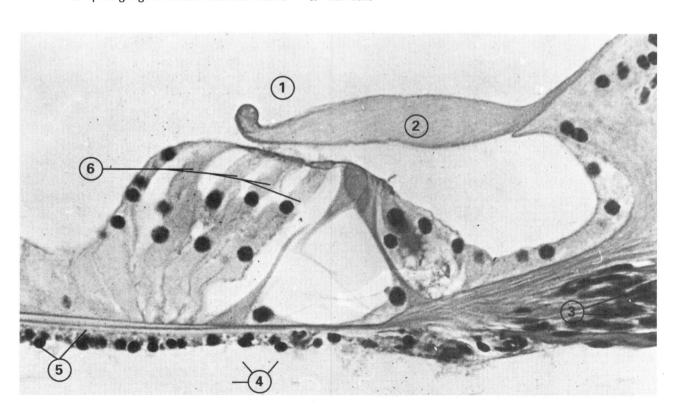

EXERCISE 3   Hearing Tests

MATERIALS

mechanical stopwatch
yardstick or meter stick
tuning fork (preferably 512 cycles
    per second)

rubber reflex hammer
chair
masking tape

PROCEDURE

The ear transmits sound vibrations and their patterns, intensities, and directions of origin, to the temporal lobe of the brain. Sound waves have two major characteristics: **frequency**, or wavelength, which determines pitch and is measured in cycles per second (cps), and **amplitude**, or intensity, which determines loudness and is measured in decibels (db).

WATCH-TICK TEST

Work with a partner. Tape a yard- or meterstick horizontally to a wall, level with your partner's ear when he or she is in a seated position. Hold a stopwatch next to your partner's ear and slowly move it away from the ear until the ticking is no longer audible. Repeat for the other ear. Record, in centimeters or inches, the distance at which the sound becomes inaudible.

WEBER AND RINNE TESTS

A tuning fork can be struck with a rubber reflex hammer and used to compare hearing in both ears. It is also possible to differentiate between conductive and perceptive (or sensorineural) deafness. To do the **Weber** test, strike the tuning fork with the hammer and place the handle on your forehead in a medial position. If the tone is heard in the middle of the head, you have equal hearing or loss of hearing in both ears. If nerve deafness is present in one ear, the tone is heard in the other ear. In conduction deafness, the sound is heard in the ear in which there is hearing loss.

In the **Rinne** test, strike the tuning fork and place the handle on the mastoid process. Then place the tines parallel to the external auditory meatus. Alternate the fork in the two positions a number of times. If the tone is heard equally well in both places, hearing loss is probably mixed. If hearing is normal, the tone is louder and heard longer at the side of the ear (air conduction). If it is louder through the mastoid process (bone conduction), a conductive hearing loss may be indicated.

Question 8B.5

At what distance from the right ear did the sound become inaudible? the left ear?

EXERCISE 4   Nystagmus

MATERIAL

swivel armchair

PROCEDURE

The semicircular canals of the inner ear function to maintain equilibrium, especially with respect to rotational or angular movement and sudden changes in body direction. **Nystagmus** refers to the involuntary, rapid rhythmic movements of the eyeball that occur with these changes in direction. It persists for a short time after the movement stops. In this exercise, working in groups of three or four, you will demonstrate nystagmus.

Be sure the back of the chair is tightened so it will not tilt backwards. Choose one student from your group to sit in the chair with his or her head flexed 30°, eyes closed, feet off the floor, and hands tightly gripping the arms of the chair.

Have another student rotate the chair 10 revolutions in no more than 20 seconds, then abruptly stop the chair. (A third student serves as a standby assistant, in case the student in the chair loses his balance.)

Immediately ask the seated student to open his eyes. Observe the movement of his eyeballs for one minute.

Question **8B.6**

In what direction did the eyeballs move after the chair's motion was stopped?

....

Question **8B.7**

How many seconds did it take for eyeball movement to stop?

....

Identification 8.1

8.1

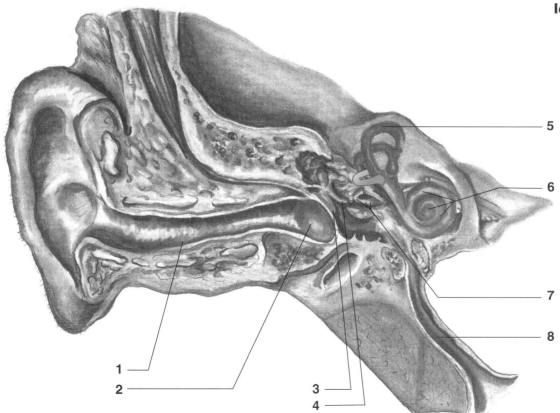

C. Olfactory, Taste, and Cutaneous Receptors

PURPOSE

Unit 8C will familiarize you with the basic anatomy and physiology of olfactory, taste, and cutaneous receptors.

OBJECTIVES

After completing Unit 8C, you will be able to

- perform physiological exercises related to olfactory discrimination.
- perform physiological tests for various taste discriminations.
- identify histological characteristics of cutaneous receptors.
- perform exercises illustrating physiological characteristics of cutaneous receptors.

EXERCISE 1   Olfactory Discrimination

MATERIALS

human skulls
midsagittal section of human head model
samples of nutmeg, onion powder, cinnamon, vanilla extract, mustard powder, ginger, garlic powder, pepper, oregano, cloves, paprika, coffee
slices of apple and potato

PROCEDURE

Observe a human skull and a midsagittal section of a human head. Identify the structures printed in boldface.

Receptors for the sense of smell are located high in the interior of the nose between the **median septum** and in the region of the **superior turbinate**. This area is referred to as the **olfactory cleft**. **Olfactory nerves** pass superiorly through the **cribriform plate** of the ethmoid, which is lateral and inferior to the **crista galli** (**Figure 5.18**), and from there pass to the olfactory bulbs.

The nasal passages are lined with mucous membranes and cilia, which moisten and warm the air during breathing. When someone has a cold, the membranes tend to swell and secrete excessive mucus that covers the lining of the cavity and therefore impairs the sense of smell.

The following exercises are designed to test your ability to discriminate different scents. The sense of smell is complex, and the ability to determine scents varies with the individual. Test your ability and compare it with that of your classmates.

Groups of students are suggested for this exercise. Your instructor has selected several different spices. You are to smell each with your eyes closed and identify the spice. Keep a list of the numbered spice samples and your responses.

Pinch your nostrils together, close your eyes, and have your partner place a piece of either potato or apple in your mouth. Is it apple or potato? Explain why it is difficult to determine which it is.

EXERCISE 2 Taste

MATERIALS

prepared slides of cross section of
    tongue
cotton-tipped applicators
PTC paper
10% sugar solution
10% NaCl solution
lemon juice

menthol eucalyptus
mustard
ice cubes
hot water (105–110°F)
paper cups
styrofoam cups
compound microscopes

PROCEDURE

Examine a prepared microscope slide of a cross section of a tongue, showing taste buds (**Figure 11.11**). The tongue, which is covered with **papillae** (**Figures 11.12** and **11.13**), is the principal organ of taste. **Taste buds**, which are located in the "trenches" of the papillae, are sensitive to substances dissolved in water.

Taste, like smell, is a result of the stimulation of sensory neurons by chemical substances. The four tastes are **salty**, stimulated by metallic cations; **sour**, stimulated by hydrogen ions; **sweet**, stimulated by a hydroxyl group in sugar or alcohol; and **bitter**, triggered by alkaloids.

Dip a separate cotton-tipped applicator into each of the following: 10% sugar solution, 10% NaCl solution, lemon juice, menthol eucalyptus, and mustard. Touch each applicator to your tongue in the following areas: the apex, or tip, the posterior medial region, anterior lateral region, and posterior lateral region. You may find it helpful to wait one minute or to rinse your mouth with water between applications.

On the following drawing of a tongue, map where you can most readily identify each substance.

Hold an ice cube on your tongue for one minute. Then apply the sugar solution. Can you taste the sugar?

Question 8C.1

Lemon juice is representative of which taste?

Question 8C.2

How does your completed map compare with those of your classmates?

Hold hot tap water that is 105–110°F in your mouth for no longer than one minute. Apply the sugar solution to your tongue. Can you taste the sugar?

Place a piece of PTC paper in your mouth. The ability to taste the chemical phenylthiocarbamide in the paper is inherited. Compare your ability to taste the chemical with that of your classmates.

EXERCISE 3   **Cutaneous Receptors**

MATERIALS

prepared slides of skin and sensory receptors	Bunsen burners (optional)
skin models	ice
powdered charcoal	centimeter rulers
hot water (105–110°F)	compound microscopes
large dissecting pins	coins
finishing nails (optional)	black or blue ballpoint pens

PROCEDURE

On a skin model or a microscope slide of the skin, review the epidermal and dermal layers.

Because various areas of the skin contain different sensory (or cutaneous) receptors and special staining techniques are often necessary to visualize these structures on a slide, use a skin model to locate the cutaneous receptors printed in boldface.

The epidermis contains receptors that have free nerve endings. That is, their dendrites are not enclosed in a capsule of connective tissue. Examples of these are three types of **free** (or **naked**) **dendritic nerve endings**, which sense pain, heat, and cold; **root hair plexuses** at the base of hair follicles, which detect hair movement; and **Merkel's discs**, which detect light touch. The dendritic endings of Merkel's discs have a characteristic disclike shape.

The dermis contains receptors whose dendrites are encapsulated, or corpuscular. **Meissner's corpuscles** are found in the dermal papillae. Their capsules are thin; they sense light to moderate touch. **Pacinian corpuscles** are located deeper in the dermis and are larger than Meissner's corpuscles. The capsules of Pacinian corpuscles appear as concentric layers of tissue surrounding a dendrite. They are sensitive to deep pressure (**Figures 4.27** and **4.28**). **Ruffini's endings** and **Krause's end bulbs** are also specialized types of touch corpuscles in the dermis.

The following exercises will familiarize you with a variety of cutaneous sensations. You will be working with a partner.

Question 8C.3

What effect do cold and heat have on taste sensation?

Question 8C.4

What percentage of the class are "tasters?"

What percentage are "nontasters?"

TOUCH

Using a ballpoint pen, draw a square 2 cm × 2 cm on the palm of your hand, and another on the ventral surface of your wrist. Subdivide your square into 0.5-cm sections to make a grid as shown below:

With your eyes closed, have your partner use a light touch to explore the marked areas with a dissecting pin, using the same stimulus intensity in each of the 16 different spots within the squares of the grid. If a sensation is felt, mark a T for touch in the corresponding grid:

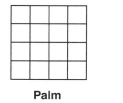

Palm   **Wrist**

Have your partner place a coin on the skin of the inner aspect of your forearm. Determine how many seconds it takes for the initial sensation of pressure to give way to an indifferent sensation. Repeat.

TEMPERATURE

With your eyes closed, have your partner map the cold and heat receptors in the same grids of your palm and inner wrist. Use a dissecting pin or finishing nail that has been cooled in ice water and wiped dry. Again explore the 16 different spots within the grids in the same manner as before. Mark a C for every cold receptor located within the grids.

Question 8C.5

Did you notice a difference in your ability to sense the stimuli between your palm and wrist? Explain.

Question 8C.6

Was there a difference in time required to reach the accommodation of stimuli? What was the time in each case?

Make sure the pin or nail is hot enough to be detected but not hot enough to burn the skin.

After mapping the areas for cold receptors, wait five minutes before exploring for heat receptors. To test for heat receptors, use a dissecting pin or finishing nail that has been passed through the flame of a Bunsen burner. Wait at least five seconds between testing each spot; mark an H in the designated grid subsection:

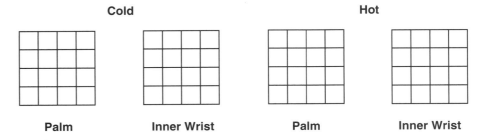

Question **8C.7**

Which area was most sensitive to temperature? the least sensitive?

Repeat the touch and temperature procedures, using the lateral surface of the mid-calf region and the ball of the foot. Compare the differences in sensation between areas of the upper and lower extremities.

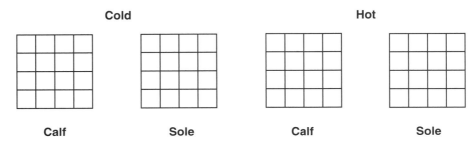

DISCRIMINATION OF STIMULI

Close your eyes and have your partner simultaneously touch the palm of your hand at two points that are close together with two dissecting pins. Then determine how far apart the two pins need to be before you can discriminate two distinct points of stimulation. Two distinct points are felt when two separate touch receptors are stimulated. Your partner should validate your response by occasionally touching only a single point to the skin.

Question 8C.8

In which area could you discriminate between the points of contact most easily? What was the distance between the points, in millimeters?

Question 8C.9

In which area could you discriminate between the points of contact least easily? What was the distance between the points, in millimeters?

Repeat this procedure in the following areas: palmar surface of the index finger, inner aspect of the forearm, back of the neck, and calf and ball of the foot. On the following graph, plot your results, in millimeters of separation, for each area that was stimulated.

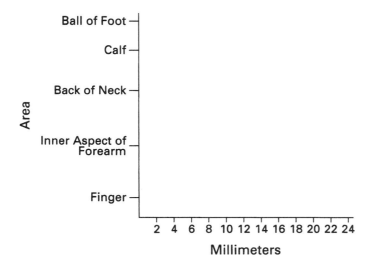

Question 8C.10

What conclusion can you draw from your graph?

DISTRIBUTION OF SWEAT GLANDS

In order to demonstrate the distribution of sweat glands, towel-dry the palm and inner wrist that were previously marked. Your partner will then lightly dust those areas with powdered charcoal. Wait 15 seconds after dusting; then, with a sharp puff, blow the excess powder from the areas. The powder will adhere to the moisture produced where sweat glands are present. Count the particles of charcoal remaining in the grids of the palm, then the inner wrist, and record those numbers below.

palm:

inner wrist:

Question 8C.11

In which of the two areas were sweat glands more numerous?

Question 8C.12

Is this the same area that perspires when you are under stress? Explain.

The Special Senses

MATCHING

Name _____

Section _____

Date _____

Column A

_____ 1. cornea

_____ 2. rhodopsin

_____ 3. iris

_____ 4. rectus muscles

Column B

a. controls the amount of light entering the eye

b. eyeball movement

c. refracts light

d. visual pigment

MULTIPLE CHOICE

_____ 5. A person with 20/30 vision would
   a. see exceptionally well.
   b. have slightly blurred vision.
   c. have average vision.
   d. see very poorly.

_____ 6. Which of these structures contains the largest number of cones?
   a. lens
   b. fovea centralis
   c. iris
   d. aqueous humor

_____ 7. Ringing in the ears (tinnitus) may indicate injury to which cranial nerve?
   a. III
   b. V
   c. VIII
   d. X

_____ 8. A cataract is an opacity of the
   a. retina.
   b. aqueous humor.
   c. cornea.
   d. lens.

_____ 9. Which of these is *not* a refractive medium of the eye?
   a. cornea
   b. vitreous humor
   c. lens
   d. retina

_____ 10. In the normal eye, images are focused
   a. on the retina.
   b. in front of the retina.
   c. behind the retina.
   d. on the blind spot.

_____ 11. Why do middle ear infections commonly follow a sore throat?
   a. Both areas are contained within the oropharynx.
   b. The mucosa of the throat is continuous with the mucosa of the auditory tube and middle ear.
   c. Sneezing spreads bacteria from the throat into the cochlea of the ear.
   d. Rupture of the tympanic membrane allows fluid to drain from the middle ear into the throat.

_____ 12. Which test determines whether hearing is equal in both ears?
- a. Weber
- b. Rinne
- c. Snellen
- d. tuning fork

_____ 13. Hearing is localized in the _____ lobe of the brain.
- a. frontal
- b. parietal
- c. temporal
- d. occipital

_____ 14. Which of these substances would be tasted at the tip of the tongue?
- a. sour and sweet
- b. salty and bitter
- c. bitter and sour
- d. salty and sweet

_____ 15. The _____ transmit the impulses for the sense of taste and smell.
- a. baroreceptors
- b. proprioceptors
- c. chemoreceptors
- d. exteroceptors

_____ 16. The middle vascular layer of the eyeball is the
- a. cornea.
- b. choroid.
- c. sclera.
- d. retina.

_____ 17. The structural portion of the inner ear that is responsible for hearing is (are) the
- a. utricle.
- b. organ of Corti.
- c. auditory ossicles.
- d. tympanic membrane.

_____ 18. The ossicles are located in the
- a. eardrum.
- b. inner ear.
- c. middle ear.
- d. external ear.

_____ 19. The primary sensory receptor for deep pressure is
- a. Pacinian corpuscle.
- b. Meissner's corpuscle.
- c. Krause's corpuscle.
- d. Ruffini's corpuscle.

_____ 20. The primary sensory receptor(s) for pain is (are)
- a. Meissner's corpuscle.
- b. free nerve endings.
- c. Ruffini's corpuscle.
- d. Pacinian corpuscle.

_____ 21. The inner ear is important for
- a. hearing.
- b. maintenance of equilibrium.
- c. both a and b.
- d. neither a nor b.

_____ 22. The lay term _white of the eye_ refers to the
- a. choroid.
- b. cornea.
- c. eyelid.
- d. sclera.

_____ 23. Conjunctiva is a membrane lining the
- a. inner surfaces of the eyelids.
- b. nasolacrimal duct.
- c. cornea.
- d. pupil.

_____ 24. The protective coating of the eye is called the
- a. sclera.
- b. retina.
- c. ciliary body.
- d. choroid.

_____ 25. The olfactory cleft is located
- a. in the cerebral cortex of the frontal lobe.
- b. around the anterior nares.
- c. in the infundibulum.
- d. in the region of the superior turbinate and median septum.

26. Which structures of the eyeball contribute to changing the size of the pupil? _____

_____

_____

_____

_____

_____

27. How are the Weber and Rinne tests diagnostic? _____

_____

_____

_____

_____

_____

28. Is the ability to taste approximately the same for all individuals?

Explain. _____

_____

_____

_____

_____

_____

29. What conditions can influence variations in the ability to sense touch?

_____

_____

_____

_____

_____

30. What types of color blindness are the most common? _____

_____

_____

_____

_____

_____

CASE STUDY

SPECIAL SENSES

Marva, now 22 years old, suffered numerous middle ear infections (otitis media) as a child. To relieve the pressure and fluid buildup, several myringotomies were performed in her right ear. Marva is now having difficulty hearing in her right ear. She sought help from an otologist who determined that she had scar tissue proliferation on her tympanic membrane.

Describe where her problem exists.

Label the appropriate structures and explain the reason why Marva is having difficulty hearing now.

Why did she seek help from an otologist?

The Endocrine System

A. The Endocrine Glands

PURPOSE

Unit 9A will enable you to identify the major endocrine glands and their secretions and to recognize different types of endocrine tissues.

OBJECTIVES

After completing Unit 9A, you will be able to
- identify the major endocrine glands in the human body.
- name the endocrine glands and the hormones each secretes.
- use a microscope to recognize sections of the pituitary, thyroid, parathyroid, pancreas, and adrenal glands.

MATERIALS

human torso
models of endocrine glands
compound microscope
dissecting microscope

microscope slides of pituitary, thyroid, parathyroid, pancreas, and adrenal glands

PROCEDURE

    Endocrine glands are ductless glands that secrete chemical agents known as **hormones.** Hormones from each endocrine gland are transported in the blood to another part of the body where they evoke systemic responses or adjustments by acting on target tissues or organs. The endocrine glands, together with the nervous system, integrate the functions of the organs and systems in the body.

Question 9A.1

Why could the pancreas be considered to be a "mixed" gland?

EXERCISE 1   **Anatomy of Endocrine Glands**

The **pituitary gland**, or **hypophysis**, is located in the sella turcica of the sphenoid bone. The pituitary is regulated by the **hypothalamus** of the brain. It is also connected to the hypothalamus by an **infundibular stalk**. The anterior lobe of the pituitary is the **adenohypophysis**, or **pars distalis**. This lobe secretes several hormones known as tropic hormones; they exert their effects indirectly by stimulating other endocrine glands or body tissues to secrete hormones.

The posterior lobe of the pituitary is known as the **neurohypophysis**, or **pars nervosa**. The neurohypophysis secretes two hormones, oxytocin and antidiuretic hormone, or ADH. These hormones are produced in the hypothalamus and then transported by nerve fibers to the neurohypophysis, where they are stored and then secreted into the blood.

The **thyroid gland** is located anterior to the trachea and inferior to the larynx. It has two **lateral lobes** that are connected by an **isthmus** of thyroid tissue. A small, elongated **pyramidal lobe** that projects superiorly from the isthmus may be visible on your model. The thyroid secretes three hormones: two of them, thyroxine (T_4) and triiodothyronine (T_3), are secreted by microscopic follicular cells; the third, calcitonin, is secreted by microscopic parafollicular cells, or C-cells, of the thyroid gland.

The four **parathyroid glands** are located on the posterior surface of the lateral lobes of the thyroid gland. The parathyroids secrete parathyroid hormone (PTH).

The **pancreas** is located just inferior to the stomach and is bordered on the right by the curve of the duodenum off the distal portion of the stomach. The larger, rounded **head** of the pancreas lies within the curve of the duodenum. The **body** is inferior to the stomach and is partially obscured by the inferior border of the stomach. The **tail** of the pancreas is the tapered end that usually extends beyond the greater curvature of the stomach on the left. The pancreas has an endocrine function in addition to its exocrine function of secreting digestive enzymes. Scattered among the acinar cells of the pancreas are the islets of Langerhans, which produce and secrete the hormones insulin, glucagon, and somatostatin.

There are two **adrenal**, or **suprarenal**, glands, one positioned on the superior border of each kidney. Each gland has an outer **cortex** that surrounds an inner **medulla**. The adrenal cortex secretes the steroid hormones aldosterone, cortisol, androgens, and smaller amounts of estrogen. The medulla secretes the hormones epinephrine and norepinephrine.

The **pineal gland** is suspended from the roof of the third ventricle of the brain. The elongated, bilobed **thymus gland** is located in the mediastinum, anterior to the superior portion of the sternum.

On a human torso or using models of endocrine glands, find the structures printed in boldface in Exercise 1. As you identify the structures, label them on **Figure 9.1**.

The accompanying tables summarize the major hormones secreted by the pituitary and other endocrine glands in the body.

Figure 9.1

Endocrine system

1. adrenals	6. thyroid
2. parathyroids	7. testes
3. ovaries	8. pancreas
4. pituitary	9. thymus
5. pineal	

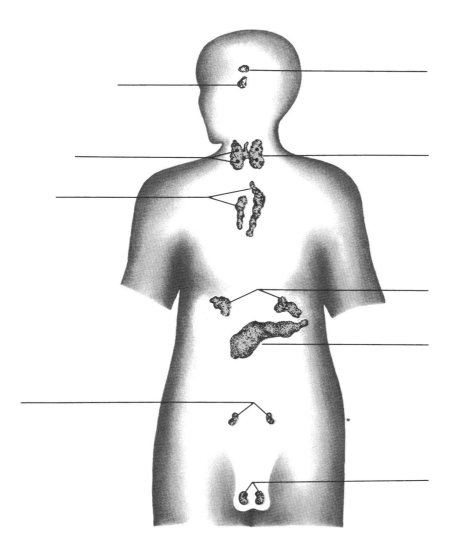

Anterior Pituitary Hormones and Their Principal Actions

Hormone	Principal Actions
Tropic Hormones Growth hormone (GH)	Influences the growth of body cells and protein anabolism through the action of somatomedins in the liver
Thyroid-stimulating hormone (TSH)	Controls the secretion of thyroid hormones
Adrenocorticotropic hormone (ACTH)	Controls the secretion of certain hormones by the adrenal cortex
Follicle-stimulating hormone (FSH)	*In the female*, initiates development of ova and induces secretion of estrogens by the ovaries *In the male*, stimulates sperm production by the testes
Luteinizing hormone (LH)	*In the female*, stimulates follicle maturation, ovulation, and formation of the corpus luteum; prepares the uterus for implantation; and initiates milk production in the mammary glands. *In the male*, hormone (also known as ICSH, or interstitial cell-stimulating hormone) stimulates interstitial cells in the testes to mature and produce testosterone
Prolactin (PRL)	Initiates and maintains milk secretion by the mammary glands
Melanocyte-stimulating hormone (MSH)	Stimulates melanin activity in melanocytes

Posterior Pituitary Hormones and Their Principal Actions

Hormone	Principal Actions
Oxytocin (OT)	Stimulates contraction of the smooth muscle cells of the pregnant (gravid) uterus during labor; stimulates the contractile cells of the mammary glands for milk ejection
Antidiuretic hormone (ADH)	Decreases urine volume; raises blood pressure during severe hemorrhage by constricting the arterioles

Major Hormones Secreted by Endocrine Glands and Their Principal Actions

Gland	Hormone	Principal Actions
Thyroid	Thyroid hormones	
	Thyroxine (T_4)	Regulates metabolism, growth and development, and activity of the nervous system
	Triiodothyronine (T_3)	Same as for T_4
	Calcitonin (CT)	Decreases blood levels of calcium by opposing the action of parathyroid hormone (PTH)
Parathyroid	Parathyroid hormone (PTH)	Increases the blood calcium level and decreases the blood phosphate level by increasing the rate of calcium absorption from GI tract into blood; increases the number and activity of osteoclasts, increases calcium absorption by the kidneys, increases phosphate excretion by the kidneys, activates vitamin D, which promotes calcium absorption from the GI tract
Adrenal Medulla	Epinephrine	Produces effects that resemble those of ANS during stress
	Norepinephrine	Similar to effects of epinephrine
Adrenal Cortex	Mineralocorticoids (e.g., aldosterone)	Enhance sodium reabsorption in the distal convoluted tubules of the kidneys, resulting in increased blood levels of sodium; enhance potassium secretion in the distal convoluted tubules of the kidneys, resulting in decreased blood levels of potassium
	Glucocorticoids (e.g., cortisol)	Promote normal metabolism and resistance to stress, and counter inflammatory response
	Sex steroids	
	Androgens	Minimal significance
	Estrogens	Minimal significance
Pancreas	Insulin (from beta cells of islets of Langerhans)	Lowers the blood sugar (glucose) level by accelerating the transport of glucose into skeletal muscle and adipose cells, converting glucose into glycogen in the liver (glycogenesis); decreases glycogenolysis and gluconeogenesis in the liver; promotes protein synthesis and lipogenesis
	Glucagon (from alpha cells of islets of Langerhans)	Elevates blood glucose level by increasing glycogenolysis and gluconeogenesis in the liver
	Somatostatin (growth hormone inhibiting factor, GHIF) (from delta cells of islets)	Inhibits insulin and glucagon secretion
Pineal	Melatonin	May inhibit reproductive activites by inhibiting gonadotropic hormones
Thymus	Thymosin	Promotes the maturation of T-lymphocytes
Ovary	Estrogens and Progesterone	Aid in the development and maintenance of female sexual characteristics; with GnRH and gonadotropic hormones, regulate the menstrual cycle, maintain pregnancy, and regulate lactation
	Relaxin	Relaxes the cartilage of pubic symphysis; helps to dilate cervix near end of pregnancy
Testis	Testosterone	Aids in the development and maintenance of the male sexual characteristics
	Inhibin	Inhibits secretion of FSH to control sperm production

Question 9A.2

How do the functions of prolactin and oxytocin differ?

Question **9A.3**

Name and give the function of a hormone stored in region 5 of Figure 9.2.

As you observe the following slides of endocrine glands, refer back to the tables listing the principal actions of hormones.

EXERCISE 2 **Microscopic Study of the Endocrine Glands**

THE PITUITARY GLAND (HYPOPHYSIS)

Examine a section of a pituitary gland under a dissecting microscope to see its entirety. Then draw the section and label the **infundibular stalk, pars distalis (anterior lobe), chromophil cells, pars nervosa (posterior lobe), and pars intermedia** (Figure 9.2 and Color Plate 37).

Figure 9.2

Hypophysis (mammal) (100×)

1. sella turcica
2. hypothalamus
3. infundibular stalk
4. pars intermedia
5. pars nervosa
6. pars distalis containing chromophil cells (acidophils and basophils)
7. dura mater

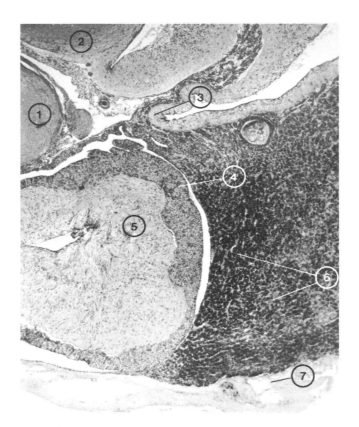

Drawing:

THE THYROID GLAND

In this section, you will see the **cuboidal epithelium** of **follicular cells** surrounding **colloid**, which stains pink and contains **thyroglobulin**, the precursor and storage form of the thyroid hormones thyroxine and triiodothyronine (**Figure 9.3**). **Parafollicular cells (C-cells)** are in the interstitial areas. These produce and secrete the hormone calcitonin. Draw and label the colloid and cuboidal epithelium.

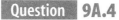

Question **9A.4**

What is the function of calcitonin?

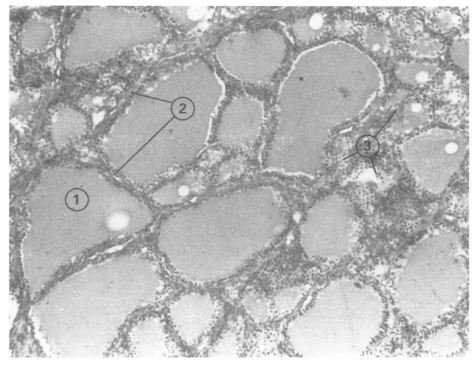

Figure 9.3

Thyroid gland (430×)

1. thyroid follicle containing thyroglobulin
2. follicular cells of cuboidal epithelium
3. parafollicular cells

Drawing:

Question **9A.5**

Name and give the function of a hormone stored in region 2 of Figure 9.4.

THE PARATHYROID GLANDS

Parathyroid tissue in the human adult is made up of two major types of cells: **chief** (or **principal**) cells and **oxyphilic** cells. Chief cells are small, stain deep purple, and occur in masses, whereas oxyphilic cells contain more cytoplasm, stain a paler blue, and occur singly or in small groups. Oxyphils are not present in young children (**Figure 9.4**). Draw a section of parathyroid tissue and label the oxyphilic cells, chief cells, and connective tissue.

Figure 9.4

Parathyroid gland (430×)

1. oxyphil (acidophil) cells
2. principal (chief) cells
3. blood vessels

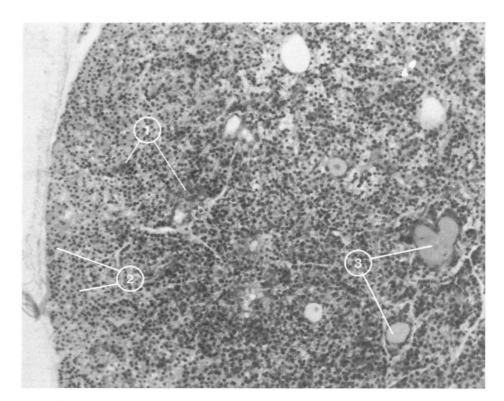

Drawing:

The Pancreas

Observe the scattered, pale, patchlike **islets of Langerhans** among the **acinar cells** (**Figures 9.5** and **9.6**). Draw a section of pancreas and label the acinar and islet portions.

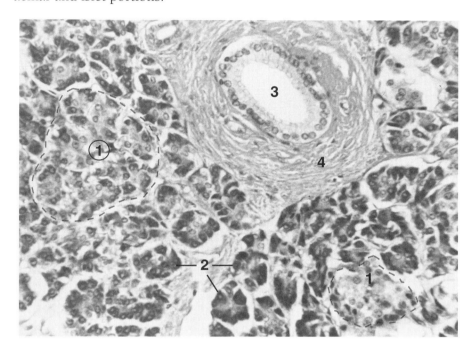

Figure 9.5

General view of pancreas (100×)

1. pancreatic islets (islets of Langerhans)
2. pancreatic acini
3. interlobular duct
4. interlobular connective tissue

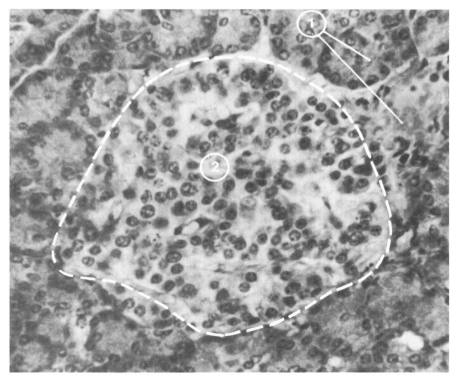

Figure 9.6

Pancreas, islet of Langerhans (1000×)

1. pancreatic acini
2. islet of Langerhans

Drawing:

Which of the adrenocortical hormones

- has an anti-inflammatory function?

- maintains sodium/potassium balance?

THE ADRENAL GLANDS

Each adrenal gland has an outer **cortex** surrounding the **medulla**. The cortex consists of three zones: the outer **zona glomerulosa**, which secretes the mineralocorticoid aldosterone, the middle **zona fasciculata**, which secretes glucocorticoids such as cortisol and the inner **zona reticularis**, which secretes sex hormones (**Figures 9.7** and **9.8**).

Figure 9.7

Human adrenal gland (whole mount, 10×)

1. cortex
2. medulla
3. vein

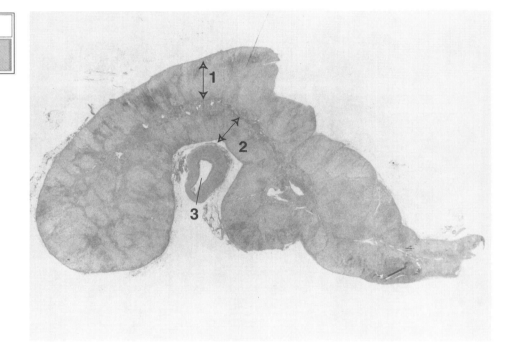

Figure 9.8

Adrenal gland, cortex and medulla (40×)

1. capsule
2. zona glomerulosa
3. zona fasciculata
4. zona reticularis
5. medulla
6. vein

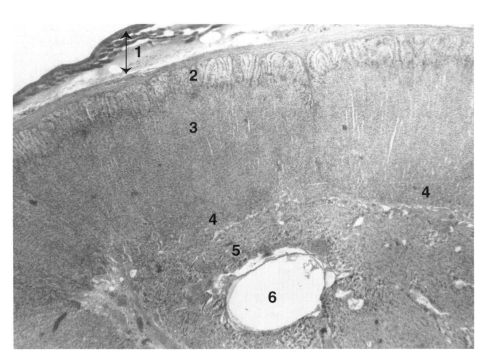

Drawing:

BLOODBORNE PATHOGENS

Use care when working with body fluids.

B. Endocrine System Physiology

PURPOSE

Unit 9B will enable you to understand selected physiological functions of the endocrine system.

OBJECTIVES

After completing Unit 9B, you will be able to

- determine your blood glucose level using an electronic method blood glucose monitor.
- demonstrate and reverse insulin shock in a fish.

Because of the possibility of infections being transmitted from one student to another during exercises involving human body fluids, it is important that you take great care in collecting and handling such fluids. ***The body fluids used in the exercises should be your own. Follow proper procedures for collecting, handling, and disposing of body fluids as directed by your instructor.*** Because of particular concern over exercises involving human blood, it may be preferable for the instructor to perform, as a demonstration, the exercises that use human blood.

EXERCISE 1 **Determination of Blood Glucose Level Using an Electronic Blood Glucose Monitor**

Regulation of the blood glucose level in the body is an endocrine function of the islets of Langerhans in the pancreas. The normal fasting blood glucose level is approximately 60–110 mg/dL, depending on the method used. An elevated level may be indicative of the metabolic disorder **diabetes mellitus**. Decreased blood glucose is characteristic of hypoglycemia. Recent developments in biomedical technology have resulted in rapid, accurate blood glucose tests that may be performed in nonclinical settings.

It is possible to determine the blood glucose level electronically, using a blood glucose monitor.

These instructions are for the Accu-Chek® Advantage blood glucose monitor. If you use another model, follow its accompanying instructions.

The Accu-Chek® Advantage blood glucose monitor is a battery-powered meter that uses Chemstrip bG® Test Strips to measure blood glucose level in a single drop of blood obtained from your finger. **Figure 9.9** illustrates the parts of the Accu-Chek® Advantage blood glucose monitor.

MATERIALS

Accu-Chek Advantage System*
containing:

Accu-Chek Advantage Meter	disposable gloves
Accu-Chek Softclix lancet device	paper towels
Accu-Chek Softclix lancets	sealable, puncture-proof container
Accu-Chek Comfort Curve test strips	large plastic garbage bag

or other blood glucose monitor

*Photo and instructions for use courtesy of Roche Diagnostics, Indianapolis, IN.

BLOODBORNE PATHOGENS

Use a new, sterile lancet or hemolet. Exercise care when working with body fluids.

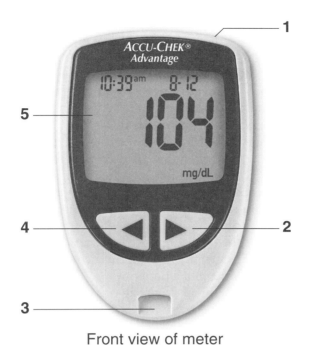

Front view of meter

Figure 9.9

Accu-Chek® Advantage
Blood Glucose Monitor

1. ON/OFF button
2. Right arrow button
3. Test strip slot
4. Left arrow button
5. Display

PROCEDURE*

If you are using the blood glucose monitor for the first time, it will be necessary for you or your instructor to insert the battery, insert the code key for each new vial of test strips you use, and if you use the meter's memory, set the time and date on the meter. You should also run a control test before using the instrument in the classroom. Specific instructions for ensuring that the meter, accessories, and test strips are working properly and applications for using the left and right arrow buttons are in the booklet that accompanies the Accu-Chek Advantage System.

To use the blood glucose monitor, it will, of course, be necessary to prick your finger to express a drop of blood. The drop will more uniformly match the amount required to run the test if you use the lancet device that accompanies the monitor.

After washing your hands, pull the cap that has the numbers 1.2.3. . . off the lancet device and insert a lancet until it clicks.

Twist off the lancet's protective cover and snap the cap of the device back on after lining up the notches. The cap easily snaps into place.

Twist the cap to set the lancet depth. Start at 2 or 3. For tougher skin, dial to a higher number.

Press the plunger at the opposite end of the device. The release button will turn yellow when it is ready. Then set aside.

Before performing the blood test, be sure you will be working over two or three layers of paper towels. Wash and dry your hands, then glove your opposite hand. Take one strip from the vial, then close the vial cap tightly.

*Photo and instructions for use courtesy of Roche Diagnostics, Indianapolis, IN.

Question **9B.1**

What is your blood glucose level?

Question **9B.2**

In what units is blood glucose measured?

Question **9B.3**

Which hormone is secreted by the alpha cells of the islets of Langerhans? What is its function?

Insert the end of the test strip with the silver-colored bars into the meter, with the yellow window facing up. The meter will turn on automatically.

Be sure the code on the meter matches the code on the test strip vial.

When you see the flashing drop of blood on the screen, hold the lancet device to the *side* of your fingertip and press the release button on the side of the device, next to the inscription "Accu-Chek Softclix." Then gently squeeze your fingertip to get a drop of blood.

Touch and hold the drop to the edge—*not* the top—of the yellow window. Make sure the yellow window fills completely. If you still see yellow, apply another drop of blood within 15 seconds, or start over with a new test strip. The hourglass will flash on the screen, then the test result appears. After recording your results, press the on/off button on the top edge of the meter to turn the meter off.

To remove the lancet, remove the lancet device cap and point the lancet end away from yourself. Slide out the ejector at the distal end of the device to discharge the lancet into a puncture-proof container. Discard the paper towels and gloves in a plastic garbage bag.

EXERCISE **2**   **Insulin Shock in Fish**

Insulin is essential for human survival. It facilitates the diffusion of glucose from the bloodstream through the cell membranes of body cells (except erythrocytes, liver, and nerve cells) and into the cell, where it is metabolized. During digestion, glucose molecules enter the blood, elevating the blood glucose level. This stimulates the beta cells of the islets of Langerhans of the pancreas to secrete insulin, which will lower the blood glucose level. An elevated blood glucose level is known as **hyperglycemia**. A decreased level is known as **hypoglycemia**. If the blood glucose continues to drop, the nervous system becomes hypersensitive to impulses, and symptoms of insulin shock—such as irritability and convulsions—appear, leading to coma and death.

In this exercise, you will experimentally induce, observe, and reverse insulin shock in a fish.

MATERIALS

regular insulin (U-100)	U-100 insulin syringes
500 mL beakers	small minnows or large guppies
20% glucose solution	puncture-resistant container
fish net	

PROCEDURE

Place 200 mL of a 20% glucose solution in a 500-mL beaker, label, and save.

Obtain from your instructor a 500 mL beaker containing a small minnow or large guppy in 200 mL of water. Observe and record the fish's gill and swimming movements for a 5-minute period.

After the observation time, add 800 units of regular insulin to the 200 mL of water. You have now experimentally created the conditions for insulin shock.

Observe for any abnormality in the gill and swimming movements of the fish. Watch for signs of insulin shock in your fish by observing rapid and irregular gill movements, or rapid and erratic swimming movements. Floating on the surface of the water (unlikely to occur) indicates that the fish is in a coma.

When you observe any of these signs, use a net to gently remove the fish from the insulin water beaker and immediately place it in the beaker containing 20% glucose solution.

Observe the fish for normal gill and swimming movements. Record the time it takes for your fish to return to its normal pattern of movements. Because of differing experimental conditions such as the fish's size and metabolism, your results may vary from those obtained by others in the class.

Dispose of the insulin syringe with the needle attached and uncapped in a puncture-resistant container.

CAUTION
BLOODBORNE PATHOGENS

Disposable sharps must be placed in puncture-resistant containers labeled as biohazardous to protect others.

The Endocrine System

MATCHING

_____	**1.** both an endocrine and an exocrine gland
_____	**2.** contains the following structures: pars distalis, pars nervosa, and pars intermedia
_____	**3.** comprised of chief and oxyphilic cell types
_____	**4.** usually four of these glands in the human body
_____	**5.** produces tropic hormones
_____	**6.** produces steroid hormones in its cortex and epinephrine and norepinephrine in its medulla
_____	**7.** produces androgens and estrogens in both sexes
_____	**8.** contains follicular and parafollicular cells

a. pituitary
b. parathyroid
c. thyroid
d. pancreas
e. adrenal

Name _____

Section _____

Date _____

MULTIPLE CHOICE

_____ **9.** Which of these hormones may be used to induce labor in a pregnant woman?
  a. luteinizing hormone   **c.** ADH
  b. oxytocin   **d.** progesterone

_____ **10.** Which of these hormones is secreted in response to an elevated blood glucose level?
  a. insulin   **c.** epinephrine
  b. glucagon   **d.** thyroxine

_____ **11.** The chemical element necessary for normal functioning of the thyroid gland is
  a. cobalt.   **c.** zinc.
  b. iodine.   **d.** calcium.

_____ **12.** When there is an insufficient amount of thyroxine in the circulating blood the anterior pituitary normally increases its output of thyroid stimulating hormone. This is an example of
  a. reverse homeostasis.   **c.** decreased metabolic rate.
  b. the all-or-none law.   **d.** negative feedback.

_____ 13. Which of these is not true of the adrenal medulla?
- **a.** It increases liver glycogenolysis.
- **b.** It elevates blood pressure.
- **c.** It is necessary for life.
- **d.** It secretes epinephrine, which increases the heart rate.

_____ 14. Calcitonin is to the thyroid gland as cortisol is to the
- **a.** neurohypophysis.
- **b.** pancreas.
- **c.** adrenal cortex.
- **d.** parathyroid gland.

_____ 15. What is the mechanism of action of insulin?
- **a.** It increases the rate of glucose oxidation within cells.
- **b.** It facilitates transport of glucose across the cell membrane in skeletal muscle and adipose tissue.
- **c.** It increases urinary excretion of excessive blood glucose.
- **d.** It increases the basal metabolic rate.

_____ 16. Which of the following would *not* develop as a result of a hypophysectomy (removal of the pituitary gland)?
- **a.** sterility
- **b.** diabetes mellitus
- **c.** hypothyroidism
- **d.** diabetes insipidus

_____ 17. Females may develop masculinization of features due to a tumor of the
- **a.** adrenal cortex.
- **b.** thyroid.
- **c.** pancreas.
- **d.** thymus.

_____ 18. Glucagon is secreted by which of these glands?
- **a.** adrenal
- **b.** pancreas
- **c.** ovary
- **d.** pituitary

_____ 19. The adrenal glands are located
- **a.** in the neck.
- **b.** in the pelvis.
- **c.** above each kidney.
- **d.** in the chest cavity.

_____ 20. _____ stimulates follicle maturation in the ovary.
- **a.** Thyroid stimulating hormone
- **b.** Insulin
- **c.** Luteinizing hormone
- **d.** Progesterone

_____ 21. _____ is a hormone that lowers serum calcium levels and functions as an antagonist to parathyroid hormone.
- **a.** Glucagon
- **b.** Calcitonin
- **c.** Insulin
- **d.** Growth hormone

_____ 22. The zona glomerulosa of the adrenal gland secretes
- **a.** sex hormones.
- **b.** aldosterone.
- **c.** glucocorticoids.
- **d.** glucagon.

_____ 23. A lack of insulin results in
- **a.** hypothyroidism.
- **b.** hyperthyroidism.
- **c.** hypoglycemia.
- **d.** hyperglycemia.

TRUE OR FALSE

_____ **24.** Hormones influence only the process of metabolism and have no functional control over other bodily processes such as growth and development.

_____ **25.** Hormones are secreted directly into the circulatory system.

DISCUSSION QUESTIONS

26. Using your text or reference books, identify the disorder that may result from:

a. removal of the parathyroid glands

_____

b. hypersecretion of GH in an adult

_____

c. hyposecretion of insulin

_____

d. hypersecretion of corticosteroids

_____

e. hyposecretion of corticosteroids

_____

f. deficiency of thyroxine in an adult

_____

g. insufficient iodine in the diet

_____

h. hypothyroidism present from birth

_____

27. Does the human pituitary have a distinct pars intermedia?

_____

28. How does thyroglobulin differ from thyroxine functionally?

_____

_____

_____

29. If a thyroid gland were overactive, would you expect the follicular epithelium to retain its cuboidal shape? Explain.

_____

_____

_____

CASE STUDY

ENDOCRINE SYSTEM

Mary Jones is 45 years old and has had insulin-dependent diabetes mellitus for 25 years. She injects herself with 20 units of human insulin each morning.

In terms of glucose transport, what is Mary's basic problem?

How does the insulin Mary injects alleviate this problem?

If Mary does not take her insulin, she becomes very thirsty. Why?

The Blood, Lymphatic, and Cardiovascular Systems

A. Anatomy of the Blood, Lymphatic System, Blood Vessels, and Heart

PURPOSE

Unit 10A will enable you to identify the characteristics of blood cells, the lymphatic system, the heart, the blood vessels, and patterns of circulation.

OBJECTIVES

After completing Unit 10A, you will be able to
- identify the components of human blood and frog blood.
- describe the anatomy of the lymphatic system.
- identify differences in structure and function among arteries, veins, capillaries, and lymph vessels.
- identify the structures of the human, cat, and sheep hearts.
- locate the major blood vessels in the cat and human.

MATERIALS

compound microscopes
blood cell models
prepared microscope slides of:
  normal human blood smear
    (Wright's stain)
  frog blood
  sickle cell anemia
  leukemia
  lymph node
  lymphatic vessel
  ileum portion of small intestine
  palatine tonsil
  spleen
  thymus gland
  artery, vein, and nerve
lancets or hemolets
alcohol wipes
disposable gloves
alcohol-cleaned microscope slides
paper towels

rectangular coverslips
model of human lymph
  node
preserved cats
staining jars
staining racks
Wright's blood stain (or dropper
  bottles of Wright's blood stain)
dropper bottles of Wright's buffer
  solution
wax pencils
wash bottles containing deionized
  water
immersion oil
disinfectant solution
puncture-resistant, sealed disposal
  container
demonstration cat with injected
  lymphatic system (optional)
human cadaver (optional)

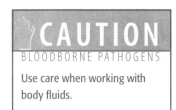

CAUTION
BLOODBORNE PATHOGENS
Use care when working with
body fluids.

Because of the possibility of infections being transmitted from one student to another during exercises involving human body fluids, it is important that you take great care in collecting and handling such fluids. *The body fluids used in the exercises should be your own. Follow proper procedures for collecting, handling, and disposing of body fluids as directed by your instructor.* Because of particular concern over exercises involving human blood, it may be preferable for the instructor to perform, as a demonstration, the exercises that use human blood.

PROCEDURE

EXERCISE 1 **Microscopic Examination of Blood**

Blood is a specialized form of connective tissue and consists of both a fluid component and formed elements. It is transported through vessels where it receives oxygen from the lungs and collects metabolic products from the tissues. Blood also contains hormones, antibodies, nutrients, ions, and other substances, which it transports to all parts of the body.

Fresh whole blood, when centrifuged, separates into two distinct portions: the clear, straw-colored fluid known as **plasma**, and cellular **formed elements.** By volume, more than 50% of blood is plasma. The formed elements include **erythrocytes**, or red blood cells; **leukocytes**, or white blood cells; and **platelets** or **thrombocytes** (**Color Plate 20**).

Erythrocytes are highly specialized for the transport of oxygen. Mature erythrocytes are anucleate: they lack a nucleus, allowing them to carry increased amounts of oxygen. The cells are also biconcave—the two surfaces curve inward toward each other. These properties enable them to accommodate more oxygen-carrying hemoglobin molecules. As they flow through capillaries, erythrocytes have a tendency to adhere to one another along their concave surfaces, a phenomenon known as Rouleaux formation. In human males, there are 5–5.5 million red blood cells per cubic millimeter of blood; in females, 4.5–5 million.

Question 10A.1

Why do males have more red blood cells than females?

Leukocytes are cells containing nuclei of various shapes. Leukocytes average 5,000–9,000 per cubic millimeter in normal blood. The count in children is higher, and variations from the normal number occur in pathological conditions. Different types of leukocytes may be identified in a differential white blood cell count. Leukocytes are classified as agranular and granular.

Agranular leukocytes have a homogeneous cytoplasm and large round or kidney-shaped nuclei. Agranular leukocytes include **lymphocytes** and **monocytes** (**Color Plates 21** and **22**).

When stained with Wright's blood stain, lymphocytes have a relatively large, round, dark-staining nucleus surrounded by a narrow rim of clear bluish cytoplasm. Lymphocytes may be of varying sizes and constitute 25%–30% of the total leukocyte count of normal blood. Lymphocytes are active in the immune response, forming antibodies and releasing chemicals that stimulate the immune response (**Figure 10.1**).

Monocytes are larger than lymphocytes and have an oval or kidney-shaped nucleus surrounded by a pale bluish-gray cytoplasm. These cells constitute

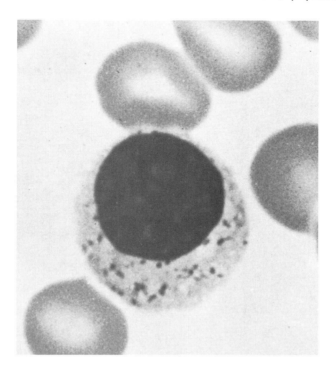

Figure 10.1

Lymphocyte (1000×)
(Color Plate 22)

3%–8% of the total leukocyte count. Monocytes are able to squeeze between the cells of capillary walls by a process known as diapedesis, and then move into tissues, where they become macrophages, which phagocytose (engulf) bacteria and cellular debris for destruction and removal from the cells of those tissues (**Figure 10.2**).

Granular leukocytes can be distinguished by the presence of cytoplasmic granules and nuclei that appear to be more lobed. Granulocytes are of three types: **neutrophils**, **eosinophils** (or **acidophils**), and **basophils** (**Color Plates 23–25**).

Neutrophils are the most numerous leukocytes and constitute 65%–75% of the total leukocyte count. The neutrophil is a polymorphonuclear leukocyte; that is, the nucleus shows a variety of forms but usually consists of three to five irregularly oval-shaped lobes connected by thin strands of nuclear material. The cytoplasm contains fine, pale lilac-colored granules. Neutrophils also function to phagocytose bacteria (**Figure 10.3**).

Eosinophils or acidophils are somewhat larger than neutrophils. The nucleus is usually irregularly shaped and partially constricted into two lobes. The cytoplasm contains coarse, round, uniformly sized, red-orange granules. They constitute 2%–4% of the normal leukocyte count. Eosinophils detoxify foreign substances and increase in number during an allergic response (**Figure 10.4**).

Basophils are difficult to find in human blood, because they constitute only 0.5%–1% of the total leukocyte count. The nucleus is irregularly shaped and may appear as two lobes. The cytoplasm contains irregular, blue-purple granules that usually obscure most of the bilobed nucleus. Basophils secrete heparin, which prevents blood from clotting within vessels (**Figure 10.5**).

Platelets, or **thrombocytes**, are tiny fragments of larger cells known as **megakaryocytes** that are found within the bone marrow. Platelets circulate

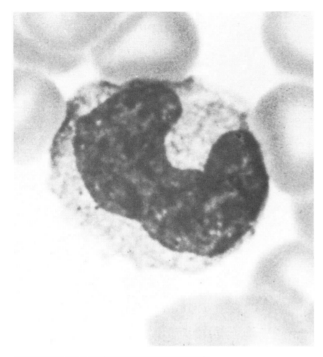

Figure 10.2

Monocyte (1000×)
(Color Plate 21)

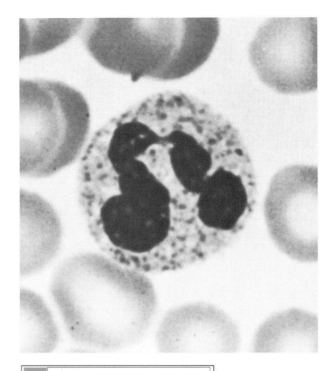

Figure 10.3

Neutrophil (1000×)
(Color Plate 23)

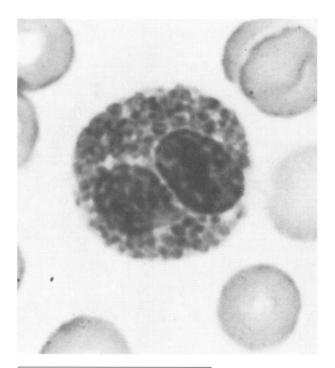

Figure 10.4

Eosinophil (1000×)
(Color Plate 24)

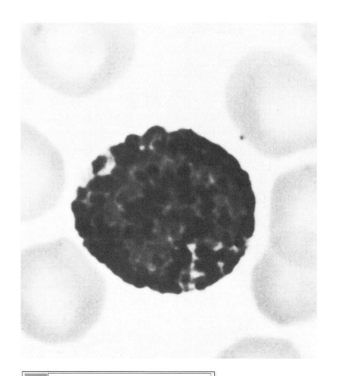

Figure 10.5

Basophil (1000×)
(Color Plate 25)

freely in the blood and function in hemostasis, the process by which platelets adhere to injured areas of blood vessels. Hemostasis then leads to the formation of a blood clot (thrombus). The normal platelet count in adults is 150,000–400,000 per cubic millimeter of blood.

PROCEDURE

Before observing slides of various types of blood cells, you may find it helpful to study models of blood cells and differentiate among **erythrocytes**, **lymphocytes**, **monocytes**, **neutrophils**, **eosinophils**, **basophils**, and **platelets**. Identify those structures printed in boldface on both the models and on slides.

HUMAN BLOOD SMEAR

Examine first under high power, a prepared slide of a human blood smear that has been stained with Wright's blood stain. Then place a drop of immersion oil on the slide and examine it under the oil immersion objective. Identify the following types of cells: **erythrocytes**, **lymphocytes**, **monocytes**, **neutrophils**, **eosinophils**, and **basophils**. Eosinophils and basophils may be difficult to identify in this preparation. You may also see much smaller **platelets** on the slide (**Figure 10.6**).

You may prepare a smear of your own blood. *Read these instructions carefully and follow them precisely; you will need to work quickly in order to attain satisfactory results.*

Wash your hands. Wearing a disposable glove on the opposite hand and working over a paper towel, wipe the tip of your third or fourth finger with an alcohol wipe; allow it to dry.

Quickly puncture the clean finger with a lancet or hemolet. Press the same finger just proximal to the puncture site in order to form a large drop of blood. Place the blood on one end of a dry, alcohol-cleaned slide. Immediately take

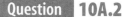
Question **10A.2**

Rank the types of leukocytes from the most to least numerous.

CAUTION
BLOODBORNE PATHOGENS
Use care when working with body fluids.

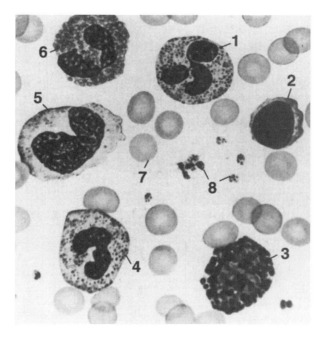

| **Figure 10.6** |
| Human blood, showing formed elements (1000×) (Color Plate 20) |

1. neutrophil
2. lymphocyte
3. basophil
4. band cell (immature neutrophil)
5. monocyte
6. eosinophil
7. erythrocyte
8. platelets

another clean slide and move its edge into the drop of blood (**Figure A**). Holding the slide at about a 30° angle, slowly and evenly pull this slide across the first (**Figures B** and **C**). This pulls a thin film of blood across the slide, resulting in a blood smear.

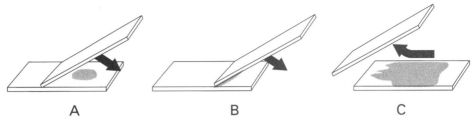

A   B   C

Allow the smear to air dry for at least 2 minutes. Place the smear in a staining jar containing Wright's blood stain, or on a staining rack. Drop Wright's stain over the entire surface of the smear. Allow the smear to be immersed in the stain for 1 minute. Pick up the slide at one end and tilt it until the excess stain runs off.

Set the smear on a staining rack and cover it with 8 drops of Wright's buffer solution; then let stand for $2\frac{1}{2}$ minutes. Rinse gently with distilled water, let air dry, and apply a coverslip. You may also use a staining jar containing buffer solution for this step.

Observe your blood under the oil immersion objective. **Erythrocytes** appear a soft pink color; **neutrophils**, **lymphocytes**, and **monocytes** have lilac-to-purple nuclei; **eosinophils** contain reddish-orange granules, and **basophils** contain dark purple granules (**Color Plates 21–25**). You may also find some very small **platelets** on your slide. You may see occasional **band cells** which are immature neutrophils. Band cells have horseshoe-shaped nuclei.

Before leaving your work area, be sure you have disposed of all hemolets and slides of blood that you prepared in a sealed, puncture-resistant container. Then wipe your work surface with a disinfectant solution. Also be sure you have wiped the excess oil from the oil immersion objective with lense paper.

FROG BLOOD SMEAR

Examine a prepared frog blood smear for eosinophils and basophils; these appear similar to those in human cells. Notice that frog **erythrocytes** are large, nucleated cells (**Figure 10.7**).

SICKLE CELL ANEMIA

Notice the sickle-shaped erythrocytes on this slide. These cells have a reduced capacity for carrying oxygen. Sickle cell anemia is an inherited condition in which the hemoglobin of the abnormally shaped cells contains the amino acid valine substituted for glutamic acid that is present in normal hemoglobin.

LEUKEMIA

Leukemia is a malignant condition characterized by an abnormal increase in the white blood cell count and the presence of immature white blood cells in the peripheral circulation. If your slide is of myelogenous leukemia, there will be an increase in **immature granular leukocytes**. If you have a lymphogenous or a monocytic leukemia slide, you will see an increase in **immature lymphocytes** or **monocytes**.

Question 10A.3

What structure in frog erythrocytes is missing in human erythrocytes?

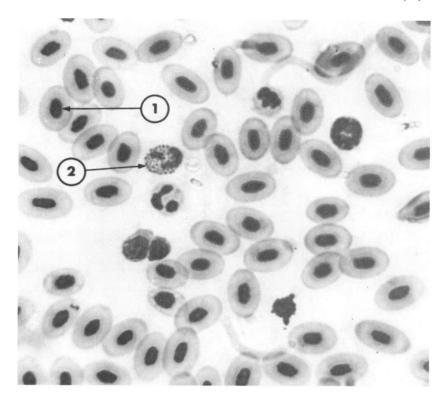

Figure 10.7

Frog blood (430×)

1. nucleus of red blood cell
2. eosinophil

EXERCISE 2   The Lymphatic System

As blood circulates through the body tissues, small quantities of blood plasma leak out of capillaries into tissues. The fluid that leaks out is known as **tissue fluid** or **interstitial fluid**. The **lymphatic system** is a specialized system of vessels, tissues, and organs for the collection, purification, and return of tissue fluid to the bloodstream. The vessels of the lymphatic system are called **lymph vessels** or **lymphatic vessels**. The tissue fluid that is contained and transported through these vessels is known as **lymph**. Lymph vessels are very similar in structure to veins. However, they have much thinner walls. Like veins, lymph vessels contain valves to prevent the backflow of fluid as it flows through them.

Lymph vessels serve as an adjunct to veins. They return lymph to the bloodstream by merging into progressively larger vessels that converge in the chest region to form two large vessels or ducts. The **right lymphatic duct** is formed by the convergence of progressively larger lymph vessels from the right shoulder and arm and the right side of the head and neck. The right lymphatic duct then descends to the junction of the right subclavian and right internal jugular veins, where it returns lymph from these areas into the bloodstream. The **thoracic duct** begins as a dilated sac (**cisterna chyli**) that is formed by the convergence of progressively larger lymph vessels from the lower extremities, abdomen, left shoulder and arm, and left side of the head and neck. The cisterna chyli is located in the abdominal cavity at the level of L2 vertebra. It then constricts as it ascends above the L1 vertebra to form the actual thoracic duct. The thoracic duct then continues into the thorax to the junction of the left subclavian and left internal jugular veins, where it returns lymph to the systemic blood circulatory system (**Figure 10.8**).

Question 10A.4

Why is it important that lymph be returned to the blood circulatory system?

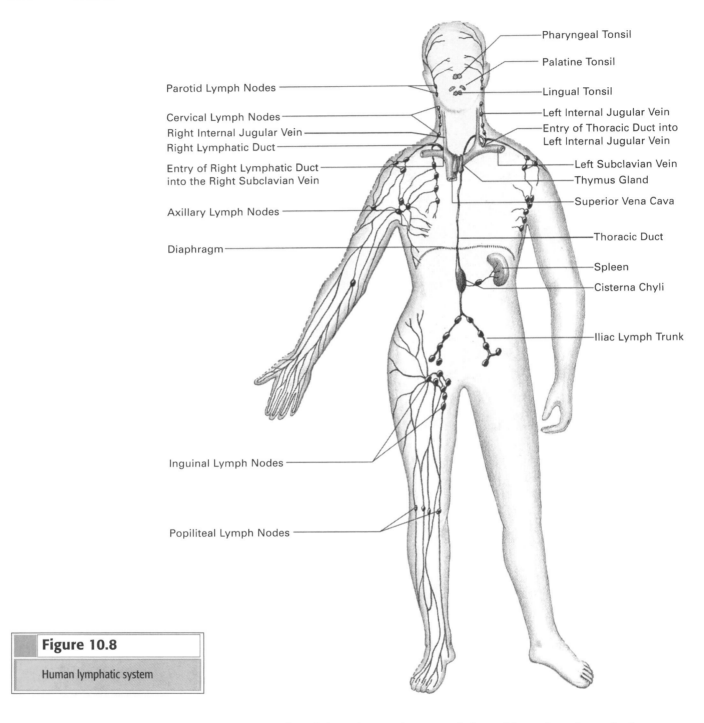

Parotid Lymph Nodes

Cervical Lymph Nodes
Right Internal Jugular Vein
Right Lymphatic Duct

Entry of Right Lymphatic Duct
into the Right Subclavian Vein

Axillary Lymph Nodes

Diaphragm

Inguinal Lymph Nodes

Popiliteal Lymph Nodes

Pharyngeal Tonsil

Palatine Tonsil

Lingual Tonsil

Left Internal Jugular Vein

Entry of Thoracic Duct into
Left Internal Jugular Vein

Left Subclavian Vein

Thymus Gland

Superior Vena Cava

Thoracic Duct

Spleen

Cisterna Chyli

Iliac Lymph Trunk

Figure 10.8

Human lymphatic system

Associated with lymph vessels are oval-shaped **lymph nodes**, which contain lymphocytes and connective tissue. Lymph nodes are found in clusters at intervals along lymph vessels. They function in immunity and also to filter and engulf bacteria, cancer cells, and other substances from lymph. Major groups of lymph nodes are found in the following body regions: cervical, axillary, thoracic, and inguinal (**Figure 10.8**). Lymphatic organs in the human body include lymph nodes, the tonsils, spleen, and thymus gland.

MATERIALS

compound microscope	dissecting pan
prepared microscope slides of lymph node, ileum, lymph vessels, tonsil, spleen, and thymus gland	newspaper
	disposable gloves
	demonstration cat injected for study of lymphatic system (if available)
model of a lymph node	cadaver (if available)
preserved cat	
dissecting kit	

PROCEDURE

EXAMINATION OF LYMPHATIC ORGANS

As you observe each slide, model, your cat, or a cadaver, find the structures printed in boldface.

On a model of a lymph node, observe the concave depression known as the **hilum**, from which two **efferent lymphatic vessels** emerge. These structures carry filtered lymph away from the node. On the opposite side of the node, you will see several **afferent lymphatic vessels**, which convey lymph into the node. Both sets of vessels contain **valves**. Within the lymph node, observe the outer **capsule** and the **trabeculae** of connective tissue extending into the node from the capsule. Note the **lymphatic nodules** containing lighter-colored **germinal centers** of the **cortex** and the **medullary cords** of cells in the inner **medulla** of the node.

Examine a slide of a lymph node under low power. You may see **adipose tissue** and small cross sections of **blood and lymphatic vessels** adjacent to a **capsule** of connective tissue. Strands of connective tissue known as **trabeculae** extend into the node from the capsule. Within the lymph node are two major regions, the cortex and medulla (**Figure 10.9**).

Question 10A.5

Are there germinal centers in the medulla of a lymph node?

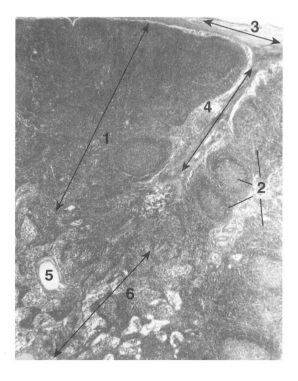

Figure 10.9

Lymph node, overview (40×)

1. cortex
2. follicles
3. capsule
4. trabeculum
5. sinusoid
6. medulla

The outer **cortex** contains dense spherical **nodules** of lymphocytes. These nodules usually stain darker at the periphery and lighter in the center. Under high power, observe these lighter central areas, known as **germinal centers**. In general, the inner **medulla** of the lymph node appears more lightly stained and less dense (**Figures 10.9** and **10.10**).

Observe a slide of the ileum, or distal portion of the small intestine, under low power. Look for dark purple staining groups of lymph nodules known as **Peyer's patches** or **aggregated lymphatic follicles**. Also observe the less dense **germinal centers** inside individual nodules under both low and high power.

Next, observe a slide of a longitudinal section of a lymphatic vessel, first under low power and then under high power. Note the thin vessel walls and locate a **valve** within the vessel. You can find valves in lymphatic vessels at points where the vessel walls appear to bulge (**Figure 10.11**).

The tonsils, the spleen, and the thymus gland are also lymphoid organs. Observe a slide of palatine tonsil under low power. Underlying the surface **epithelium** you will see typical purple-staining, spherical **lymphatic nodules**. If you follow the surface epithelium, you will note that it descends into **crypts** that are lined with the same epithelium. Deeper regions of the slide may contain **striated muscle fibers** and small **blood vessels**.

Under low power, observe a slide of the spleen. It has an outer covering known as the **capsule**, **trabeculae** extending inward from the capsule, and round **lymphatic nodules** (or **splenic corpuscles**). Each splenic corpuscle contains a **central artery**.

Observe a slide of thymus gland under low power. Like the other slides of lymphatic organs, the connective tissue **capsule** covers the surface. In the thy-

Question   10A.6

Name three lymphoid organs that contain a capsule.

Figure 10.10

Lymph node, cortical region (100×)

1. capsule
2. follicle with germinal center
3. afferent lymphatic vessels

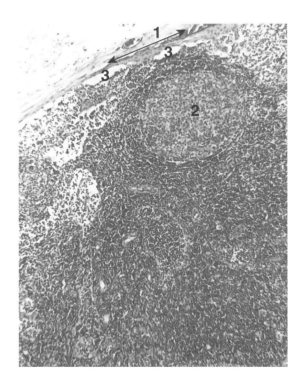

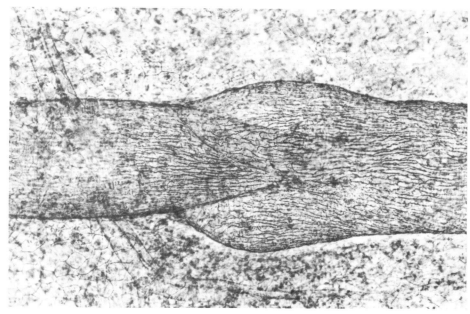

Figure 10.11

Valve of lymphatic vessel (100×)

mus, connective tissue extensions of the capsule are called **septa** rather than trabeculae. The septa divide the thymus into **lobules**, which are shaped like polygons. Each lobule contains a darker staining outer **cortex** and a lighter staining **medulla**. In the center of the medulla of each lobule is one or more cystlike structures known as a **thymic** (or **Hassal's**) **corpuscle**. These corpuscles usually stain pink or red and are composed of nests of epithelial cells.

If a cadaver is available, observe the head region. The paired **palatine tonsils** protrude from the lateral walls of the posterior oral cavity. The paired **lingual tonsils** are elevations on the posterior aspect of the tongue. The single **pharyngeal tonsil**, or **adenoid**, is found within the posterior wall of the nasopharynx (**Figures 11.5** and **10.8**).

The **spleen** is about 12 cm in length and is located in the left hypochondriacal region between the superior portion of the stomach and the diaphragm. It has a soft consistency (**Figures 10.8** and **10.12**).

The **thymus gland** is an elongated bilobed structure in the superior mediastinum, dorsal to the sternum and ventral to the ascending aorta (**Figure 10.8**).

Examination of the Cat Lymphatic System

You will need to open the thoracic cavity in order to see its contents. Because you will later examine the abdominal cavity, you may now open both cavities. Using your scissors (blunt blade down) or a scalpel, make a vertical incision in the chest wall beginning at the midline of the neck and extending inferiorly through the sternum, all the way to the level of the pubic symphysis. As you are cutting, you may meet resistance and hear a cracking sound. Be careful not to cut too deeply, to avoid injuring internal organs.

Figure 10.12

Human spleen

| 1. spleen | 3. small intestine |
| 2. liver | 4. reflected abdominal wall |

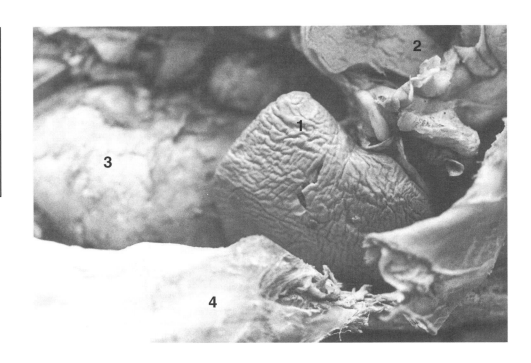

CAUTION

BLOODBORNE PATHOGENS

Be careful when you are making the incision that you do not cut yourself. In the event that you do cut yourself, exercise proper cleanup and disposal of body fluids.

Question 10A.7

Which two body cavities are separated by the diaphragm?

Question 10A.8

In which quadrant(s) of the abdomen is the spleen located?

At the level of the xiphoid process in the thorax, extend the vertical incision laterally and inferiorly to the right and left, cutting through the intercostal muscles to the midaxillary line. Retract the ribs laterally to expose the thoracic cavity. Use your fingers to pull the thin, brown-colored **diaphragm** away from the lateral and posterior body walls. Identify the structures printed in boldface in the following instructions.

Return to the ventral neck region of your cat. Follow the left common carotid artery and left external jugular vein superiorly. Just superior to the point at which the left external jugular vein turns medially and crosses over the left common carotid artery you will see a large, roughly triangular-shaped **lymph node**.

Observe the abdominal region of your cat. In the left lumbar region, locate a firm, dark-brown, slightly roughened, elongated organ that resembles a large tongue. This is the **spleen**.

Also remove the excess fat or connective tissue from the soft, cream-colored, elongated **thymus gland** that is found in the anterior mediastinum.

If a demonstration cat that has been injected with green latex to color the lymphatic system is available, locate the lymphatic structures printed in boldface. The **right lymphatic duct** is in the inferior neck region at the junction of the right subclavian and internal jugular veins. The **cisterna chyli** is in the abdomen, just inferior to the diaphragm and to the right of the second lumbar vertebra. Follow the cisterna chyli superiorly to where it constricts, forming the **thoracic duct**. Note that the thoracic duct enters the circulating blood at the junction of the left internal jugular and subclavian veins. You may also see aggregates of **lymph nodes** in the cervical, axillary, and inguinal regions.

Question **10A.9**

Can an adult human being live without a spleen? Explain.

EXERCISE 3 Microscopic Examination of an Artery, Vein, Capillary and Nerve

The circulatory system transports blood to and from tissues. **Arteries** are large vessels that carry blood from the ventricles of the heart to the tissues in progressively smaller vessels called **arterioles** and **metarterioles**, and then to the **capillaries.** The exchange of oxygen, carbon dioxide, and other metabolic substances occurs between the capillaries and cells of various tissues.

The arterial wall is composed of three distinct layers: the **tunica intima** or innermost layer, which consists of a thin layer of **endothelium** resting on an **internal elastic membrane**; the **tunica media** or middle layer, composed of a layer of thick smooth muscle fibers and elastic connective tissue; and the **tunica adventitia**, or **tunica externa**, which is the outermost layer, made up of loose collagen fibers and adipose tissue. The elastic tissue in the tunica media allows for the pulsation of arteries as blood flows through them. The thick walls of arteries also prevent them from collapsing under periods of low blood pressure and accommodate expansion under high blood pressure.

Question **10A.10**

Why are the layers of the wall of a vein thinner than those of an artery?

Capillaries are thin-walled, single-layer vessels; their single layer of endothelium allows respiratory gases and metabolic substances to diffuse through it easily. Capillaries usually form networks in tissues to allow for complete perfusion of cells with blood.

After passing through a capillary network, blood flows through **veins**. Like arteries, veins are composed of three layers. In veins the layers are thinner and the lumen may be larger and more elongated or irregular in shape.

The tunica media of veins is composed of several layers of smooth muscle, and the tunica adventitia is a layer of collagen fibers. The walls of veins contain valves that assist the return of blood to the heart against gravity. The valves consist of semilunar flaps that open to permit the flow of blood to the heart and close to prevent the backflow of blood. The opening in the center of a vessel through which blood flows is the **lumen**.

Larger arteries and veins also contain minute vessels, the **vasa vasorum**, that nourish the vessel walls themselves. **Nerves** appear as solid structures with threads of connective tissue within them.

Question **10A.11**

What would you expect a capillary model to look like?

MATERIALS

models of arteries, veins, and
  capillaries
compound microscope

prepared slides of arteries, veins,
  nerves, and capillaries

PROCEDURE

You may find that studying models of blood vessels first will help orient you to what you will be looking for on slides. As you observe these structures, identify the parts printed in boldface.

On the artery model, find the **tunica intima, internal elastic membrane, tunica media,** and **tunica adventitia.** Find the **tunica intima, tunica media, tunica adventitia,** and a **valve** on the model of the vein. Also look for **vasa vasorum** in both models.

Examine a prepared slide of a cross section of an artery under low power. Identify the **tunica intima, internal elastic membrane, tunica media,** and **tunica adventitia** (Color Plate 40).

To see the detail of the arterial wall, observe it under low and high power (**Figures 10.13** and **10.14**). Observe the innermost layer, the **tunica intima.** Note the darkly staining, wavelike **internal elastic membrane,** which separates

Figure 10.13

Cross section of artery, vein, and nerve (neurovascular bundle) (40×)

1. lumen of artery
2. lumen of vein
3. nerve fiber
4. lymphatic vessel
5. internal elastic membrane of tunica intima

6. tunica media
7. tunica externa (adventitia)
8. skeletal muscle
9. arteriole

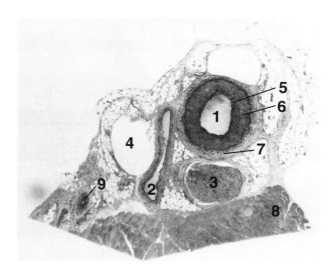

the tunica intima from the tunica media. The **tunica media** contains thick smooth muscle and elastic connective tissue. Look for the **external elastic membrane** that separates the tunica media from the **tunica adventitia.** You may also see **vasa vasorum** in the adventitia layer. Draw and label a section through the arterial wall in Box A (page 312).

Examine a prepared slide of a cross-section of a vein under low and high power (**Figures 10.13** and **10.14**). Identify the **tunica intima, tunica media,** and **tunica adventitia** layers. You may also see **vasa vasorum** in the tunica adventitia. Draw and label a section through the venous wall in Box B (page 312).

If your slide contains a cross section of **nerves,** they appear as lightly stained solid structures surrounded by a sheath of connective tissue.

Under high power, examine a prepared slide of a capillary network. Note the single layer of **endothelial cells** of the capillary wall. Draw the capillary network in Box C (page 312).

Figure 10.14

Artery, arteriole, lymphatic vessel, and vein (400×)

1. artery containing red blood cells
2. arteriole
3. lymphatic vessel
4. vein containing red blood cells

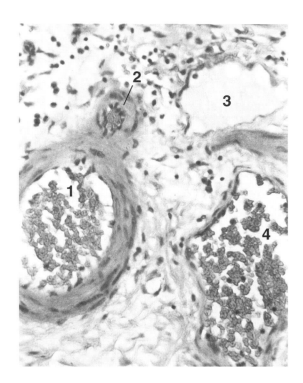

Identification 10.1

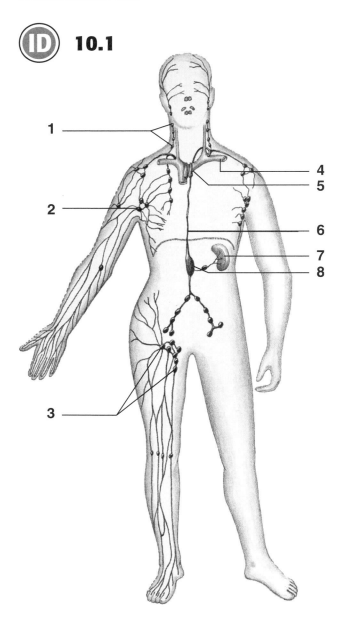

ID **10.1**

Drawings for Exercise 3

A. Labeled drawing of artery (cross section)

B. Labeled drawing of vein (cross section)

C. Drawing of capillary network

EXERCISE **4** The Human Heart

MATERIALS

human heart model
cadaver (optional)

PROCEDURE

Position the heart model so the **apex**, or inferior tapered end, is directed toward the left and the broader **base** of the heart is superior. When looking for various structures, remember that references made to right and left are according to the anatomical position of the organ.

The human heart consists of four separate chambers: two smaller superior **atria** and two larger inferior **ventricles**. Find the atria and ventricles on the model (**Figures 10.15** and **10.16**). Externally, the atria appear flaplike, with the right positioned more dorsally and the left more ventrally. Also find the structures printed in boldface in the following descriptions.

Blood from regions of the body superior to the heart empty into the **superior vena cava**, which empties into the **right atrium**. Blood from regions inferior to the heart empty into the **inferior vena cava**, which also empties into the right atrium. By opening the model and observing the interior aspect of the right atrium, you may be able to find openings or areas representing entry of the venae cavae. From the right atrium, blood flows into the **right ventricle**. Between the right atrium and right ventricle is the **tricuspid valve**, which consists of three flaps, or cusps. From the right ventricle, blood is pumped superiorly and toward the left of the body through the **pulmonary artery**. The pulmonary artery branches into the **right** and **left pulmonary arteries**. These arteries enter the right and left lungs, respectively. In the lungs, carbon dioxide is removed and blood is oxygenated.

Oxygenated blood returns from the lungs to the **left atrium** through four **pulmonary veins,** two from each lung. By observing the inside of the left atrium in the model, you may be able to see four openings representing entry of the pulmonary veins. From the left atrium, blood flows to the left ventricle. Between the left atrium and left ventricle is the **bicuspid**, or **mitral**, **valve**.

Attached to the cusps of the bicuspid and tricuspid valves are slender fibers known as **chordae tendineae**, which are attached to the muscular wall of the ventricles by **papillary muscles**. The bicuspid and tricuspid valves normally close during ventricular contraction, or ventricular systole. Chordae tendineae prevent the cusps of these valves from opening in a reverse direction into the atria when pressure in the ventricles is increased during ventricular systole. From the left ventricle, blood is pumped to the aorta, the largest artery in the body.

Observe the aorta on your model. The **ascending aorta** is the point at which the aorta exits the heart in a superior direction. This large vessel then forms a loop known as the **aortic arch**, which extends posteriorly and inferiorly toward the ventral surface of the heart.

Your model may include the bases of three vessels that arise from the aortic arch. These vessels, in the order from which they emerge from the aortic arch from right to left, are the **brachiocephalic** (or **innominate**) **artery**, which

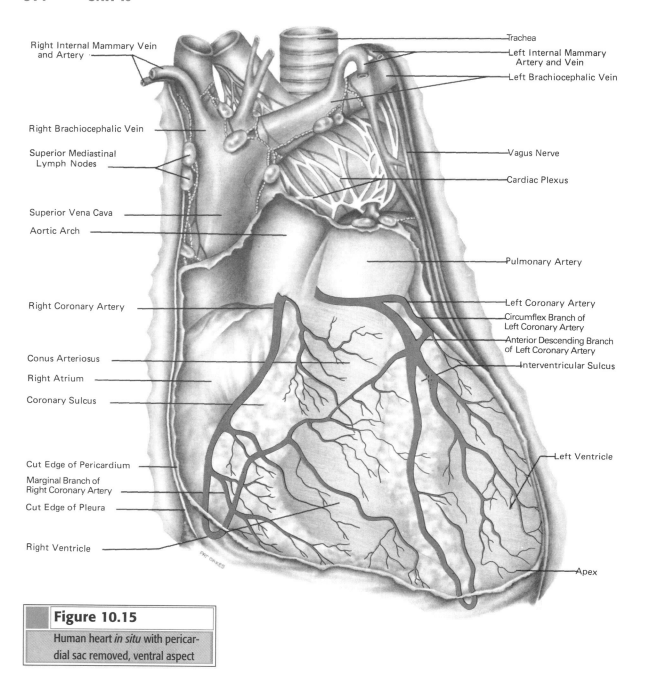

Right Internal Mammary Vein and Artery

Right Brachiocephalic Vein

Superior Mediastinal Lymph Nodes

Superior Vena Cava

Aortic Arch

Right Coronary Artery

Conus Arteriosus

Right Atrium

Coronary Sulcus

Cut Edge of Pericardium

Marginal Branch of Right Coronary Artery

Cut Edge of Pleura

Right Ventricle

Trachea

Left Internal Mammary Artery and Vein

Left Brachiocephalic Vein

Vagus Nerve

Cardiac Plexus

Pulmonary Artery

Left Coronary Artery

Circumflex Branch of Left Coronary Artery

Anterior Descending Branch of Left Coronary Artery

Interventricular Sulcus

Left Ventricle

Apex

Figure 10.15

Human heart *in situ* with pericardial sac removed, ventral aspect

will subdivide into vessels serving the right arm and the right side of the head and brain; the **left common carotid artery,** which serves the left side of the head and brain; and the **left subclavian artery,** which distributes blood to the left arm (**Figures 10.22** and **10.23**).

Return to the pulmonary artery. This vessel contains the **pulmonary semilunar valve** as it exits the right ventricle. This valve opens to allow blood to flow through the pulmonary artery as the right ventricle is contracting. Likewise, the aorta contains the **aortic semilunar valve** as it exits the left ventricle. This valve opens to allow blood to flow through the aorta as the left ventricle is contracting. The pulmonary and aortic semilunar valves normally open at the same time, because the two ventricles contract simultaneously. Note that there are no valves at the entrance of the superior and inferior venae cavae to the right atrium, nor at the entrance of the pulmonary veins to the left atrium.

Question 10A.12

Why are there no valves at the entrance of vessels into the atria?

Figure 10.16

Human heart and major blood vessels

1. ascending aorta
2. aortic arch
3. brachiocephalic artery
4. left common carotid artery
5. right atrium
6. pulmonary artery (trunk)
7. right ventricle
8. left ventricle
9. right lung
10. left lung
11. diaphragm

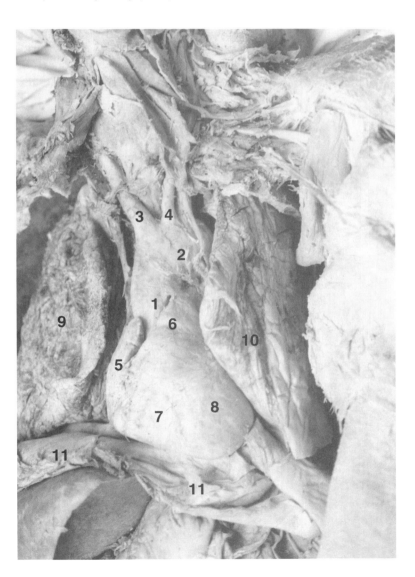

Question 10A.13

Which valves are known as atrioventricular valves, and what is their function?

Question **10A.14**

Why is the interventricular septum thicker than the interatrial septum?

The tough muscular wall of the heart is the **myocardium.** Observe that the myocardium of the atria is much thinner than that of the ventricles. Also note that the left ventricle has the thickest myocardium, a characteristic that is related to its function of pumping blood throughout the systemic circulation (**Figure 10.17**).

The left and right sides of the heart are also separated by myocardium. The thin **interatrial septum** is between the right and left atria. The thick **interventricular septum** separates the left and right ventricles (**Color Plate 47**).

Close the heart model by realigning the ventral portion with the dorsal aspect and observe the anterior portion of the model. Observe the blood vessels on the surface of the heart (**Figure 10.15**). The **left** and **right coronary arteries**, which are probably colored red, branch off the ascending aorta. Follow the left coronary artery as it extends under the left atrium and then divides into two branches. The **anterior descending artery** turns inferiorly down the left ventricle. You may see a groove, the **anterior interventricular sulcus**, into which this vessel fits. The **circumflex branch** of the left coronary artery lies in the **coronary sulcus**, which is a groove that separates the atria from the ventricles. The circumflex artery extends more horizontally to the left of the left coronary artery, curving around to the posterior aspect of the heart. The **right coronary artery** is longer than the left and extends under the right atrium. It also divides into two branches. The **marginal branch** emerges near

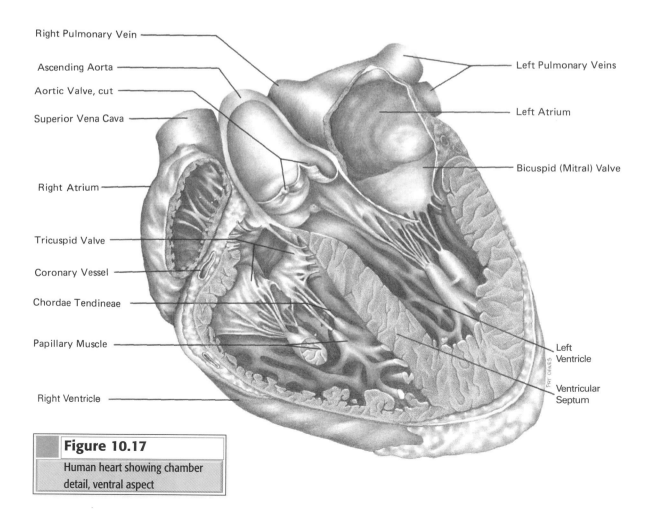

Right Pulmonary Vein

Ascending Aorta

Aortic Valve, cut

Superior Vena Cava

Right Atrium

Tricuspid Valve

Coronary Vessel

Chordae Tendineae

Papillary Muscle

Right Ventricle

Left Pulmonary Veins

Left Atrium

Bicuspid (Mitral) Valve

Left Ventricle

Ventricular Septum

Figure 10.17

Human heart showing chamber detail, ventral aspect

the right lateral border of the right ventricle. The **posterior descending branch** may be seen on the posterior surface of the heart, extending inferiorly in the **posterior interventricular sulcus** between the ventricles to the apex of the heart.

Venous blood from the heart drains into the **coronary sinus**, a large vein on the posterior surface of the heart, before returning to the right atrium. There are two main tributaries transporting blood to the coronary sinus. The **great cardiac vein** begins at the apex of the anterior aspect of the heart and then ascends in the anterior interventricular sulcus to the coronary sinus. The **middle cardiac vein** drains the posterior aspect of the heart beginning at the apex and then ascends in the posterior interventricular sulcus to the coronary sinus.

If a cadaver specimen is available, note that the heart lies obliquely in the chest with the apex pointing inferiorly and to the left. As it lies in the **mediastinum**, or middle of the thoracic cavity, the heart is surrounded by a protective fibrous sac known as the **pericardium**. You will need to remove the pericardium before you can view the heart itself. The external heart wall, or **epicardium**, is covered with varying amounts of yellowish fat that may obscure the coronary blood vessels. By lifting the apex of the heart, you can observe the underlying **cardiac notch**, the concavity in which the heart lies on the medial aspect of the left lung. Return the apex of the heart to its normal position.

Question 10A.15

A large vessel tapers off the superior portion of the right ventricle; which vessel is this?

Question 10A.16

Another large vessel arises from the left ventricle, forms an anterior to posterior loop, and extends inferiorly behind the heart; which vessel is this?

Identification 10.2

(ID) **10.2**

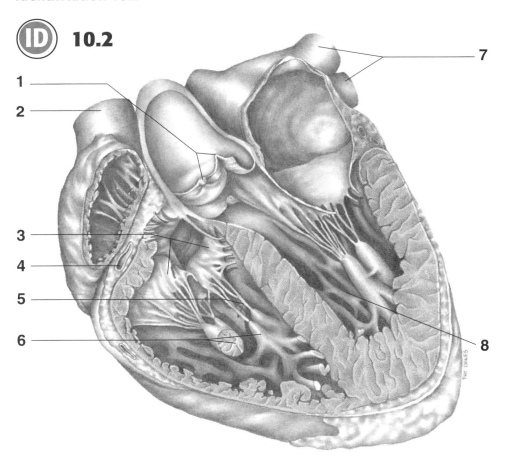

EXERCISE 5   Dissection of the Sheep Heart

MATERIALS

preserved sheep heart   scalpel
dissecting instruments   blunt probe
dissecting pan   straws or wooden probes
disposable gloves

PROCEDURE

Rinse the sheep heart with cold water in order to remove excess preservative or clotted blood. To irrigate the heart's chambers, allow a slow stream of water to run through the large vessels extending into the heart. You will need to remove the pericardium, or pericardial sac to see the heart itself. You may find it helpful to review the structures of the human heart (Exercise 4) before finding comparable structures on the sheep heart.

Using **Figure 10.18** as a guide, identify the following structures on the ventral aspect of your specimen: **trachea, pulmonary artery, pulmonary veins, left atrium, left ventricle, apex, coronary sulcus, right ventricle, right atrium, superior vena cava,** and **aorta.** You may find it helpful to insert straws or wooden probes into the openings leading into the heart. This will help you identify each vessel, based on the chamber it supplies or drains.

Question   10A.17

What would be the appropriate directional term to describe the location of the apex in a sheep heart?

Figure 10.18

Sheep heart *in situ* with pericardial sac removed, ventral aspect

1. trachea
2. pulmonary arteries
3. pulmonary trunk (artery)
4. left atrium
5. left ventricle
6. apex
7. coronary sulcus
8. right ventricle
9. right atrium
10. superior vena cava
11. aorta

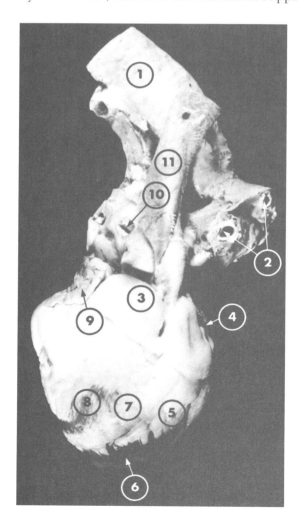

Turn the heart over so that its dorsal aspect is visible. Using **Figures 10.18** and **10.19** as a guide, identify the **trachea**, **pulmonary veins**, **inferior vena cava**, **right ventricle**, and **left ventricle**. Also observe the whitish-yellow **adipose tissue** and any **coronary vessels** on the surface of the heart and in the coronary sulcus. The aorta has a thick wall, the pulmonary veins have thinner walls, and the atria appear to be smaller, grayish flaps of tissue superior and lateral to the ventricles. The right atrium is on the ventral surface and the left is on the dorsal surface of the heart.

Cut the sheep heart in order to view the internal structures. Position the heart on its dorsal surface with the apex pointing inferiorly. Using a sharp scalpel, make a coronal incision through the apex of the ventricles and extend it superiorly until you have cut through both ventricles and the right atrium. Place the two halves of the heart side by side with the apices inferior. Using a blunt probe and **Figure 10.19** as a guide, find the following chambers and valves: **left ventricle**, **right ventricle**, **myocardium**, **interventricular septum**, **chordae tendineae**, **right atrium**, **aortic semilunar valve**, **tricuspid valve**, and **mitral (bicuspid) valve**.

Question **10A.18**

Which ventricular myocardium is thicker? Why?

Figure 10.19

Sheep heart, frontal section

1. aortic arch
2. superior vena cava
3. brachiocephalic artery
4. ligamentum arteriosum
5. pulmonary trunk
6. left pulmonary artery
7. myocardium of left atrium
8. right atrium
9. left atrium
10. tricuspid valve
11. mitral valve
12. chordae tendineae
13. right ventricle
14. left ventricle
15. myocardium of left ventricle

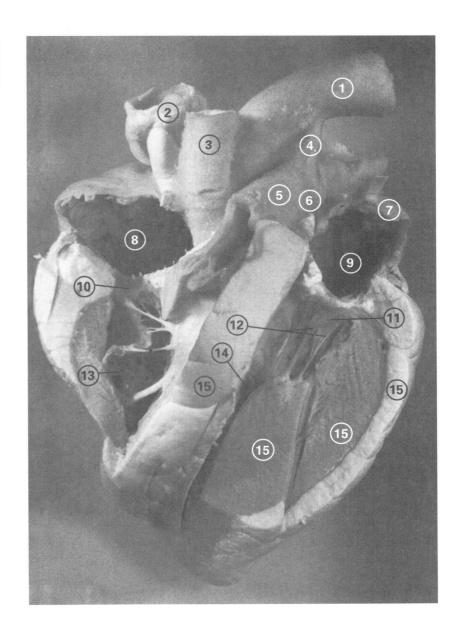

EXERCISE 6 Dissection of the Cat Heart

MATERIALS

preserved cats
dissecting instruments
dissecting pans
disposable gloves

newspaper
wash bottles containing deionized
 water

PROCEDURE

Using your blunt probe, tease away any connective tissue that may obscure your view of the cartilaginous **trachea** and the lungs within the thoracic cavity. Remove the **thymus gland** and any connective tissue that may obscure your view of the cat heart inside its **pericardium**.

The pericardium itself consists of an outer, fibrous **parietal pericardium**, or **pericardial sac**, which surrounds the heart, and an inner, transparent **visceral pericardium**, or **epicardium**, that closely adheres to the myocardium of the heart itself. The **pericardial cavity** is the space between the two layers of pericardium. Using your scissors and forceps, gently cut through the parietal pericardium and remove it from around the heart. Notice that the pericardial sac is not attached to the heart itself but to the superior blood vessels behind the heart.

Locate the superior **base** of the heart, from which blood vessels enter and emerge. The inferior end of the heart that is pointed and located just above the diaphragm is the **apex**.

The **atria** are two irregular, thin-walled chambers at the base of the heart. The two atria are separated from the ventricles by a horizontal groove, the **coronary sulcus**. The scalloped border of each atrium is known as the **auricle**. The large, muscular, posterior chambers of the heart are the **ventricles**. The left ventricle is more muscular and therefore firmer than the right ventricle, which correlates with its function of pumping blood to the peripheral circulation. Externally, the left and right ventricles are separated by a shallow groove or depression, the **anterior interventricular sulcus**, that extends obliquely across the ventral surface from the base to the apex. On the surface of the heart, you may be able to see tiny, threadlike, red **coronary arteries** and blue **coronary veins**. You will probably be able to see the **anterior descending branch** of the left coronary artery, which runs obliquely between the left and right ventricles. Also, note the **coronary sulcus**, a groove that separates the right atrium from the right ventricle.

Without removing the heart from the cat, use your scalpel to make a coronal incision into the heart from apex to base, dividing it into dorsal and ventral halves. Do not totally sever the two halves of the heart at the base. Remove any latex that may be present in the chambers with a blunt probe, and rinse with distilled water. Identify the following internal features of the cat heart, referring to **Figure 10.20** as needed.

The inner surface of the right and left atria is composed of **endocardium**. Note the openings for the **pulmonary veins** in the dorsal wall of the left atrium. The inner surface of the ventricular walls is corrugated because of the presence of muscular cords known as **trabeculae carnae**. Identify the **tricuspid**

Question 10A.19

Is it an advantage or a disadvantage for the pericardial sac to be detached from the heart itself? Explain.

Question 10A.20

Which chambers of the heart receive blood from the anterior descending branch?

Figure 10.20

Detail of cat heart *in situ* with pericardium removed, ventral aspect

1. anterior vena cava
2. brachiocephalic (innominate) artery
3. left subclavian artery
4. aortic arch
5. left coronary artery
6. left ventricle
7. right ventricle
8. posterior vena cava
9. right atrium
10. pulmonary (trunk) artery

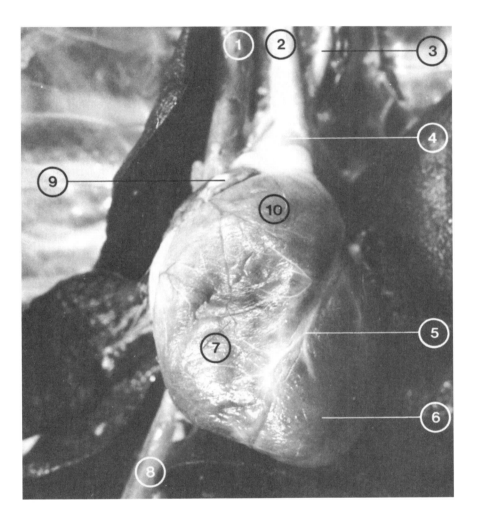

Question **10A.21**

How many cuspid valves
are there in the heart?
Name them.

Question **10A.22**

How many semilunar
valves are there in the
heart? Name them.

valve between the right atrium and right ventricle. Each of the three cusps is
connected to thin fibers, **chordae tendineae**, which in turn are anchored to the
ventricular wall of the myocardium by **papillary muscles**. The **pulmonary
semilunar valve** is located at the base of the **pulmonary artery**. Identify the
bicuspid, or mitral, valve between the left atrium and left ventricle. This valve
is also attached to the ventricular wall by chordae tendineae and papillary
muscles. Also locate the **aortic semilunar valve** that opens into the **ascending
aorta**.

EXERCISE 7   Dissection of the Cat Circulatory System with Reference to the Human

Your specimen has been prepared for the dissection of the arterial and venous blood vessels by the injection of colored latex. Latex simulates the natural texture of blood vessels and aids in their identification. Arteries have been colored red and veins blue. Arteries are injected through the carotid or femoral arteries. Veins are injected through the jugular vein in the neck or the femoral vein in the leg. The skin has been cut at the injection site and you may see a piece of string or thread protruding from that area.

In order for you to identify blood vessels clearly, all arteries and veins must be teased free from the adjacent supporting tissue using a blunt probe. Be careful to avoid severing vessels as you expose them. As you dissect your specimen, identify the vessels printed in boldface in the following discussions.

Circulation is a closed system in that the heart is in the center, and blood vessels form a continuous circuit around the body. Blood vessels are usually named according to direction and location. Identifying blood vessels will reinforce your knowledge of medical terminology because the names of many blood vessels are derived from body regions, bones, or proximity to major organs. As a further aid in identifying blood vessels, remember that arteries carry blood away from the heart; therefore, you will be working from the heart to the peripheral organs. Veins carry blood toward the heart, so you may find it easier to study their circulation moving from the periphery toward the heart.

MATERIALS

double-injected, preserved cats
dissecting instruments
dissecting pan
glycerol-Lysol preservative solution
newspaper
paper towels

large plastic bag
name tags with fastening wires
disposable gloves
human torso
cadaver (if available)

PROCEDURE

DISSECTION OF ARTERIES WITHIN THE THORACIC CAVITY (FIGURES 10.21–10.24)

The heart itself is covered with a protective membrane, the pericardium. The pericardium was removed when you dissected or exposed the cat heart.

The **aorta** is the largest artery in the body. Connective tissue and a gray-to-brown, smooth **lymph node** at the site of exit of the aorta from the heart should be carefully removed in order to expose portions of the aorta. The aorta emerges from the left ventricle of the heart as the **ascending aorta** and then forms the **aortic arch** prior to descending along the dorsal body wall of the thoracic and abdominal cavities. The **descending thoracic aorta** is the portion of the vessel anterior to the diaphragm, a flat sheet of skeletal muscle that separates the thoracic from the abdominal cavities. The **descending abdominal aorta** is the portion of the vessel posterior to the diaphragm. It will be easier to work with blood vessels of the abdomen if you first lift and trim the **greater omentum**, a large apron of connective tissue that covers the ventral surface of

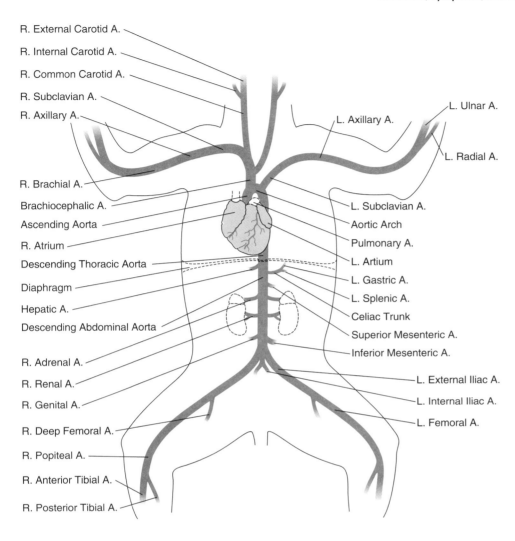

R. External Carotid A.
R. Internal Carotid A.
R. Common Carotid A.
R. Subclavian A.
R. Axillary A.
R. Brachial A.
Brachiocephalic A.
Ascending Aorta
R. Atrium
Descending Thoracic Aorta
Diaphragm
Hepatic A.
Descending Abdominal Aorta
R. Adrenal A.
R. Renal A.
R. Genital A.
R. Deep Femoral A.
R. Popiteal A.
R. Anterior Tibial A.
R. Posterior Tibial A.

L. Ulnar A.
L. Axillary A.
L. Radial A.
L. Subclavian A.
Aortic Arch
Pulmonary A.
L. Artium
L. Gastric A.
L. Splenic A.
Celiac Trunk
Superior Mesenteric A.
Inferior Mesenteric A.
L. External Iliac A.
L. Internal Iliac A.
L. Femoral A.

Figure 10.21

Arterial System of the Cat

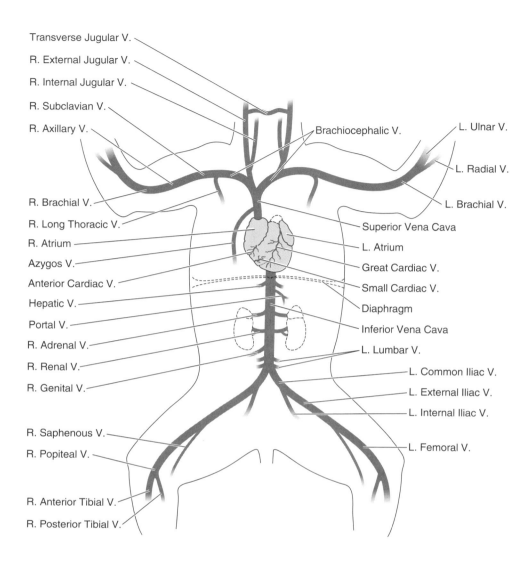

Transverse Jugular V.
R. External Jugular V.
R. Internal Jugular V.
R. Subclavian V.
R. Axillary V.
Brachiocephalic V.
L. Ulnar V.
L. Radial V.
R. Brachial V.
L. Brachial V.
R. Long Thoracic V.
Superior Vena Cava
R. Atrium
L. Atrium
Azygos V.
Great Cardiac V.
Anterior Cardiac V.
Small Cardiac V.
Hepatic V.
Diaphragm
Portal V.
Inferior Vena Cava
R. Adrenal V.
L. Lumbar V.
R. Renal V.
L. Common Iliac V.
R. Genital V.
L. External Iliac V.
L. Internal Iliac V.
R. Saphenous V.
L. Femoral V.
R. Popiteal V.
R. Anterior Tibial V.
R. Posterior Tibial V.

Figure 10.22

Venous System of the Cat

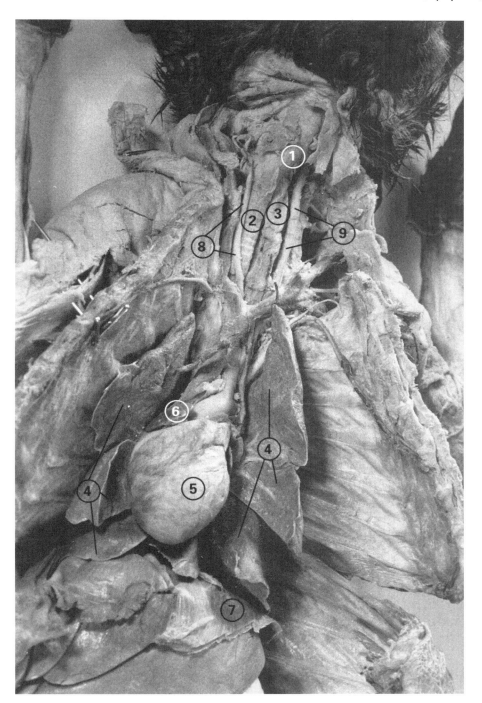

Figure 10.23

Major organs of the thoracic cavity, cat

1. larynx
2. trachea
3. esophagus
4. left lung
5. left ventricle of heart
6. right atrium of heart
7. diaphragm
8. right common carotid artery
9. left common carotid artery

Figure 10.24

Major blood vessels of the neck and thoracic cavity

1. cartilaginous rings of trachea
2. esophagus
3. apex of heart
4. left ventricle
5. right atrium
6. pericardium
7. pulmonary artery
8. left atrium, edge
9. ascending aorta
10. aortic arch
11. brachiocephalic (innominate) artery
12. anterior vena cava
13. right innominate vein
14. left innominate vein
15. right external jugular vein
16. left external jugular vein
17. right common carotid artery
18. left common carotid artery
19. right subclavian artery
20. right axillary artery
21. right brachial artery

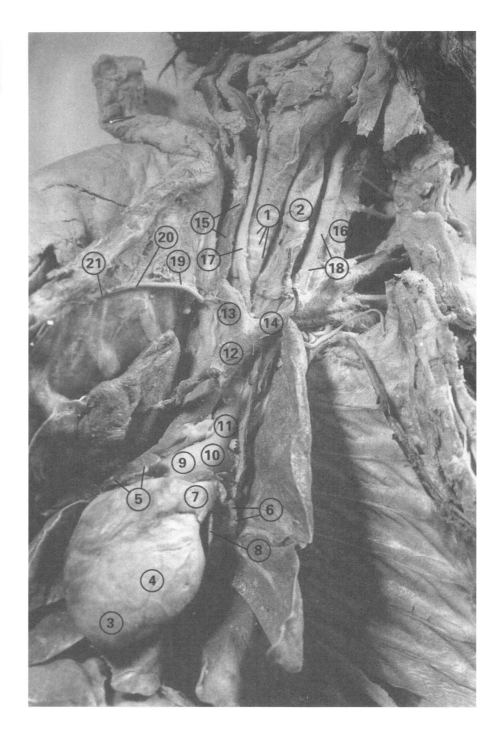

the abdominal cavity. The greater omentum also contains a considerable amount of adipose tissue. Shift the contents of the abdominal cavity to the cat's own right and identify the descending abdominal aorta; then tease away the connective tissue that anchors it to the dorsal body wall.

The **pulmonary artery**, or **pulmonary trunk** tapers superiorly and to the left from the right ventricle of the heart and carries deoxygenated blood to each lung. This vessel is usually injected with blue latex. Follow this vessel and note that it branches into the **right** and **left pulmonary arteries** that enter each lung.

DIVISIONS OF THE AORTIC ARCH AND BRACHIOCEPHALIC ARTERY, AND THEIR BRANCHES   (FIGURE 10.25)

Locate the **brachiocephalic (innominate) artery**, which is the first and largest branch emerging from the aortic arch. In the cat, this vessel divides into a **right subclavian artery**, a **right common carotid artery**, and a **left common**

Question   10A.23

Can an artery transport deoxygenated blood? Explain.

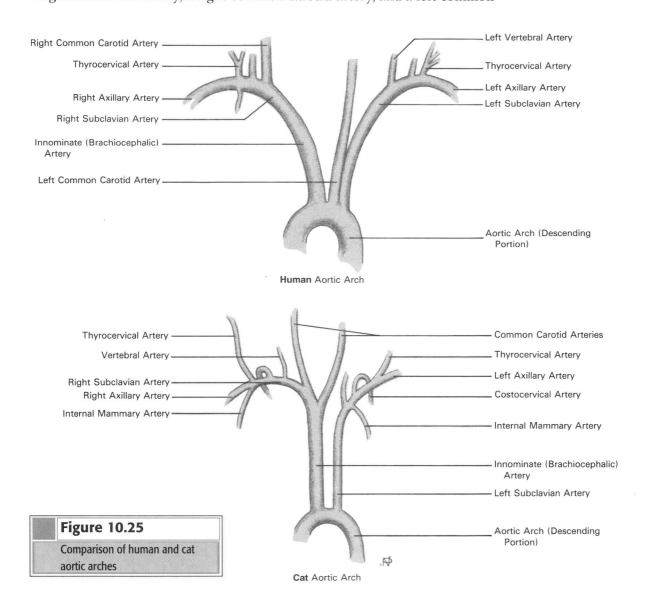

Right Common Carotid Artery
Thyrocervical Artery
Right Axillary Artery
Right Subclavian Artery
Innominate (Brachiocephalic) Artery
Left Common Carotid Artery

Left Vertebral Artery
Thyrocervical Artery
Left Axillary Artery
Left Subclavian Artery

Aortic Arch (Descending Portion)

Human Aortic Arch

Thyrocervical Artery
Vertebral Artery
Right Subclavian Artery
Right Axillary Artery
Internal Mammary Artery

Common Carotid Arteries
Thyrocervical Artery
Left Axillary Artery
Costocervical Artery
Internal Mammary Artery
Innominate (Brachiocephalic) Artery
Left Subclavian Artery
Aortic Arch (Descending Portion)

Cat Aortic Arch

Figure 10.25

Comparison of human and cat aortic arches

Question 10A.24

What are the origins of the terms axillary and brachial?

carotid artery, respectively. The second branch emerging from the aortic arch is the **left subclavian artery.** In the human being, there are three branches off the aortic arch. The first is the brachiocephalic (innominate) artery, which divides into the right subclavian and right common carotid arteries. The second branch is the left common carotid artery. The third vessel branching off the aortic arch is the left subclavian artery.

At the superior border of the larynx, a cartilaginous, boxlike structure at the superior end of the trachea, the common carotid arteries divide into **internal** and **external carotid arteries.** The internal carotid artery supplies blood to the brain. The external carotid serves the scalp.

Blood vessels serving the forelegs and hindlegs will be easier to locate if you expose them on the same side on which you located the muscles of your cat.

As the right (or left) subclavian artery leaves the thoracic cavity, it is termed the **axillary artery.** As the axillary artery extends into the foreleg, it becomes the **brachial artery.** Distal to the elbow, the brachial artery divides into the **radial** and **ulnar arteries.**

Return to the descending thoracic aorta. Note the 10 pairs of small **intercostal arteries** that supply the intercostal muscles, branching from this vessel. Also observe the **esophageal arteries,** which travel along the length of the esophagus, a flattened muscular tube anterior to the descending thoracic aorta.

ARTERIES OF THE ABDOMEN

Now lift the small intestine and find the **descending abdominal aorta** immediately beneath the diaphragm. The **celiac trunk artery** is the first major branch of the descending abdominal aorta near the liver. To expose it, move the viscera to the cat's own right; then cut or tear the diaphragm away from the left peritoneal wall of the cat (if you did not do so in Exercise 2). Using a blunt probe, remove the tough, brown connective tissue from the dorsal abdominal wall, just inferior to the diaphragm. The celiac artery divides into three main branches: the superior **hepatic,** the middle **left gastric,** and the inferior **gastrosplenic artery.** The hepatic artery serves the liver, but before reaching the liver, the **gastroduodenal artery** branches off toward the distal pyloric portion of the stomach. The left gastric artery supplies the lesser curvature and ventral surface of the stomach. In order to see this vessel, remove a small portion of the greater and lesser omentum. The gastrosplenic artery supplies blood to the spleen, stomach, and pancreas.

The next artery branching off the descending abdominal aorta is the **superior mesenteric artery,** which emerges from the ventral surface of the descending abdominal aorta, just posterior to the celiac artery. This large vessel supplies the small intestine and parts of the colon. Fan out the mesentery of the small intestine and observe how the superior mesenteric artery branches to serve the small intestine and parts of the colon.

The **renal arteries** arise about 4 cm posterior to the superior mesenteric artery and emerge laterally to supply each kidney. To see this vessel clearly, remove adipose and connective tissue from around a kidney. Branching may occur before the renal artery enters the kidney.

The **genital arteries** are paired arteries that arise posterior to the renal artery. In a male cat, trace the **internal spermatic artery** from its origin, through the inguinal canal, and down to the scrotum. In a female, trace the **ovarian artery** along the dorsal body wall to the corresponding ovary and uterine horn. The ovarian artery is usually easier to locate than the internal spermatic artery.

Moving in a posterior direction, you will see the **inferior mesenteric artery**, a single vessel that arises from the ventral surface of the abdominal aorta to supply the descending colon and rectum (**Figures 10.21** and **10.26**).

ARTERIES OF THE HINDLEG

Continue following the descending abdominal aorta posteriorly until it bifurcates into the **external iliac arteries**, which are large vessels that extend into the hindlegs. The **deep femoral artery** branches medially off the external iliac artery in the proximal thigh. The external iliac artery continues as the **femoral artery**. Behind the knee, the femoral artery becomes the **popliteal artery**. This artery bifurcates into the **anterior** and **posterior tibial arteries**, which serve the distal hindleg.

The **internal iliac artery** is found immediately posterior to the origin of the external iliac artery. This vessel supplies the uterus and rectum (**Figures 10.21** and **10.26**).

VEINS ASSOCIATED WITH THE HEART, THORACIC CAVITY, NECK, AND FORELEGS

The veins in your specimen may not be as well injected as the arteries, because veins have thinner walls and may not withstand the pressure under which latex is injected. Most veins have the same names as arteries serving corresponding parts of the body (**Figures 10.22** and **10.24**).

Beginning with the heart, observe the two large blue veins entering the right atrium. The **anterior vena cava** enters the right atrium from the superior portion of the body, where it returns blood from the head and forelegs. The **posterior vena cava** enters the right atrium from the posterior portion of the body. The **pulmonary veins**, which return oxygenated blood from the lungs to the left atrium can best be seen by retracting the heart laterally.

Return to the anterior vena cava and reflect the heart and right lung to the cat's own left to see the area of the **azygos vein** immediately to the right of the vertebral column. Tease away connective tissue with your blunt probe to expose it. Branches of the azygos are the **intercostal** and **esophageal veins**.

The **right** and **left brachiocephalic (innominate) veins** drain into the superior vena cava from the head and forelegs.

The **external jugular** and **subclavian veins** drain into the brachiocephalic veins. The largest vein in the neck is the external jugular vein, which drains blood from the cranial region. The **internal jugular vein** from the base of the skull descends and empties into the external jugular vein.

The forelegs are drained by the **ulnar** and **radial veins**, which converge into the brachial vein. This vessel, in turn, drains into the **axillary vein** in the axillary region, which drains into the **subclavian vein**, the more inferior of the two veins entering the brachiocephalic vein.

You may be able to see the **thoracic duct** of the lymphatic system, which enters the junction of the left subclavian and left external jugular veins behind the peritoneum. Recall that this vessel drains lymph from the left superior and both inferior regions of the body.

Question **10A.25**

Can veins transport oxygenated blood? Explain.

Question **10A.26**

How many brachiocephalic arteries are there?

Figure 10.26

Major blood vessels of the
abdominopelvic cavity, cat

1. diaphragm
2. liver
3. stomach (reflected)
4. descending abdominal
   aorta
5. celiac trunk (artery)
6. superior mesenteric
   artery
7. spleen (reflected)
8. left renal artery
9. left kidney
10. left renal vein
11. small intestine
12. posterior vena cava
13. inferior mesenteric
    artery
14. right internal iliac artery
15. right external iliac artery
16. left external iliac artery
17. urinary bladder
18. femoral vein
19. saphenous vein

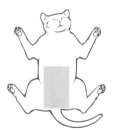

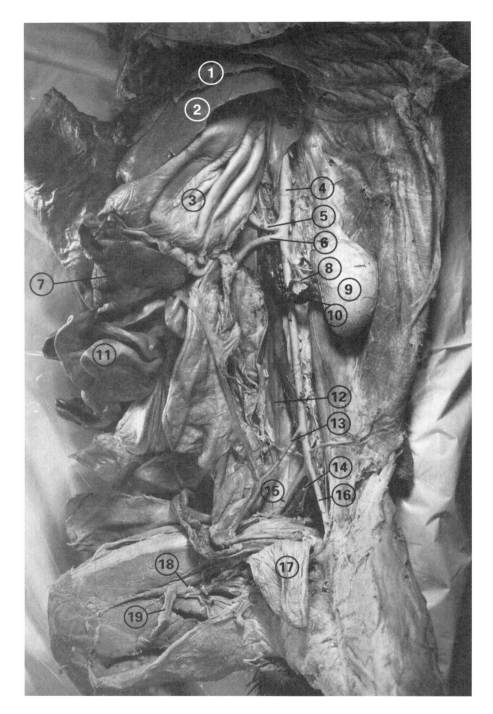

Veins of the Abdomen (Figures 10.22 and 10.26)

You have already identified the posterior vena cava, which enters the right atrium from the posterior portion of the body. This vessel proceeds through the diaphragm where it lies adjacent to the thoracic aorta, then continues along the abdominal portion of the aorta. Tease away any connective tissue along the length of the posterior vena cava through its bifurcation at its inferior end. Abdominal tributaries of the posterior vena cava include the **hepatic veins**, which are best viewed by cutting and removing a wedge of liver tissue.

The **renal veins** drain blood from the kidneys. If you are observing the right kidney, you may see two renal veins emptying into the posterior vena cava.

The **right spermatic** (or **ovarian**) **vein** drains directly into the posterior vena cava. The left empties into the left renal vein.

The **hepatic portal vein** and other lesser tributaries compose the hepatic portal system. Blood from the digestive organs and spleen is carried to the liver through the hepatic portal vein, the largest vessel entering the posterior surface of the liver. To expose the hepatic portal vein, lift up the larger ventral lobes of the liver. This vessel is in the medial region, adjacent to the common bile duct. The **superior mesenteric**, **gastrosplenic**, and **inferior mesenteric veins** drain the small intestine, stomach and spleen, and colon, respectively. It will be difficult to identify the veins of the hepatic portal system unless you have a triple-injected specimen, in which the vessels of this system are colored yellow.

Veins of the Hindleg

Return to the bifurcation at the inferior end of the posterior vena cava and remove any excess tissue using a blunt probe. The vessels forming this bifurcation are the **right** and **left common iliac veins**, which are counterparts of the external iliac arteries. The hindlegs are drained distally by the **anterior** and **posterior tibial veins** which empty into the **popliteal vein** behind the knee. This vein becomes the **femoral vein** in the thigh, which drains into the **external iliac vein**, which in turn drains into the **common iliac vein**. The **greater saphenous vein**, which carries blood from the posterior surface of the calf, also drains into the femoral vein (**Figure 10.22**).

Arteries and Veins of the Human

Using the torso and cadaver, if available, find the human arteries and veins labeled on **Figures 10.27** and **10.28**. The vessels present in the human that you also found in your cat are printed in boldface type.

Question 10A.27

Is there a common iliac artery in the cat?

Figure 10.27

Human circulatory system: major arteries

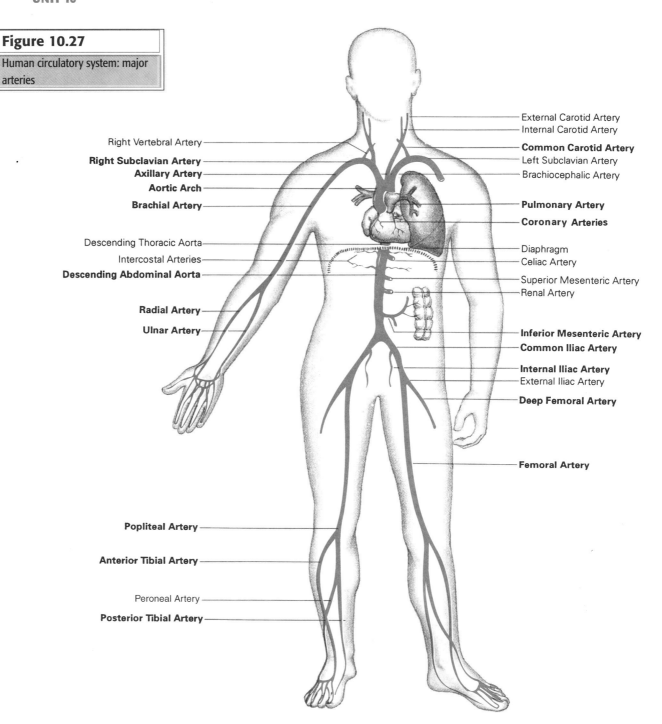

Right Vertebral Artery

Right Subclavian Artery

Axillary Artery

Aortic Arch

Brachial Artery

Descending Thoracic Aorta

Intercostal Arteries

Descending Abdominal Aorta

Radial Artery

Ulnar Artery

Popliteal Artery

Anterior Tibial Artery

Peroneal Artery

Posterior Tibial Artery

External Carotid Artery

Internal Carotid Artery

Common Carotid Artery

Left Subclavian Artery

Brachiocephalic Artery

Pulmonary Artery

Coronary Arteries

Diaphragm

Celiac Artery

Superior Mesenteric Artery

Renal Artery

Inferior Mesenteric Artery

Common Iliac Artery

Internal Iliac Artery

External Iliac Artery

Deep Femoral Artery

Femoral Artery

Question 10A.28

Using arrows, trace a drop of blood from the heart to the hand.

Figure 10.28

Human circulatory system: major veins

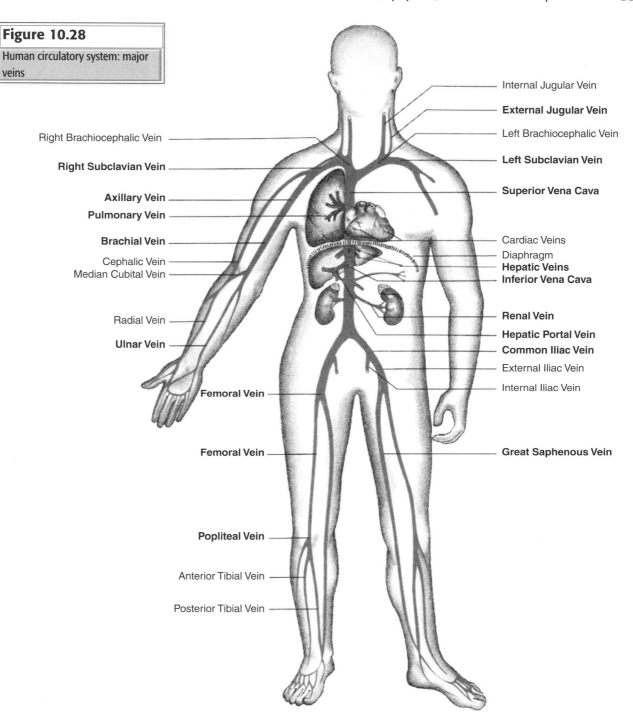

Right Brachiocephalic Vein

Right Subclavian Vein

Axillary Vein

Pulmonary Vein

Brachial Vein

Cephalic Vein

Median Cubital Vein

Radial Vein

Ulnar Vein

Femoral Vein

Femoral Vein

Popliteal Vein

Anterior Tibial Vein

Posterior Tibial Vein

Internal Jugular Vein

External Jugular Vein

Left Brachiocephalic Vein

Left Subclavian Vein

Superior Vena Cava

Cardiac Veins

Diaphragm

Hepatic Veins

Inferior Vena Cava

Renal Vein

Hepatic Portal Vein

Common Iliac Vein

External Iliac Vein

Internal Iliac Vein

Great Saphenous Vein

Question 10A.29

Using arrows and the names of blood vessels, trace a drop of blood from the right ankle to the right atrium.

Identification 10.3

10.3

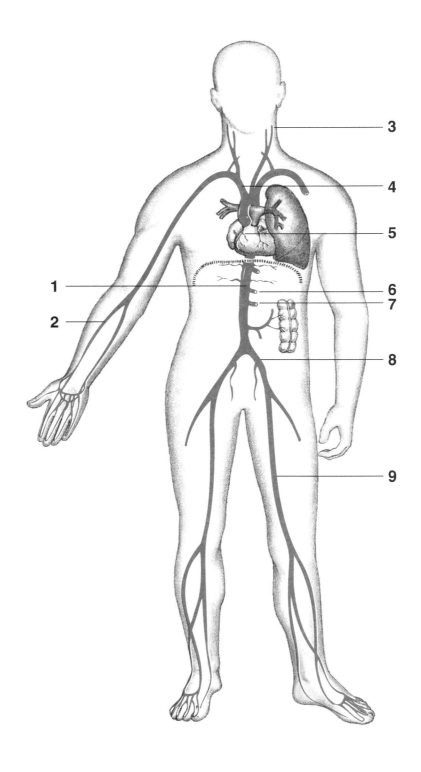

Identification 10.4

 10.4

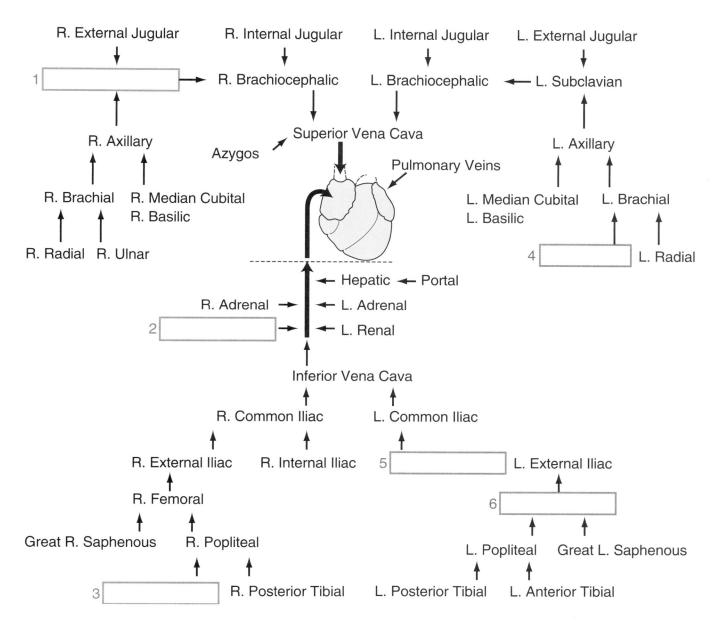

R. External Jugular   R. Internal Jugular   L. Internal Jugular   L. External Jugular

1 [ ] → R. Brachiocephalic   L. Brachiocephalic ← L. Subclavian

R. Axillary   Superior Vena Cava   L. Axillary

Azygos

R. Brachial   R. Median Cubital   Pulmonary Veins

R. Basilic   L. Median Cubital   L. Brachial

L. Basilic

R. Radial   R. Ulnar   4 [ ]   L. Radial

Hepatic ← Portal

R. Adrenal → ← L. Adrenal

2 [ ] → ← L. Renal

Inferior Vena Cava

R. Common Iliac   L. Common Iliac

R. External Iliac   R. Internal Iliac   5 [ ]   L. External Iliac

R. Femoral   6 [ ]

Great R. Saphenous   R. Popliteal   L. Popliteal   Great L. Saphenous

3 [ ]   R. Posterior Tibial   L. Posterior Tibial   L. Anterior Tibial

EXERCISE 8   **Human Fetal Circulation**

Circulation in the **fetus** differs from circulation following birth. This is because the fetal lungs and digestive system are not yet functional; therefore, oxygen and nourishment must be obtained from the mother's blood. Using **Figure 10.29** or a model, find the fetal structures listed in the table titled Human Fetal Structures and Functions.

Question 10A.30

Using the table Human Fetal Structures and Functions, identify structures by their definitions:

1. _____

2. _____

a. atrophies and becomes ligamentum venosum of the liver

b. constricts and becomes ligamentum arteriosum

Human Fetal Structures and Functions

Fetal Structure	Function	Changes Occurring at Birth
Placenta (pla-SEN-ta)	A structure that provides an indirect connection between mother and fetus, through which oxygen, nutrients, and waste products in the blood are exchanged	Shed from lining of mother's uterus immediately after delivery of baby
Umbilical (um-BIL-i-kal) **arteries**	Two vessels that are extensions of the fetal iliac arteries; carry deoxygenated blood from the fetus to the placenta	External portion is shed from umbilicus after umbilical cord is cut; internal portion becomes **lateral umbilical ligaments**
Umbilical vein	A vessel that carries oxygenated blood from the placenta to the fetus, emptying into the inferior vena cava	External portion is shed from umbilicus after umbilical cord is cut; internal portion becomes **round ligament** of liver
Ductus venosus (ve-NŌ-sus)	A continuation of the umbilical vein inside the body of the fetus; passes under the liver, draining blood into the inferior vena cava. Only a small percentage of blood enters the fetal liver	Atrophies, becomes **ligamentum venosum** of liver
Ductus arteriosus (ar-tē′-rē-Ō-sus)	A small vessel that connects the pulmonary artery with the aorta, bypassing the fetal lungs	Constricts and gradually becomes occluded with connective tissue over a period of several weeks, becoming **ligamentum arteriosum**
Foramen ovale (fō-RĀ-men ō-VĀ-le)	An opening in the atrial septum through which blood passes from the right to the left atrium, bypassing fetal lungs	Valve closes over foramen when newborn begins to breathe; complete closure occurs over several months, becoming the **fossa ovalis**

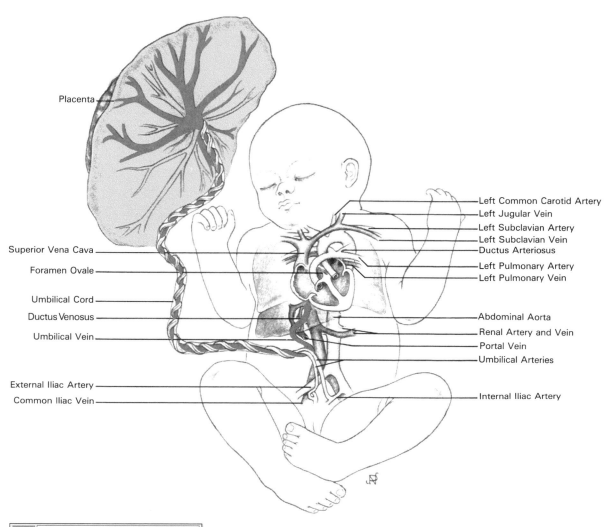

Placenta

Superior Vena Cava
Foramen Ovale

Umbilical Cord
Ductus Venosus
Umbilical Vein

External Iliac Artery
Common Iliac Vein

Left Common Carotid Artery
Left Jugular Vein
Left Subclavian Artery
Left Subclavian Vein
Ductus Arteriosus
Left Pulmonary Artery
Left Pulmonary Vein

Abdominal Aorta
Renal Artery and Vein
Portal Vein
Umbilical Arteries

Internal Iliac Artery

Figure 10.29

Fetal circulation

B. Human Blood Physiology

PURPOSE

Unit 10B will familiarize you with basic blood tests and cardiovascular measurements.

OBJECTIVES

After completing Unit 10B, you will be able to

- demonstrate an understanding of and ability to determine ABO and Rh blood types.
- understand the purpose of measuring hemoglobin and hematocrit levels and perform these tests on a blood sample.
- demonstrate proficiency in pulse and blood pressure determinations and understand their medical significance.

PROCEDURE

Use caution when working with human blood. Read all the blood exercises you will be doing *before* attempting to perform them. Organize and set up the materials for all the exercises you will be doing in sequence *before* puncturing your finger for a blood sample. After finishing these exercises, wipe your work surface with a sponge or paper towel that has been dipped in a disinfectant solution.

EXERCISE 1   Determination of ABO and Rh Blood Types

Red blood cells contain proteins known as **antigens (agglutinogens)** that are located on the surface of erythrocytes. Research has identified more than 100 erythrocyte antigens, each an expression of an inherited gene. Common antigens found in human beings that are important for determining basic blood types are A, B, AB (both A and B), O, and Rh. Your blood type corresponds to the particular antigens that are present. See the table entitled Antigens and Antibodies in ABO Blood Groups.

Antigens and Antibodies in ABO Blood Groups

Blood Group	Antigens on Red Cells	Antibodies in Serum	Frequency Percentage in U.S. Population			
			Whites	African Americans	Native Americans	Asian Americans
A	A	Anti-B	40	27	16	28
B	B	Anti-A	11	20	4	27
O	Neither A nor B	Anti-A Anti-B	45	49	79	40
AB	A and B	Neither A nor B	4	4	<1	5

An individual who is exposed to antigens that are different from those in his or her own tissues will produce proteins known as **antibodies** that will inactivate the foreign antigens. Antibodies are specific; they will react only with antigens that are identical to those that stimulated their production.

In order to determine your ABO blood type, anti-A and anti-B sera will be used. Anti-A serum contains antibody A, anti-B serum contains antibody B. If your blood type is A, there will be **agglutination,** or clumping, with anti-A serum, because antibody A in the serum will react with antigen A on the ery-throcyte surface. Likewise, if blood-type B is present, agglutination will take place with anti-B serum. If you are type AB, agglutination by both anti-A and anti-B sera will occur. If type O is present, neither anti-A nor anti-B sera will result in agglutination. Agglutination is usually visible to the naked eye.

Human erythrocytes are also classified according to the Rh factor as either Rh positive (Rh+) or Rh negative (Rh–), depending on whether or not the Rh antigen is present. This antigen is also known as the D antigen. In order to determine the presence or absence of the Rh antigen, a sample of blood is tested with anti-D serum. If agglutination takes place, the individual is Rh positive; if no agglutination occurs, he or she is Rh negative.

MATERIALS

alcohol wipes	rectangular coverslips
compound microscope	disposable gloves
sterile lancets (hemolets)	puncture-resistant, sealed
flat toothpicks	disposal container
anti-A, anti-B, and anti-D sera	paper towels
(at room temperature)	sponges
light warming box (optional)	disinfectant solution
clean microscope slides	

PROCEDURE

Place three clean microscope slides on a paper towel. Label the slides A, B, and D, respectively.

Wearing a glove on your opposite hand, wipe the pad of your third or fourth finger with an alcohol wipe, then puncture the finger with a hemolet. Place one drop of blood on each slide.

Add one drop of anti-A serum to the drop of blood marked A, a drop of anti-B serum to the slide marked B, and a drop of anti-D (anti-Rh) serum to the slide marked D. Mix each slide of blood and antiserum with a flat tooth-pick. Use a separate toothpick for each slide to avoid contamination. Likewise, *do not allow the droppers containing the sera to touch the slides.*

Allow the slides to stand for 2 minutes. You may then observe for aggluti-nation. You may find it helpful to pick up each slide and gently rock it back and forth while observing for agglutination. To confirm your findings, place a coverslip on the slides marked A and B and observe under a microscope at low power.

Place the slide marked D on the light warming box at 40°C and mix by rocking the slide back and forth for 2 additional minutes. Examine for aggluti-

Question **10B.1**

What would be the blood type of an individual who has neither A nor B antigens on his or her erythrocytes?

Question **10B.2**

Why should someone who is blood type B not receive blood type A in a transfusion?

CAUTION
BLOODBORNE PATHOGENS

Use care when working with body fluids.

nation during this period. If in doubt, apply a coverslip and observe under a microscope at low power.

Record your results in the tables entitled ABO Blood Types and Rh Blood Types. Mark agglutination as $(+)$ and no agglutination as $(-)$.

ABO Blood Types

Anti-A	Anti-B	Your Blood Type	Number of Individuals	Class Percentage	Theoretical Percentage in Your Population Group
					(A)
					(B)
					(AB)
					(O)

Rh Blood Types

Anti-D	Your Rh Type	Number of Individuals	Class Percentage	Theoretical Percentage*	
				Caucasian	African American
				85	88
				15	12

*Research statistics are not available at this time for Native American and Asian-American populations.

EXERCISE **2**   Hemoglobin Estimation

Two methods of determining the hemoglobin level of blood are the older Tallquist method and a more recent and accurate method that uses a **hemoglobinometer.**

TALLQUIST METHOD

The amount of hemoglobin present in red blood cells is an indicator of the oxygen-carrying capacity of the blood. A simple, semiquantitative method of measuring hemoglobin is to compare a small piece of Tallquist paper that has been saturated with a drop of blood with a Tallquist color scale.

MATERIALS

Tallquist scale and test paper
alcohol wipes
sterile hemolets
disposable gloves

paper towels
puncture-resistant, sealed disposal
    container

PROCEDURE

Working over a paper towel, remove one square of paper from the Tallquist booklet of test paper. Wipe your third or fourth finger with an alcohol wipe. Holding the tip of the hemolet with your gloved hand, quickly puncture your finger. Wipe away the first drop, and place the second large drop of blood in the center of the test paper. Wait 15 seconds and then compare your sample with the Tallquist scale. Record the scale estimate.

HEMOGLOBINOMETER METHOD   (FIGURE 10.30)

A more accurate method of determining the hemoglobin content of a blood sample is by using a hemoglobinometer, an instrument that compares the absorption of light by the hemoglobin of a hemolyzed sample of blood with a color standard within the instrument.

Question 10B.3

In what units per volume is hemoglobin measured?

MATERIALS

hemoglobinometer, with charged
    batteries
hemolysis applicator
sterile lancets (hemolets)
sterile alcohol wipes

disposable gloves
paper towels
puncture-resistant, sealed disposal
    container

Question 10B.4

In general, how do hemoglobin values of males and females compare?

PROCEDURE

After identifying the sample chamber, cover plate, and clip on the hemoglobinometer, remove the sample chamber from the side of the hemoglobinometer and open it (**Figure 10.30**).

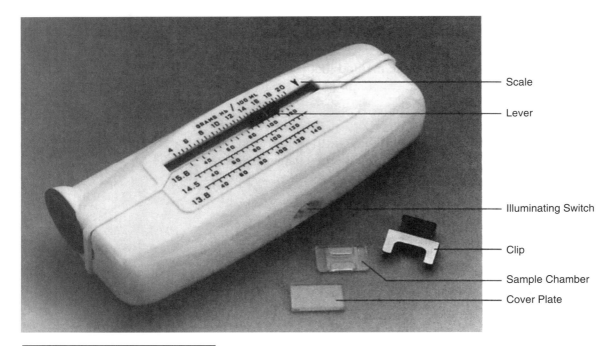

Figure 10.30

Hemoglobinometer

Question 10B.5

What actually happens to red blood cells when they are hemolyzed?

Cleanse your third or fourth finger with a sterile alcohol wipe. Obtain a drop of fresh blood using a sterile lancet held in your gloved hand. Drop the blood from your finger into the sample chamber.

Using a hemolysis applicator, hemolyze the blood in the sample chamber by stirring it for 40–45 seconds. Hemolysis has occurred when the blood changes from a cloudy to a clear red solution.

Slide the sample chamber into the side of the hemoglobinometer, raise the eyepiece to eye level, turn on the illuminating switch on the bottom of the hemoglobinometer, and observe the split field through the eyepiece. Slowly move the lever on the side of the instrument until the colors of the two fields match. Then read and record the hemoglobin concentration from the scale on the side of the hemoglobinometer.

Hemoglobin concentration for adult males averages 14–16 grams/100 mL, for adult females 12–14 g/100 mL.

Question 10B.6

What is your hemoglobin concentration?

Question 10B.7

How does this compare with your result using the Tallquist scale?

EXERCISE 3 Pulse Determination

In addition to height and weight, there are four measurements, or vital signs, that are used to evaluate an individual's general health. These are temperature, pulse rate, respiratory rate, and blood pressure. In this exercise, you will determine your pulse rate by both palpation (feeling pulses through the skin) and electronically.

As the chambers of the heart contract, they exert pressure, which results in blood flow. Because the atria pump blood only to the ventricles, pressure within the atria is low. The right ventricle pumps blood to the lung capillaries, which are also under low pressure. The left ventricle, which pumps blood through the arteries, capillaries, and veins of the systemic circulation and thus exerts the greatest pressure against the blood in this chamber, has the thickest wall. Because the aorta exits directly from the left ventricle, it needs to withstand the high pressure generated from the contraction of the left ventricle.

As blood leaves the left ventricle and enters the arterial system, it flows in spurts. The spurting action results in waves of pressure known as **pulses**, which correspond to the contraction of the ventricles. Pulses are normally felt or palpated through the skin where an artery runs over a bony prominence. The usual pulse rate in an adult is 60–100 per minute, with 72–80 being average.

PROCEDURE

DETERMINATION OF THE RADIAL PULSE

Use the index and middle fingers of one hand to palpate your pulse in the opposite wrist. Do not use your thumb because a pulse is present in the thumb itself. Place your forearm with the palm up on the laboratory table. Reaching over your wrist, feel the radial pulse posterior to your thumb, between the large tendon and bony prominence (styloid process of the radius) at the lateral aspect of the wrist. Too much pressure will obliterate the pulse. You also need to concentrate, screening out other distractions or sensations. Practice until you can feel the pulse without its fading in and out. If the beats occur regularly, count them for 15 seconds, then multiply by 4 to obtain beats per minute (bpm).

Record your radial pulse rate

while sitting quietly _____ bpm

immediately after running in place for 1 minute _____

immediately after hyperventilating (breathing rapidly and deeply)
for 30 seconds _____

DETERMINATION OF CAROTID AND DORSALIS PEDIS PULSES

To find your carotid pulse, place the second and third fingers of one hand on one side of the larynx ("Adam's apple") in your neck. Gently press medially and dorsally until you feel the pulse. Count the beats for 15 seconds and multiply by 4.

Rate _____

The dorsalis pedis pulse can be found in the midline of your foot at the junction of the tarsals and metatarsals. Apply moderate pressure, and count the beats for 15 seconds and multiply by 4.

Rate _____

Question 10B.8

If an individual's heart rate is 84, what would you expect his or her pulse rate to be?

Question 10B.9

How did your radial pulse rate compare with your other pulse rates?

Question 10B.10

Why should you avoid palpating both carotid pulses at the same time?

ELECTRONIC DETERMINATION OF PULSE RATE AND OXYGEN SATURATION

It is possible to measure the pulse rate electronically. In clinical situations, the radial pulse and another determination, oxygen saturation (SpO_2), are often performed together in this manner. SpO_2 refers to the percentage of oxygen-carrying sites on hemoglobin molecules of arterial blood that are saturated with oxygen. In most individuals, this figure is between 95–100% and closely parallels the measurement of pO_2, or the partial pressure of oxygen in arterial blood.

The instrument that measures both pulse rate and oxygen saturation is a pulse oximeter. Two wavelengths of light, one red, the other infrared, are passed through body tissue (typically a finger), to a photoelectric sensor. Each time the heart beats, capillaries dilate, recording a pulse. The sensor also measures the amount of light that is absorbed by oxygen-saturated blood, which is recorded as percent saturation.

MATERIALS

Digit™ Finger Pulse Oximeter® (**Figure 10.31**) or similar device

PROCEDURE

Remove the oximeter from its carrying case and verify that it contains two AA batteries. Insert your index finger into the built-in clip at the end of the instrument designated with the inscription DIGIT. If you are wearing nail polish, your results may not be accurate. Wait until %SpO_2 and pulse beats per minute, BPM, are displayed. The strength of your pulse is proportionate to the number of horizontal bars that are displayed to the left of the %SpO_2 readout.

You may also follow the same protocol after running in place for 1 minute, and after hyperventilating for 30 seconds.

Record your results:

At rest: *After running in place:*   *After hyperventilating:*

%SpO_2 _____   %SpO_2 _____   %SpO_2 _____

BPM _____   BPM _____   BPM _____

Question 10B.11

How did your results at rest compare with typical adult values?

Question 10B.12

What changes occurred in oxygen saturation and pulse after running in place? hyperventilating?

Figure 10.31

Digit™ Finger Pulse Oximeter®

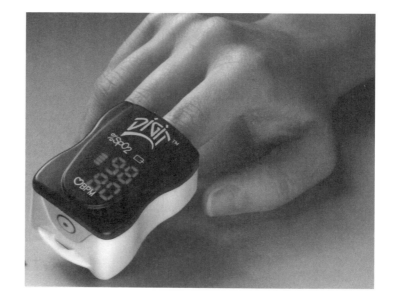

Photo courtesy of Smiths Medical PM, Inc.

EXERCISE 4   ## Blood Pressure Determination

As blood flows from the heart, it exerts pressure against the arterial walls. This pressure is measured in millimeters of mercury. **Systolic blood pressure** is the pressure exerted against the wall of the artery in which the pressure is being measured when the ventricles of the heart are contracting. **Diastolic blood pressure** is the pressure exerted against the wall of the same artery when the ventricles are relaxed. Blood pressure is written as systolic pressure over diastolic pressure. For example, if systolic pressure is 120 and the diastolic reading is 82, arterial blood pressure is ___/___.

DETERMINATION OF ARTERIAL BLOOD PRESSURE BY AUSCULTATION

Extend your partner's forearm across a table with the palm up. Either arm may be used, but the left is preferable. Use the same arm for Exercises 3 and 4 of this unit.

The instrument routinely used to measure blood pressure is a **sphygmomanometer**. It consists of a mercury or aneroid manometer and a cuff that contains an inflatable rubber bladder, tubing, and a squeeze bulb with a screw type valve attached to it.

MATERIALS

mercury or aneroid (dial) sphygmomanometer
stethoscope

PROCEDURE

Wrap the cuff of the sphygmomanometer around your lab partner's arm 2″ above the elbow. The inflatable area should be 2″ above the antecubital fossa. Some cuffs are marked with arrows to assist you in its proper positioning. The cuff should fit snugly and be secured with the velcro tabs or clips on the cuff itself. Slip the dial into the tab on the side of the cuff or set the mercury column vertically on the lab table. Place the stethoscope in your ears with the earpieces facing forward. Place the flat diaphragm of the stethoscope over the medial portion of the antecubital fossa.

Inflate the cuff by turning the screw next to the top of the bulb clockwise to its fullest extent. Then squeeze the bulb until the dial or mercury column reaches 180. Slowly open the valve by turning the screw in a counterclockwise direction. Listen carefully as you let air out of the cuff at a rate of about 4 mm between sounds. The point at which the first tapping sound is heard is the systolic pressure. The point at which the sounds assume a muffled quality represents diastolic pressure. These regular tapping sounds of blood pulsating through the artery are the **sounds of Korotkoff.** If you cannot detect the change to the muffled sound, the point at which the tapping ceases is considered to be diastolic pressure. It is important that one method be used consistently with any given subject. If you did not distinctly hear the sounds, repeat the procedure, allowing the pressure in the cuff to return to zero between readings. After completing the readings, refer to the table entitled Mean Blood Pressure Values for Men and Women.

Question 10B.13

How did the pulse rates you obtained through palpation compare with those determined electronically?

Question 10B.14

Why is it preferable to measure blood pressure in the left arm?

CAUTION

Be sure not to cut off circulation to the arm and hand. Monitor your partner carefully for signs of faintness.

Question 10B.15

What is the anatomical advantage of placing the earpieces of the stethoscope in your ears in a forward direction?

Question 10B.16

What is your blood pressure?

DETERMINATION OF ARTERIAL BLOOD PRESSURE ELECTRONICALLY

Electronic blood pressure measurement allows for the assessment of blood pressure without using a stethoscope. A microphone in the cuff detects the sounds of Korotkoff as the cuff is deflated. These signals are then processed electronically. The internal portion of the instrument panel contains an aneroid manometer and a battery-powered electrical system. Depending on the model used, the actual blood pressure will be displayed on a digital read-out or by a flashing light on a panel. The light will begin to flash at the point of systolic blood pressure and stop at the point of diastolic pressure. This particular exercise describes the model utilizing the flashing light.

Mean Blood Pressure Values for Men and Women

Age (Years)	Men Systolic	Men Diastolic	Women Systolic	Women Diastolic	Age (Years)	Men Systolic	Men Diastolic	Women Systolic	Women Diastolic
½	89	60	93	62	18	120	74	116	72
1	96	66	95	65	19	122	75	115	71
2	99	64	92	60	20–24	123	76	116	72
3	100	67	100	64	25–29	125	78	117	74
4	99	65	99	66	30–34	126	79	120	75
5	92	62	92	62	35–39	127	80	124	78
6	94	64	94	64	40–44	129	81	127	80
7	97	65	97	66	45–49	130	82	131	82
8	100	67	100	68	50–54	135	83	137	84
9	101	68	101	69	55–59	138	84	139	84
10	103	69	103	70	60–64	142	85	144	85
11	104	70	104	71	65–69	143	83	154	85
12	106	71	106	72	70–74	145	82	159	85
13	108	72	108	73	75–79	146	81	158	84
14	110	73	110	74	80–84	145	82	157	83
15	112	75	112	76	85–89	145	79	154	82
16	118	73	116	72	90–94	145	78	150	79
17	121	74	116	72	95–106	145	78	149	81

From Diem, K., and Lentner, C., eds. *Documenta Geigy Scientific Tables.* 7th ed. Basle, Switzerland: J. R. Geigy S. A., 1970. With permission.

MATERIALS

electronic blood pressure cuff and instrument panel

PROCEDURE

The cuff should be securely wrapped around your partner's upper arm with the microphone over the brachial artery rather than in the antecubital fossa, as with the auscultatory method. The microphone is marked with a circle on the cuff. The correct position of the microphone is about 2″ above the elbow, on the flat, medial portion of the arm where the biceps muscle appears to flatten out. When using the cuff on a heavier arm, move the microphone more laterally and toward the antecubital fossa.

Plug the cuff into the instrument panel. Turn on the instrument and close the air valve by rotating the screw on the bulb in a clockwise direction. Inflate the cuff to 180 mm Hg by squeezing the bulb. Disregard any flashing of light that may occur while the cuff is being inflated.

Momentarily tighten the screw when the dial reads 180, then slowly deflate the cuff by turning the screw counterclockwise. Deflate the cuff at the rate of about 4 mm per second. The point on the dial at which the first regular flash of light occurs is the systolic pressure. The light will continue to flash synchronously with the pulse until the diastolic pressure is reached, as indicated by the last flash of light.

Measure and record two consecutive blood pressure readings on your lab partner, allowing 15–20 seconds between readings.

1: _____   2: _____

Question 10B.17

How does your blood pressure compare with the value given in the table?

CAUTION

Be sure not to cut off circulation to the arm and hand. Monitor your partner carefully for signs of faintness.

Question 10B.18

How did the blood pressures obtained by auscultation compare with those measured electronically?

Question 10B.19

Describe two sources of error that may occur using the auscultatory method.

C. Cardiac Muscle Physiology

PURPOSE

Unit 10C will demonstrate the action of the heart and properties of cardiac muscle when stimulated by various agents.

OBJECTIVES

After completing Unit 10C, you will be able to
- identify the normal wave configurations of the human electrocardigram.
- identify variations from normal wave configurations of the human electrocardiogram.
- use a computer to recognize the normal cardiac cycle of the frog heart.
- use a computer to observe the effect of vagal nerve inhibition on the cardiac cycle.
- use a computer to determine the effect of adrenalin on the cardiac cycle and its influence on the cardiovascular system.

EXERCISE 1   **The Human Electrocardiogram**

An **electrocardiogram** (ECG) is a tracing of the heart's electrical activity as impulses are conducted through the myocardium.

The conduction system of the heart (**Figure 10.32**) is composed of tissues that are specialized for conducting electrical impulses that initiate each heartbeat. The conduction system consists of four parts: (1) the sinoatrial node, (2) the atrioventricular node, (3) the atrioventricular bundle, and (4) the Purkinje fibers.

Each impulse normally begins in an area of specialized muscle tissue in the superior portion of the right atrium known as the **sinoatrial (SA) node**. The SA node is the "pacemaker" of the heart. It initiates a steady rhythm of impulses at a rate of 60–100 times per minute, which is normally faster than any impulses that would be initiated by other areas of the myocardium. From the SA node, each impulse spreads through conduction pathways in the two atria. The atria contract as a unit immediately following this electrical excitation. The impulse then normally continues to the **atrioventricular (AV) node**, located in the inferior medial wall of the right atrium. From the AV node, each impulse travels to the **atrioventricular (AV) bundle (bundle of His)** at the superior portion of the interventricular septum. The impulse continues through the AV bundle and down the **right** and **left bundle branches**, ending with the **Purkinje fibers** within the myocardium of the ventricles. The ventricles contract as a unit immediately following ventricular excitation.

The recording of each heartbeat contains three units: the P wave, the QRS complex, and the T wave. The **P wave** represents electrical impulse conduction (depolarization) through the atria. Contraction of the atria immediately follows the P wave. The **QRS complex**, which includes Q, R, and S waves, represents depolarization of the ventricles. This depolarization is immediately followed by contraction of the ventricles. The **T wave**, which is the third com-

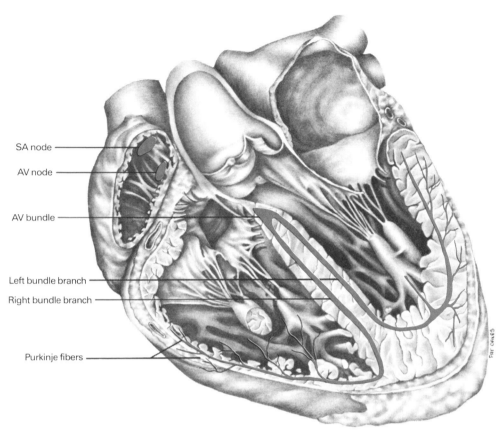

SA node

AV node

AV bundle

Left bundle branch

Right bundle branch

Purkinje fibers

Figure 10.32

The conduction system of the heart

ponent of the tracing of each normal heartbeat, represents electrical recovery (repolarization) of the ventricles and is immediately followed by ventricular relaxation (**Figure 10.33**). It is not possible to see the wave that represents atrial repolarization in a normal ECG, because it is masked by the QRS complex.

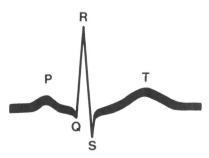

Figure 10.33

ECG of a normal single heartbeat

Figure 10.34	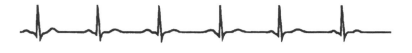
Normal sinus rhythm	

Figure 10.34 illustrates an ECG of a heart that is in **normal sinus rhythm**, in that impulses are originating in the SA node at a steady rate of 60–100 beats per minute.

The ECG may be used to detect variations from normal sinus rhythm. These variations are known as **arrhythmias**. Arrhythmias stemming from the SA node primarily affect heart rate and regularity of rhythm. P waves, QRS complexes, and T waves are all present. **Sinus tachycardia** is an arrhythmia in which the heart rate is increased to 100–150 beats per minute. Another is **sinus bradycardia**, in which the heart rate is decreased to fewer than 60 beats per minute.

A more serious arrhythmia is **atrial fibrillation**, in which impulses are initiated in atrial tissue other than the SA or AV nodes. The atria beat rapidly and asynchronously, rather than together as a unit. The ECG appears to have indistinct, wavy P waves and there are fewer QRS complexes than P waves.

Ventricular fibrillation is a life-threatening arrhythmia in which impulses are initiated in portions of the ventricular myocardium rather than in the usual conduction system of the heart. As a result, the ventricles are unable to contract as a unit and the heart loses its pumping ability. This arrhythmia is fatal unless the heart is defibrillated. The ECG appears totally disorganized, with no distinct P waves, QRS complexes, or T waves.

Question 10C.1

Match the arrhythmias in Column A with the ECG tracings in Column B. Refer to Figure A for comparison.

Column A	Column B
_____ 1. ventricular fibrillation	a.
_____ 2. atrial fibrillation	b.
_____ 3. sinus tachycardia	c.
_____ 4. sinus bradycardia	d.

The electrocardiographic (ECG) signal occurring during each heartbeat is detected by skin electrodes and is then amplified. Figure 10.35 shows the necessary assemblies for this mode of operation: the basic Cardio Tach instrument, ECG cable with grabbers for connection to electrodes, and three reusable skin electrodes. In addition to identifying P waves, QRS complexes, and T waves, the ECG can be used to determine heart rate. The time interval between consecutive R waves is measured and converted into frequency. The readout is either instantaneous or a four-beat average.

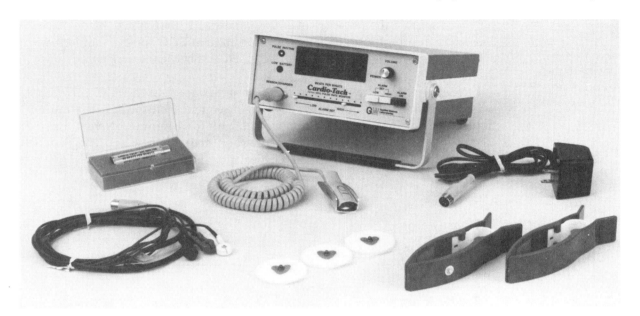

Figure 10.35

Cardio Tach Series 4600 with ECG module

The purpose of this exercise is to demonstrate ECG wave configurations and to determine heart rate measurement using the Cardio Tach electronic monitor. Your instructor will provide you with operating instructions for any other electronic monitor you use.

MATERIALS

Cardio Tach with ECG module
ECG cable with grabber connections
reusable or disposable skin electrodes

PROCEDURE

Plug the ECG cable into the sensor/charger socket. Attach each of the three color-coded grabber electrodes to three button electrodes on your partner as follows:

- black grabber—the indifferent electrode that can be attached anywhere on the body
- white grabber—the upper electrode that is attached to a spot above the heart
- green grabber—the lower electrode that is attached to a spot below the heart

If wrist-clamp electrodes are used, attach the white grabber to the right clavicle area, the green grabber to the left wrist, and the black grabber to the right wrist or right ankle. Switch on the power by turning the volume control in a clockwise direction and observe your partner's ECG on the monitor.

Question 10C.2

If your ECG equipment has a recording device, attach a strip below and label a P wave, QRS complex, and T wave.

Occasionally, the presence of an irregular heart rhythm may be displayed. This is the result of interference by less prominent ECG waves. Should this occur, interchanging the white and the green grabber leads may eliminate such difficulties.

Record the heart rate for your partner and yourself in the table provided while performing the following activities in sequence: sitting at rest, running in place for 2 minutes, and holding your breath for 1 minute. (*Note:* Wrist electrodes must be firmly attached while running in place. Electrodes may be taped to the wrists in order to reduce electrode movement.)

Heart Rates		
	Partner	You
Sitting at rest		
Running in place		
Holding your breath for 1 minute		

Question 10C.3

What were the effects of exercise and holding your breath on heart rate?

Question 10C.4

List and describe three factors that may have caused this change in heart rate.

1)

2)

3)

MULTIPLE CHOICE

_____ 1. Hematocrit may be defined as
    a. a measure of the volume of RBCs in a given volume of blood.
    b. an estimate that can be used to determine the blood volume of a person.
    c. a measure of blood clotting time.
    d. a hormone excreted in the kidneys in response to low numbers of RBCs.

_____ 2. A clot that freely floats in blood vessels is known as a(n)
    a. thrombus.
    b. infarction.
    c. embolus.
    d. thromboblast.

_____ 3. Which of the following layers is normally thicker in arteries than in veins?
    a. tunica media
    b. tunica intima
    c. tunica adventitia
    d. serosa

_____ 4. The first blood vessel to branch from the aortic arch is the
    a. innominate artery.
    b. carotid artery.
    c. subclavian artery.
    d. coronary artery.

_____ 5. The period of time in which the ventricles of the heart are filling with blood while the myocardium is relaxed is
    a. a heart attack.
    b. a stroke.
    c. the diastolic phase.
    d. the systolic phase.

_____ 6. A person who is blood typed as B positive would have which type of antigens and antibodies?
    a. B antigen, D antigen, anti-D antibodies
    b. B antigen, D antigen, anti-A antibodies
    c. A antigen, anti-B antibodies
    d. A antigen, anti-D antibodies, anti-B antibodies

_____ 7. The QRS wave of the electrocardiogram represents
    a. atrial depolarization.
    b. ventricular depolarization and relaxation.
    c. atrial relaxation.
    d. ventricular depolarization.

_____ 8. An implanted electronic pacemaker serves to augment or replace the normal functioning of the
    a. vagus nerve.
    b. glossopharyngeal nerve.
    c. SA node.
    d. AV node.

_____ 9. Why is it advantageous for red blood cells to be anucleate?
a. Oxygen-carrying capacity is increased.
b. It prevents excessive numbers of erythrocytes from being formed through mitotic division.
c. It makes it easier to distinguish them from leukocytes.
d. It enables them to pass through capillaries more easily.

_____ 10. In the presence of acute bacterial infection, the leukocyte count
a. increases.
b. decreases.
c. remains the same.
d. increases, then decreases.

_____ 11. Which of these vessels would have the thickest wall?
a. inferior vena cava
b. internal carotid artery
c. renal artery
d. aorta

_____ 12. The distal colon receives its blood supply from the
a. superior mesenteric artery
b. celiac artery.
c. inferior mesenteric artery.
d. iliolumbar artery.

_____ 13. Blood goes from the heart to the lungs through the
a. aorta.
b. ductus arteriosus.
c. pulmonary artery.
d. pulmonary vein.

_____ 14. Under which of these circumstances would counting the pulse for 15 seconds and multiplying by 4 be inaccurate?
a. if the pulse is rapid
b. if the pulse is slow
c. if there is background noise
d. if the pulse is irregular

_____ 15. A decrease in vagal impulses to the heart results in
a. increased heart rate.
b. decreased heart rate.
c. no effect on heart rate.
d. sinus bradycardia.

_____ 16. What does diastolic blood pressure represent?
a. the pressure of blood against the wall of the aorta when the ventricles contract
b. the pressure of blood against the wall of the aorta when the ventricles relax
c. the pressure of blood against the wall of the brachial artery when the ventricles relax
d. the pressure of blood against the wall of the left ventricle when the ventricles contract

_____ 17. Which of these is (are) not included in the "formed" elements of blood?
a. erythrocytes
b. monocytes
c. plasma
d. platelets

_____ 18. The mitral valve is found
a. between the left atrium and left ventricle.
b. between the right atrium and right ventricle.
c. between the two atria.
d. in the pulmonary artery.

_____ 19. If one were to centrifuge fresh whole blood, the clear liquid resulting would be
   a. red and white blood cells.　　c. serum.
   b. prothrombin.　　d. plasma.

_____ 20. Heart valves controlling the flow of blood from the atria to the ventricles are generally termed
   a. myocardial.　　c. sinoatrial.
   b. semilunar.　　d. cuspid.

_____ 21. Blood from the intestines going to the heart must first pass through the
   a. inferior vena cava.　　c. portal vein.
   b. superior vena cava.　　d. common iliac vein.

_____ 22. Mature, anucleate blood cells are
   a. normoblasts.　　c. agranulocytes.
   b. granulocytes.　　d. erythrocytes.

_____ 23. Large arteries carrying blood to the head are
   a. mesenteric.　　c. brachial.
   b. jugular.　　d. carotid.

_____ 24. The saphenous vein drains blood from the
   a. leg.　　c. brain.
   b. arm.　　d. sacral portion of the vertebral column.

_____ 25. In the human, the left common carotid artery branches off from the
   a. left subclavian artery.　　c. brachiocephalic artery.
   b. aortic arch.　　d. innominate artery.

DISCUSSION QUESTIONS

26. Using a flow chart, describe the scheme of human blood clotting.

27. Where is hemoglobin produced and what is its purpose? Can you suggest reasons why normal hemoglobin values differ in males and females?

_____

_____

_____

28. Identify the place of formation of each of the following and the function each performs in the coagulation of blood:

a. prothrombin

_____

_____

b. fibrinogen

_____

_____

c. thromboplastin

_____

_____

29. Trace the path of a drop of blood from the rectum to the heart and back to the rectum, naming all the blood vessels and structures of the heart through which the blood will pass.

30. Identify the structure of the heart that is described by each of these statements:

    a. chamber that receives blood from lungs

    _____

    b. valve between the left atrium and the left ventricle

    _____

    c. muscle tissue that makes up the heart

    _____

    d. tissue that lines the heart

    _____

    e. covering of serous tissue that surrounds the heart

    _____

    f. heart's own blood vessels

    _____

    g. left ventricle exit valve

    _____

    h. chamber that forces blood into systemic circulation

    _____

    i. ventricle partition

    _____

31. Write the term that is described by each of the following:

    a. liquid portion of the blood

    _____

    b. phagocytic cells

    _____

    c. cells that transport oxygen

    _____

    d. noncellular irregular elements essential to clotting

    _____

    e. protein molecules that confer protective immunity

    _____

    f. low hemoglobin level

    _____

    g. increased number of white blood cells

    _____

CASE STUDY

CARDIOVASCULAR SYSTEM–HEMATOLOGY

Rosa is pregnant with her second child. Her present pregnancy as well as her first were totally normal. She is concerned, however, because she recalls her own mother saying that when Rosa was a newborn she turned blue and almost died.

1. **Why is Rosa worried?**

2. **What was the problem Rosa suffered as a neonate?**

3. **Do Rosa's children have anything to fear?**

4. **Explain your answers relative to blood type.**

CASE STUDY

LYMPH SYSTEM

Kevin's mom was amazed. Kevin had very enlarged tonsils, in fact, "kissing tonsils." He didn't seem to be bothered by them and had few sore throats. He had no problem breathing and no other enlarged lymphoid tissue elsewhere in his body. She remembered that she too had very large tonsils, but when she was 6 years old her tonsils had been removed. When she asked the pediatrician about removing Kevin's tonsils, he advised against the procedure.

Why?

CASE STUDY

CARDIOVASCULAR SYSTEM

Stan, a 54-year-old dentist, suffers from hypercholesterolemia. Six years ago he had a heart attack requiring valve replacement because he had lost blood flow to the papillary muscle. His wife brought him into the ER because she said that for the past two days Stan has been confused, lethargic, and slurring his speech.

What do you think is causing Stan's present symptoms?

How does his condition relate to his corrective surgery six years ago?

The Respiratory System

A. Respiratory System Anatomy

PURPOSE

Unit 11A will familiarize you with the histological and gross anatomical features of the respiratory system.

OBJECTIVES

After completing Unit 11A, you will be able to

- demonstrate knowledge of basic lung and trachea histology.
- identify major structures and organs of the human and cat respiratory systems.
- identify the respiratory organs and related circulatory structures of sheep pluck.

MATERIALS

preserved cat	prepared slides of lung and trachea
sheep pluck	models of lung, trachea, and human
dissecting instruments	torso
compound microscope	human skulls
dissecting pan	model of sagittal section of human
dissecting pins	head
disposable gloves	cadaver (optional)
model of human larynx	

PROCEDURE

The main function of the respiratory system is to deliver oxygen to cells and remove carbon dioxide from them. For this to occur, three processes are necessary: ventilation, external respiration, and internal respiration.

Ventilation, or breathing, exchanges gases between the atmosphere and the lungs. Conduction zone structures, which do not allow for diffusion, transport air to and from the **alveoli** of the lungs, where gaseous exchange occurs. Oxygen moves from the alveoli to the blood, and carbon dioxide moves in the opposite direction, from the blood to the alveoli. This process is referred to as **external respiration**. **Internal respiration** is the exchange of gases between the blood and the body cells and the subsequent use of oxygen, with the production of carbon dioxide.

EXERCISE 1 Examination of Prepared Slides of the Lung and Trachea

Because diffusion is essential to respiratory function, the **alveolus**, a saclike structure composed primarily of simple squamous epithelium, is the main structural and functional unit of the lung. Structures associated with the alveoli (**Figure 11.9**) are the **respiratory bronchioles**, which divide into the **alveolar ducts** and **alveolar sacs**. Locate these structures on a prepared slide of the lung. Also note the **simple squamous epithelium** (**Figure 11.1**).

Obtain slides of cross-sections of bronchus and trachea and find the **ciliated pseudostratified epithelium**, **hyaline cartilage**, and **smooth muscle** (**Figures 11.2** through **11.4** and **Color Plates** 7 and 15). Sketch and label the tissues that you observe on these slides. On the slide of the trachea, you may also see **tracheal glands** (**Figure 11.3**).

Figure 11.1

Alveolar detail of lung (400×)

1. alveoli
2. simple squamous epithelium of alveolar walls

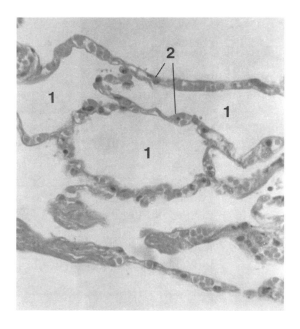

Figure 11.2

Detail of bronchus in lung (430×)

1. pseudostratified columnar epithelium
2. smooth muscle

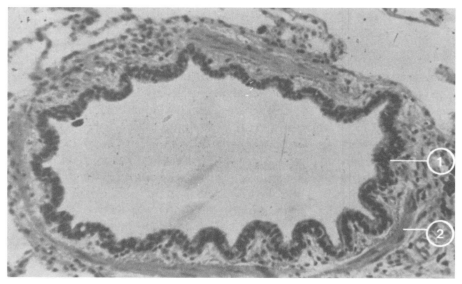

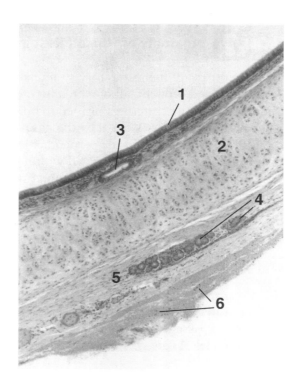

Figure 11.3

Trachea (40×)

1. pseudostratified, ciliated columnar epithelium
2. hyaline cartilage
3. duct of tracheal gland (from deeper gland)
4. tracheal glands
5. smooth muscle
6. adventitia

Question 11A.1

What is the function of ciliated pseudostratified epithelium in the upper respiratory tract?

Figure 11.4

Detail of tracheal mucosa (100×)

1. pseudostratified, ciliated columnar epithelium
2. lamina propria
3. alveolar secretory cells of tracheal glands
4. smooth muscle

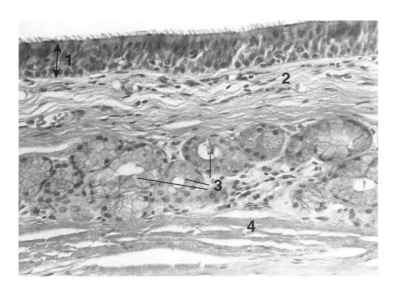

Question 11A.2

What is the function of hyaline cartilage in the trachea and bronchi?

EXERCISE 2 Anatomy of Human Respiratory Organs

The main organs of the respiratory system are the **nasal cavities**, **pharynx**, **larynx**, **trachea**, and **bronchi**, all of which function in the conduction of air, and the **lungs**. Find the structures printed in boldface on the models and torso. The nasal cavities are formed by the **nasal septum**, which consists of hyaline cartilage anteriorly and the perpendicular plate of the ethmoid posteriorly. The **paranasal sinuses** are located within the frontal, sphenoid, ethmoid, and maxillary bones. Using a human skull, identify the paranasal sinuses. See **Figures 5.5** and **5.8** to review these bones and bone markings. The **superior** and **medial conchae** are markings of the ethmoid. These and the **inferior conchae**, which are bones, project medially from the lateral walls of the nasal cavities.

The **pharynx**, or throat, is a muscular channel that serves as a passageway for both air and food. It is divided into three regions: the superior **nasopharynx**, middle **oropharynx**, and inferior **laryngopharynx**. Observe the openings in the lateral wall of the nasopharynx. These are the **eustachian**, or **auditory tubes**, which are passageways from the middle ear. Using a sagittal section of a human head model or cadaver, if available, locate these structures (**Figure 11.5**).

Observe the **larynx**. Note the boxlike shape of the organ and its sturdy cartilaginous construction. The largest of the nine cartilages is the **thyroid cartilage**, which forms the anterior surface. The thyroid cartilage is commonly known as the "Adam's apple." The inferior **cricoid cartilage** is C-shaped and extends posteriorly (**Figures 11.6** and **11.7**). The lidlike flap of cartilage at the top of the larynx is the **epiglottis**, which covers the **glottis** when swallowing. The glottis is the opening into the larynx between the vocal cords. On a model you may be able to identify the smaller, paired cartilages: the **arytenoids**, the **cuneiforms**, and the **corniculates**. Observe the internal surface of the larynx. Note the **vocal cords**, which are extensions of mucous membranes that line the laryngeal walls. These cords vibrate upon exhalation and produce sound (**Figures 11.5** and **11.7**).

The **trachea**, or "windpipe," is approximately 4.5″ in length and is easily recognizable by its cartilaginous rings, which help prevent collapse of this structure. The C-shaped rings are anteriorly positioned. Note the **trachealis muscle**, which runs longitudinally on the posterior surface of the trachea, immediately anterior to the esophagus. Inferiorly, the trachea subdivides into the **right primary** (or **mainstem**) and **left primary** (or **main stem**) **bronchi** (**Figures 11.7** and **11.9**). The bronchi repeatedly branch into smaller **secondary** and **tertiary bronchi**, then into **bronchioles**. The bronchi have histological characteristics similar to those of the trachea. Note their cartilaginous rings or plates (**Figures 11.7** and **11.9**).

On the model of the human torso, note the **larynx**, **laryngeal cartilages**, the **trachea** with its characteristic **cartilage rings**, and the **bronchi**. Observe that the **right primary bronchus** branches off higher, is slightly larger in diameter, and is more vertical than the **left primary bronchus**.

Question 11A.3

What is the function of cartilaginous rings in the trachea?

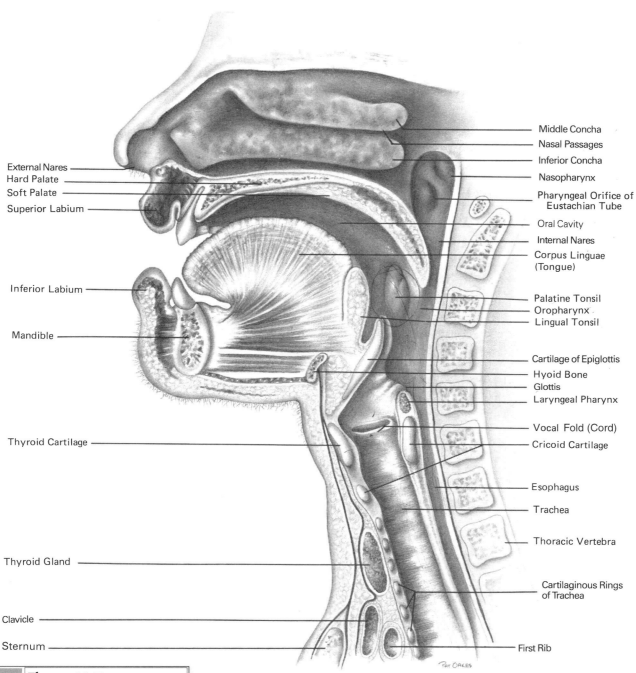

External Nares
Hard Palate
Soft Palate
Superior Labium

Inferior Labium

Mandible

Thyroid Cartilage

Thyroid Gland

Clavicle

Sternum

Middle Concha
Nasal Passages
Inferior Concha
Nasopharynx
Pharyngeal Orifice of Eustachian Tube
Oral Cavity
Internal Nares
Corpus Linguae (Tongue)
Palatine Tonsil
Oropharynx
Lingual Tonsil
Cartilage of Epiglottis
Hyoid Bone
Glottis
Laryngeal Pharynx
Vocal Fold (Cord)
Cricoid Cartilage
Esophagus
Trachea
Thoracic Vertebra
Cartilaginous Rings of Trachea
First Rib

Figure 11.5

Sagittal section of human head

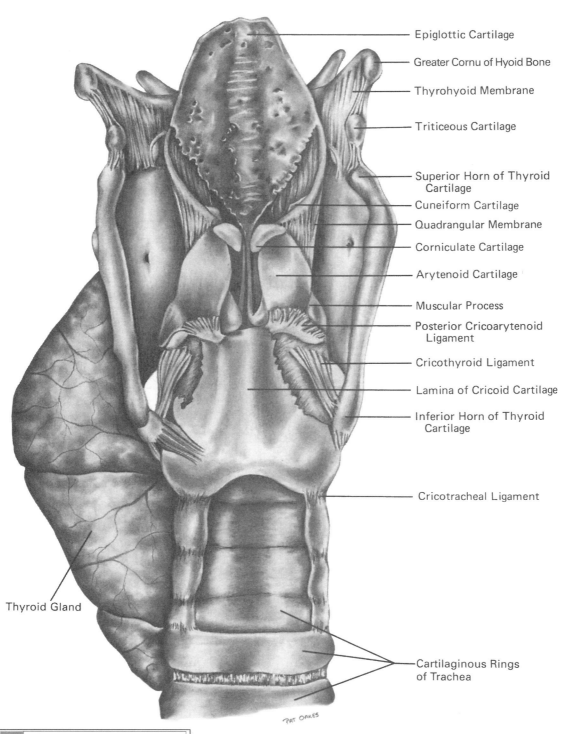

Epiglottic Cartilage

Greater Cornu of Hyoid Bone

Thyrohyoid Membrane

Triticeous Cartilage

Superior Horn of Thyroid Cartilage

Cuneiform Cartilage

Quadrangular Membrane

Corniculate Cartilage

Arytenoid Cartilage

Muscular Process

Posterior Cricoarytenoid Ligament

Cricothyroid Ligament

Lamina of Cricoid Cartilage

Inferior Horn of Thyroid Cartilage

Cricotracheal Ligament

Thyroid Gland

Cartilaginous Rings of Trachea

PAT OAKES

Figure 11.6

Human larynx viewed from the dorsal and superior aspects

Redrawn from B. J. Anson and C. B. McVay, *Surgical Anatomy.* Philadelphia: W. B. Saunders Co., 1971

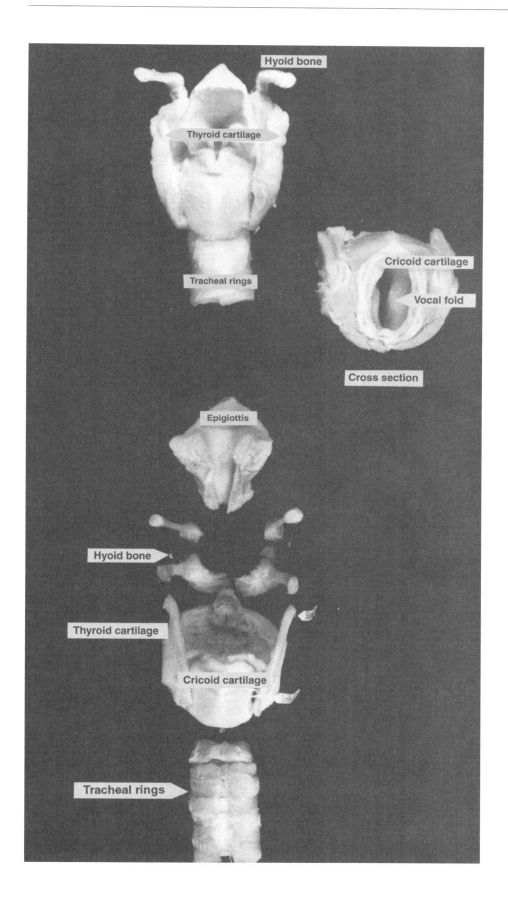

Figure 11.7

Structures of human larynx, dorsal aspect

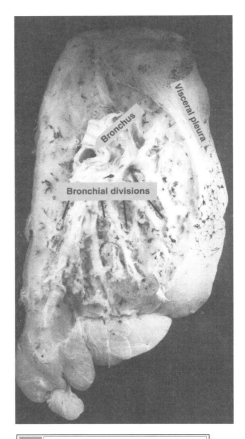

Figure 11.8

Human lung, sagittal section

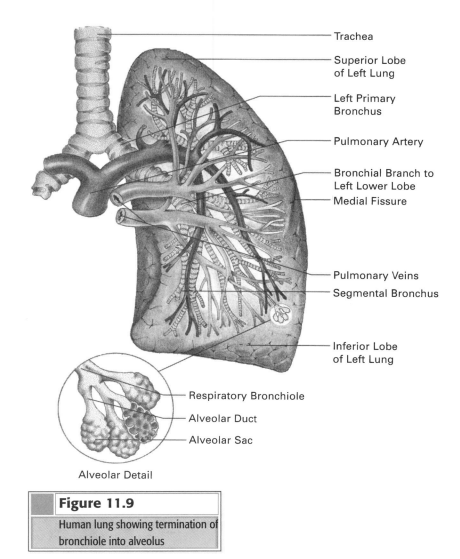

- Trachea
- Superior Lobe of Left Lung
- Left Primary Bronchus
- Pulmonary Artery
- Bronchial Branch to Left Lower Lobe
- Medial Fissure
- Pulmonary Veins
- Segmental Bronchus
- Inferior Lobe of Left Lung
- Respiratory Bronchiole
- Alveolar Duct
- Alveolar Sac

Alveolar Detail

Figure 11.9

Human lung showing termination of bronchiole into alveolus

Question 11A.4

How many lobes are in the right lung of the human being?

Question 11A.5

How many lobes are in the left lung of the human being?

On the human lung model, torso, and cadaver, if available, note the **fissures** that divide the lungs into **lobes.** On the right, there are the **oblique** and **horizontal fissures,** on the left the **medial fissure.**

On the human respiratory models and cadaver, if available, note the **hilus** (or **hilum**), the indentation on the medial surface of each lung. This is where the bronchi, blood and lymph vessels, and nerves enter and exit the lungs. The **cupola** is that portion of the superior lobes that extends above the clavicles. The **bases** of the lungs are the indentations on the inferior lobes where the lungs approach the diaphragm.

Review the muscles that function in the mechanics of breathing. The **diaphragm,** which divides the ventral cavity into the superior thoracic cavity and the inferior abdominopelvic cavity, is the major muscle of inspiration. Others are the **external intercostals, scalenes, sternocleidomastoids,** and **pectoralis minor** muscles. All contract to increase the size of the thoracic cavity during inspiration.

In quiet unlabored breathing, expiration is a passive process in which there are no muscular contractions. In forced or labored expiration, the **internal intercostals** and **abdominal muscles** contract, forcing air upward and out of the thoracic cavity.

Question **11A.6**

List the muscles that increase the size of the thoracic cavity during inspiration.

Question **11A.7**

Which muscles passively relax in quiet expiration?

Identification 11.1

 11.1

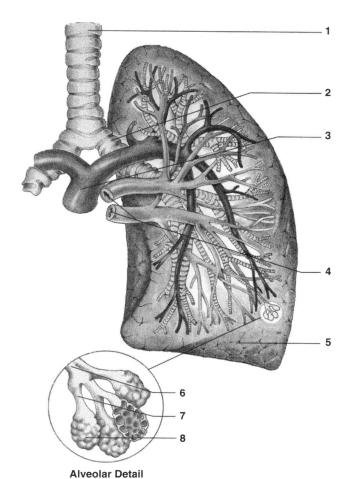

Alveolar Detail

Question **11A.8**

How does the arrangement of lung lobes differ in the cat as compared to the human being?

EXERCISE 3   Sheep Pluck

Your instructor will have a sheep pluck available for organ identification. Included will be the **lungs, heart,** and **pericardium,** as well as the **aorta, pulmonary artery,** and **pulmonary veins.** Parts of the trachea and esophagus may also be intact. Observe the structures, noting similarities and differences among the sheep, cat, and human.

EXERCISE 4   Cat Respiratory Organs

In the cat, if you have not already done so, continue superiorly the midventral incision you made when studying the cardiovascular system. This incision will allow you to expose and identify respiratory structures in the neck region. Now cut (transversely) the sternothyroid, thyrohyoid, and sternohyoid muscles (**Figure 6.17**) and gently pull the cartilaginous **larynx** at the superior end of the trachea anteriorly until it is free of its muscle attachments. Observe the **vocal folds** (**Figure 11.7**), the **epiglottis** covering the **glottis,** and the thyroid and cricoid cartilages. Tracing the **trachea** inferiorly, observe the cartilaginous rings on the surface and the **trachealis muscle** on the dorsal surface. As in the human, the trachea bifurcates, forming the **right** and **left primary bronchi,** which further branch into the **bronchial tree,** consisting of smaller secondary and segmental bronchi and bronchioles in the **lungs** (**Figure 11.11**).

Observe the soft, smooth covering of the lung lobes. The **pleura,** a double-layered serous membrane, is composed of the **parietal pleura,** which adheres to the parietal peritoneum lining the thoracic cavity, and the **visceral pleura,** which directly covers the lungs. Lubricating fluid fills the **intrapleural cavity,** the space between the pleura. Note that the cat has more lung lobes than the human being (**Figure 11.12**).

Figure 11.10

Cat respiratory organs

1. right lung lobes
2. cut diaphragm
3. left lung lobes
4. right ventricle
5. right atrium
6. left ventricle
7. left atrium

8. trachea
9. esophagus
10. superior vena cava
11. right external jugular vein
12. left external jugular vein
13. aortic arch
14. left internal jugular vein

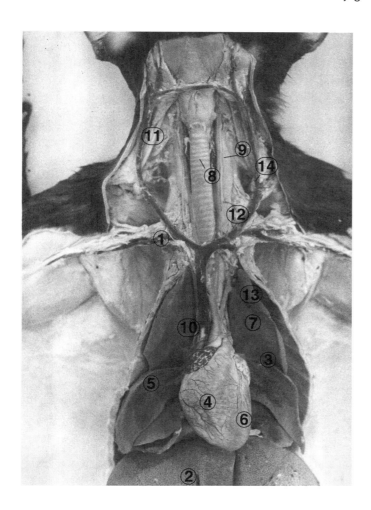

Figure 11.11

Bronchial tree of cat

1. trachea
2. bronchial branching
3. segmental bronchus

Figure 11.12

Ventral aspect of isolated heart and lungs of cat

1. right anterior lobe of lung
2. right medial lobe
3. right posterior lobe
4. caudate lobe
5. left anterior lobe
6. left medial lobe
7. left posterior lobe
8. brachiocephalic artery
9. anterior vena cava
10. aortic arch
11. left subclavian artery
12. coronary vessel
13. descending thoracic aorta
14. posterior vena cava
15. esophagus

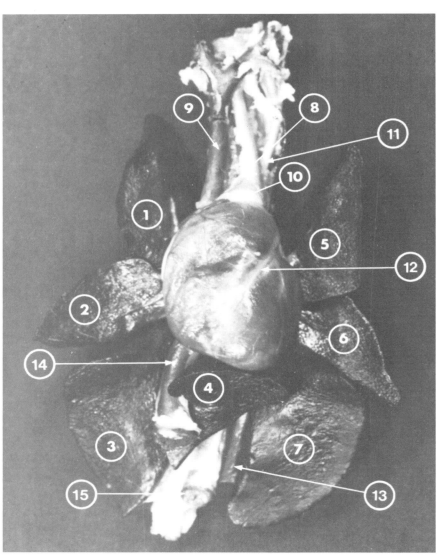

B. Respiratory Measurement

PURPOSE

Unit 11B will enable you to understand the use of the spirometer and pneumograph in measuring standard respiratory volumes.

OBJECTIVES

After completing Unit 11B you will be able to

- determine your vital capacity using a spirometer and compare it with a standard.
- calculate your total lung capacity, tidal volume, residual volume, and inspiratory and expiratory reserve volumes.
- investigate selected respiratory variations using a pneumograph.

MATERIALS

dry spirometers and disposable mouthpieces
pneumograph and recording attachments
paper cups

PROCEDURE

| EXERCISE **1** | **Spirometric Measurement and Calculation of Standard Respiratory Volumes** |

> **CAUTION**
>
> These exercises involve the interruption of regular breathing.

Variations in the size, age, and sex of an individual account for variations in respiratory volumes. Normal breathing usually moves about 500 mL of air in and out of the lungs and is referred to as **tidal volume (TV)**. Even after a normal inspiration, an individual can forcibly inhale approximately 3000 mL of air into the lungs. This forced inspiration is called the **inspiratory reserve volume (IRV)**. Also, after a normal expiration, an individual can forcibly exhale approximately 1100 mL of air, the **expiratory reserve volume (ERV)**. The **vital capacity** is the sum total of the inspiratory reserve, tidal, and expiratory reserve volumes, or approximately 4600 mL of air; thus the vital capacity is the total exchangeable air of the lungs. In addition, approximately 1200 mL of air remain in the conducting tubules. This air is not exchangeable, and is referred to as **residual volume (RV)**. **Total lung capacity** *(TLC)* is the sum total of the **tidal volume** *(TV)*, **inspiratory** and **expiratory reserve volumes** *(IRV* and *ERV)* that compose the **vital capacity**, and the **residual volume** *(RV)*. Therefore,

$$TLC = (TV + IRV + ERV) + RV$$

or

$$TLC = VC + RV$$

Use a clean mouthpiece. Monitor your partner for lightheadedness. This exercise involves the interruption of regular breathing.

You can directly measure your vital capacity using a spirometer. In order to compare your *VC* to a standard, perform the following exercise.

Set the needle of the spirometer to zero at the beginning of each use. Practice by taking three deep breaths and exhaling with force. Now inhale. Then exhale forcibly into the spirometer. You may want to hold your nose to prevent air loss. Repeat this procedure three times, recording each *VC* reading from the spirometer. Be sure to reset the dial at zero after each use. Record your readings below.

mL air

First *VC*   _____

Second *VC*   _____

Third *VC*   _____

Total *VC*   _____   mL air

$$\text{Your average } VC = \frac{\text{Total } VC}{3}$$

Using the table Normal Vital Capacity of Adults, find the standard vital capacity for someone of your sex, age, and height. Then calculate your *VC* as a percentage of the standard *VC*:

$$\frac{VC_{yours}}{VC_{standard}} \times 100 = \underline{\hspace{1cm}} \%$$

Question 11B.1

What is your average vital capacity?

Question 11B.2

What is the standard you used?

Question 11B.3

What percentage of the standard vital capacity is your vital capacity?

Normal Vital Capacity of Adults* (in cubic centimeters)

| | Height in Inches | Age in Years | | | | | |
		20	30	40	50	60	70
Males	60	3885	3665	3445	3225	3005	2785
	62	4154	3925	3705	3485	3265	3045
	64	4410	4190	3970	3750	3530	3310
	66	4675	4455	4235	4015	3795	3575
	68	4940	4720	4500	4280	4060	3840
	70	5206	4986	4766	4546	4326	4106
	72	5471	5251	5031	4811	4591	4371
	74	5736	5516	5516	5076	4856	4636
Females	58	2989	2809	2629	2449	2269	2089
	60	3198	3018	2838	2658	2478	2298
	62	3403	3223	3043	2863	2683	2503
	64	3612	3432	3252	3072	2892	2710
	66	3822	3642	3462	3282	3102	2922
	68	4031	3851	3671	3491	3311	3131
	70	4270	4090	3910	3730	3550	3370
	72	4449	4269	4089	3909	3729	3549

Adapted with permission from Propper Manufacturing Co., Inc., New York.

*Variations must be at least 20% below predicted normal to be considered subnormal. Variations can also exist depending upon size and body structure.

Using the spirometer, you can also measure your tidal volume *(TV)*. Reset the dial to zero, take in a normal breath, and exhale into the spirometer. Repeat this procedure three times, recording your *TV* each time. Be sure to reset the dial to zero after each use. Your *TV* is approximately 9% of your vital capacity.

mL air

TV_1 _____

TV_2 _____

TV_3 _____

Total *TV* _____ mL air

$$\text{Your average } TV = \frac{\text{Total } TV}{3}$$

Calculate your *IRV* and *ERV.*

Given:
IRV is approximately 51% of *VC.*
ERV is approximately 20% of *VC.*
Your *VC* × 0.51 = your *IRV.*
Your *VC* × 0.20 = your *ERV.*

Calculate your *TLC* or *x* by setting up a ratio, as follows:

Given:
VC is approximately 80% of *TLC.*
TLC is the maximum capacity (100%) of the lungs.
TLC(x) is your total lung capacity.

$$\frac{TLC(x)}{100} :: \frac{VC}{80}$$

$$80x = VC \times 100$$

$$x = \frac{VC \times 100}{80}$$

Calculate your *RV:*

$$TLC - VC = RV$$

Now calculate your **respiratory minute volume**. Have your partner count the number of breaths that you take in one minute. This is your **respiratory rate**. Record your respiratory rate _____. An individual usually breathes 14–20 times per minute. The respiratory minute volume equals the respiratory rate times the tidal volume.

Question 11B.4

What is your average tidal volume?

....

Question 11B.5

What is your inspiratory reserve volume?

....

Question 11B.6

What is your expiratory reserve volume?

....

Question 11B.7

What is your total lung capacity?

....

Question 11B.8

What is your residual volume?

....

Question 11B.9

What is your respiratory minute volume?

....

EXERCISE 2 Determination of Respiratory Variations

Human respiratory variations are determined by use of a **pneumograph**, an instrument that records variations in breathing patterns. Your instructor will advise you how to set up the instrument so you can perform various exercises and read the results. You should be actively involved in these exercises and should not be giving attention to recordings until completion. It is best to work with a partner. One partner can operate the pneumograph and identify the activities the other person performs. The pneumography tubing should be attached firmly but not tightly around the thoracic cavity, allowing space for chest expansion.

Breathe normally for approximately two minutes in a sitting position. Inspiration should be recorded by a downward deflection of the stylus. Upon inspiration, the pneumograph tubing lengthens, thus increasing the volume but decreasing the pressure within the chest cavity, allowing atmospheric air to enter.

Vary your activities in the following manner, and have your partner record the results:

Question 11B.10

How long was your recovery period?

a. talking
b. standing
c. talking and standing
d. coughing
e. contracting biceps
f. deep breathing
g. shallow breathing

h. running in place
i. concentrating
j. sitting and extending legs forward
k. drinking water
l. holding your breath

▼ CAUTION

Carefully monitor your partner for timely recovery.

Breathe normally for two minutes. Record the number of respirations. Now hold your breath for as long as you can and record your recovery period. The recovery period is reached when alveolar air contains 7%–10% CO_2. During the recovery period, respirations are usually 14–20 per minute and regular in rate.

The average recovery period is slightly longer than one minute. Now hyperventilate for 45 seconds; then time how long you can hold your breath.

Question 11B.11

Can you hold your breath for a longer or shorter period of time after hyperventilating? Why?

The Respiratory System

PICTURE IDENTIFICATION
Use this diagram for questions 1–4:

Name _____

Section _____

Date _____

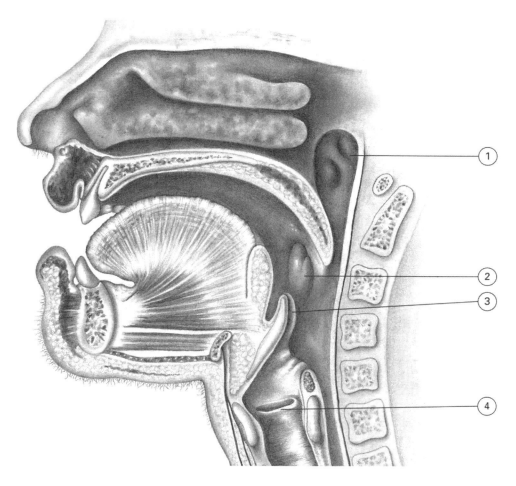

MULTIPLE CHOICE

_____ **1.** Chamber 1 is the
  a. pharyngeal orifice.
  b. oral cavity.
  c. nasopharynx.
  d. oropharynx.

_____ **2.** Structure 2 is the
  a. lingual tonsil.
  b. palatine tonsil.
  c. oropharynx.
  d. laryngeal pharynx.

_____ 3. Structure 3 is the
   a. corpus linguae.   c. lingual tonsil.
   b. epiglottis.   d. hyoid bone.

_____ 4. Structure 4 is the
   a. laryngeal pharynx.   c. thyroid cartilage.
   b. epiglottis.   d. vocal fold.

_____ 5. The anatomical structure that prevents food from entering the trachea when you swallow is the
   a. larynx.   c. epiglottis.
   b. esophagus.   d. glottis.

_____ 6. The volume of air that moves into the lungs with each inspiration is termed
   a. vital capacity.   c. residual volume.
   b. tidal volume.   d. inspiratory reserve volume.

_____ 7. The serous, a protective saclike membrane that encases the lungs is known as the
   a. pericardium.   c. pleural cavity.
   b. peritoneum.   d. pleura.

_____ 8. The anatomical site where the pulmonary artery, vein, and bronchi enter and exit the lung is the
   a. apex.   c. mediastinum.
   b. base.   d. hilum.

_____ 9. Which of the following tissue types is *not* found in the respiratory system?
   a. pseudostratified ciliated   c. osteoblasts
      epithelium   d. lamina propria with mucous
   b. goblet cells   glands

_____ 10. If an individual's tidal volume is 500 cc, expiratory reserve volume is 1200 cc, and vital capacity is 4500 cc, what is the individual's inspiratory reserve volume?
   a. 1700 cc   c. 6200 cc
   b. 2800 cc   d. cannot be calculated from the
                                         information given

_____ 11. Oxygen and carbon dioxide cross the alveolar epithelium by means of
   a. osmosis.   c. diffusion.
   b. active transport.   d. filtration.

_____ 12. What is the advantage of the trachea's lining of ciliated epithelium?
   a. to trap dust particles before   c. to aid in the movement of
      they enter the lungs   oxygen and carbon dioxide
   b. to prevent food from   molecules in and out of the
      becoming lodged in the   lungs
      trachea   d. to protect the tracheal
                                         epithelium against irritation

_____ 13. Which laryngeal cartilage forms the "Adam's apple"?
a. corniculate   c. thyroid
b. cricoid   d. epiglottis

_____ 14. Vital capacity in the adult tends to _____ with age.
a. increase   c. remain the same
b. decrease   d. there is no definite pattern

_____ 15. Vital capacity _____ in a patient with pneumonia.
a. increases   c. remains the same
b. decreases   d. doubles

_____ 16. Under which of these conditions would breathing be most rapid?
a. while asleep   c. immediately after holding
b. during normal quiet   one's breath
   breathing   d. immediately following
                             hyperventilation

_____ 17. Air remaining in the lungs even after a forced expiration is
a. tidal volume.   c. expiratory reserve volume.
b. inspiratory reserve volume. d. residual volume.

_____ 18. The nasal septum consists of which of the following bones?
a. mandibular and zygomatic   c. frontal and temporal
b. occipital and parietal   d. ethmoid and vomer

_____ 19. Structurally, the larynx is primarily composed of
a. cartilage.   c. membranes.
b. bone.   d. stratified epithelium.

_____ 20. The lower end of the trachea divides into
a. alveoli.   c. bronchi.
b. bronchioles.   d. pleural cavities.

_____ 21. The spirometer cannot directly measure
a. vital capacity.   c. tidal volume.
b. residual volume.   d. expiratory reserve volume.

_____ 22. Which of the following tissues would *not* be observed in a cross
section of the trachea?
a. trachealis muscle   c. adipose tissue
b. osseous tissue   d. ciliated pseudostratified
                             epithelium

_____ 23. Normal tidal volume is approximately how many milliliters of
air?
a. 500 mL   c. 300 mL
b. 1000 mL   d. 750 mL

_____ 24. An alveolus is composed of which tissue type?
a. simple squamous epithelium c. cuboidal epithelium
b. ciliated columnar epithelium d. hyaline cartilage

_____ 25. Which of the following terminates in an alveolar duct?
a. bronchiole   c. alveolar capillary
b. bronchus   d. pleura

DISCUSSION QUESTIONS

26. a. My tidal volume is _____

   _____

   b. My inspiratory reserve volume is _____

   _____

   c. My expiratory reserve volume is _____

   _____

   d. My vital capacity is _____

   _____

   e. My approximate residual volume is _____

   _____

   f. My total lung capacity is approximately _____

   _____

27. How do the cat and human lungs differ with respect to number and
   arrangement of lobes? _____

   _____

   _____

   _____

   _____

28. Why is the epiglottis cartilaginous? _____

   _____

   _____

   _____

   _____

29. Why are the alveolar sacs composed of simple squamous epithelium?

   _____

   _____

   _____

   _____

CASE STUDY

RESPIRATORY SYSTEM

Jack Nelson suffered from asthma in childhood, but now he is a high school swimming star. His VC is 7000 milliliters of air. He is 17 years old and 5′ 10″.

How does his VC compare to others his age and height? Explain.

What is Jack's tidal volume? How does it compare with the average TV?

Do you think that the childhood asthma had any effect on his present respiratory volumes? Explain.

The Digestive System

A. Digestive Anatomy

PURPOSE

Unit 12A will enable you to locate and state the functions of the major digestive organs and accessory glands.

OBJECTIVES

After completing Unit 12A, you will be able to
- identify the major divisions and subdivisions of the digestive tract.
- state the functions of the structures of the alimentary canal.
- briefly describe the histology of the digestive tract.

MATERIALS

models of human digestive organs
human cadaver (if available)
human skulls
human torso
dissecting instruments
disposable gloves
preserved cats
newspapers

prepared microscope slides of the
   following:
    developing tooth
    tongue
    dried tooth
    parotid gland
    esophagus
    stomach
    small instestine
    large intestine
    appendix
    rectum
    anus (anal canal)
    liver
    gallbladder

PROCEDURE

EXERCISE 1 **Human Digestive Anatomy**

On a human skull, identify the four types of teeth: the **incisors**, the **canines** (**cuspids**), the **premolars** (**bicuspids**), and the **molars** (**Figure 5.15**).

On a human torso cadaver, and models, find the structures listed in the following table, using **Figures 12.1** through **12.7** for reference.

Structures of the Human Digestive Tract

Division	Subdivision
Mouth	Teeth
	Incisors (in-SĪ-zorz)
	Canines (KĀ-nīnz) or cuspids
	Premolars (prē'-MŌ-lerz) or bicuspids
	Molars
	Tongue
Pharynx (FAR-inks)	Nasopharynx (nā'-zō-FAR-inks)
	Oropharynx (or'-ō-FAR-inks)
	Laryngopharynx (la-ring'-gō-FAR-inks)
Esophagus (e-SOF-a-gus)	
Stomach	Cardiac sphincter (SFINK-ter)
	Cardiac
	Greater and lesser curvatures
	Fundus
	Body
	Pylorus (pī-LOR-us)
	Pyloric sphincter
Small Intestine	Duodenum (doo'-ō-DĒ-num)
	Jejunum (je-JOO-num)
	Ileum (IL-ē-um)
Large Intestine (Colon)	Ileocecal (il'-ē-ō-SĒ-kal) valve (sphincter)
	Cecum (SĒ-kum)
	Appendix (not present in the cat)
	Ascending colon (KŌ-lon)
	Hepatic (he-PAT-ik) flexure
	Transverse colon
	Splenic (SPLEN-ik) flexure
	Descending colon
	Sigmoid colon (not present in the cat)
	Rectum
	Internal anal sphincter
	Anus (Ā-nus)
	External anal sphincter
Accessory Organs and Structures	Salivary (sal'-i-ver-ē) glands
	Parotid (pa-ROT-id)
	Submaxillary
	Sublingual (sub'-LING-gwal)
	Pancreas (PAN-krē-us)
	Pancreatic duct
	Liver
	Gallbladder
	Cystic (SIS-tik) duct
	Hepatic duct
	Common bile duct
	Gastrohepatic (gas'-trō-he-PAT-ik) ligament

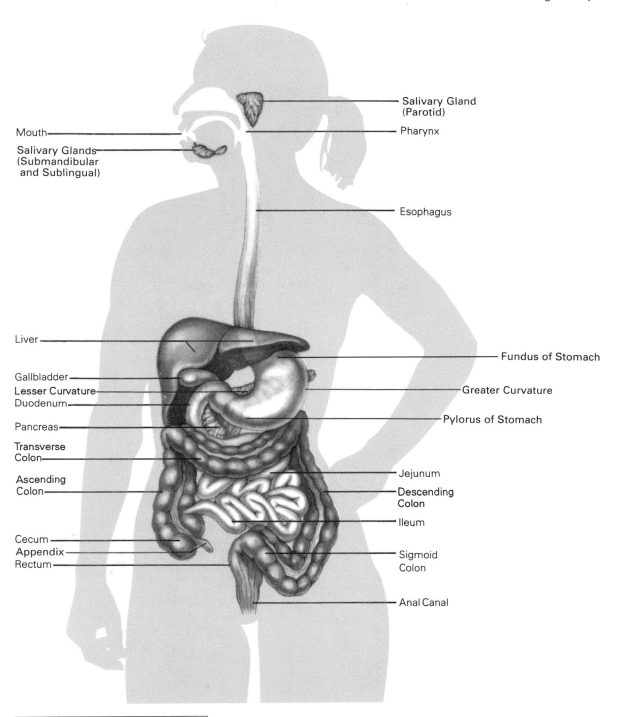

Salivary Gland
(Parotid)

Pharynx

Mouth

Salivary Glands
(Submandibular
and Sublingual)

Esophagus

Liver

Gallbladder
Lesser Curvature
Duodenum

Pancreas

Transverse
Colon

Ascending
Colon

Cecum
Appendix
Rectum

Fundus of Stomach

Greater Curvature

Pylorus of Stomach

Jejunum

Descending
Colon

Ileum

Sigmoid
Colon

Anal Canal

Figure 12.1

Human digestive system

Figure 12.2

Human digestive organs

1. liver–right lobe proper
2. liver–left lobe proper
3. gallbladder
4. stomach
5. mesentery
6. small intestine

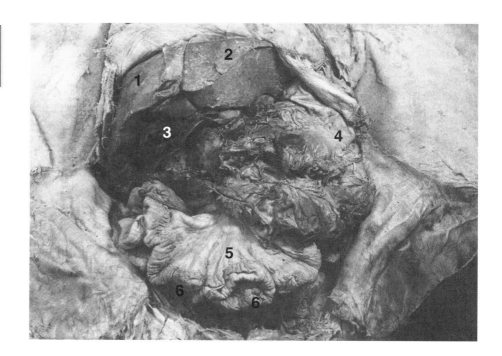

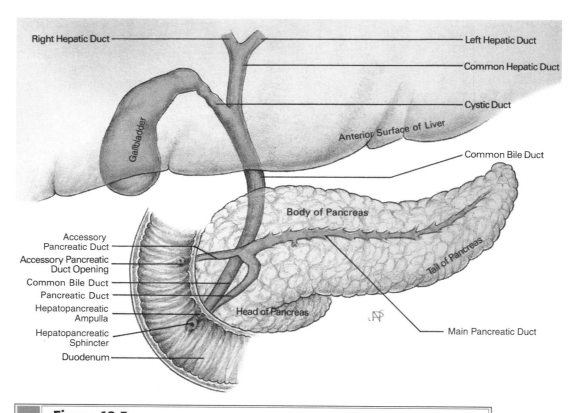

Figure 12.3

Relationship of the human liver, gallbladder, and pancreas to the duodenum

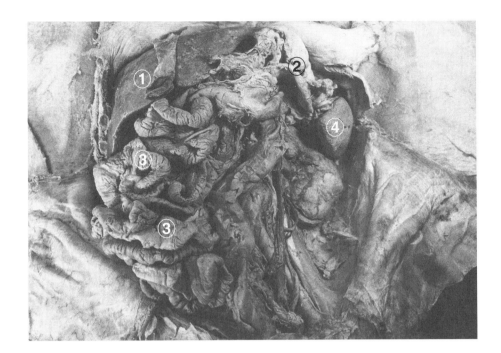

Figure 12.4

Human small intestine

1. liver–right lobe proper
2. stomach (retracted)
3. small intestine
4. spleen

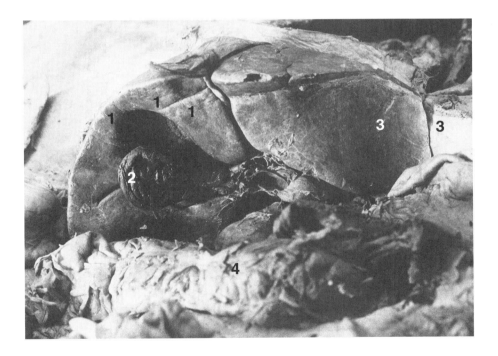

Figure 12.5

Human liver, gallbladder, and greater omentum (lateral view)

1. liver–right lobe proper
2. gallbladder
3. liver–left lobe proper
4. greater omentum

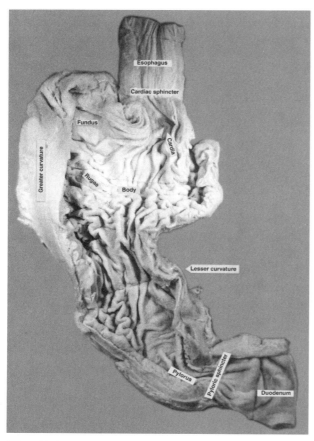

Figure 12.6

Human stomach, reflected

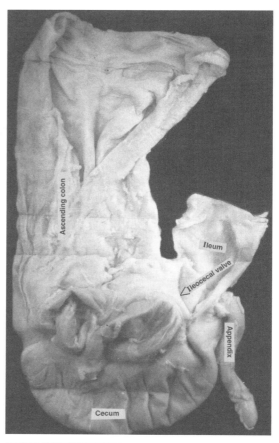

Figure 12.7

Human ileocecal junction

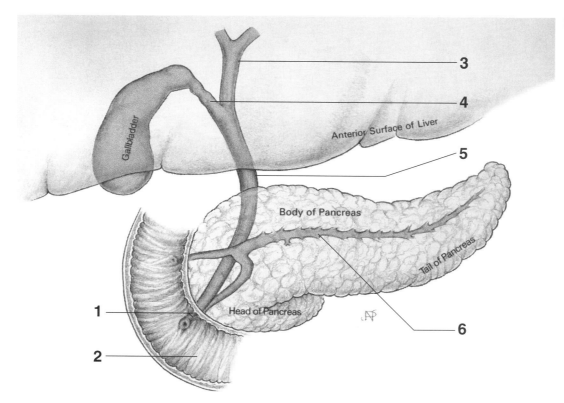

Identification 12.1

 12.1

EXERCISE 2   Cat Digestive Anatomy

On your cat, identify the **hard** and **soft palates**, the **esophagus**, and the **tongue**. The tongue is attached ventrally by the **frenulum linguae** and posteriorly to the hyoid bone. Note the **papillae**, or elevations, on the tongue.

The **pharynx** is the throat cavity and is divided into three regions. The area behind the nose is the **nasopharynx**. The region in which the **larynx**, or voice box, is located is referred to as the **laryngopharynx**. The region behind the mouth, known as the **oropharynx**, can be identified by inserting your probe into the **isthmus of the fauces**, the aperture of the throat, and moving the probe anteriorly or posteriorly.

Observe the **parotid gland** beneath the ear in the neck. The **parotid duct**, or **Stenson's duct**, extends anteriorly and empties into the posterior oral cavity. Beneath the parotid gland and posterior to the angle of the jaw is the **submaxillary gland**. Its duct, **Wharton's duct**, leads into the floor of the oral cavity. Last, identify the **sublingual glands** if possible. They are small and located at the base of the tongue and anterior to the submaxillary glands. Their ducts also empty into the floor of the oral cavity.

Laterally reflect the walls of the thoracic and abdominal cavities. The smooth membrane lining the body cavity is **parietal peritoneum**. If you have not already done so, extend your incision laterally at the level of the **diaphragm**. Carefully, remove any brownish debris or latex that may have leaked out of the blood vessels during preservation of the vessels.

Using the photographs on the following pages, you will locate the **viscera**—that is, all the organs in the abdominal cavity. Lying over most of the organs within the abdominal cavity is a connective tissue membrane, the **greater omentum** (**Figure 12.8**), which contains a considerable amount of yellow adipose tissue. The greater omentum is a large double fold of peritoneum. In order to see the organs beneath the greater omentum, carefully loosen it and move it to one side (**Figures 12.8** and **12.9**).

In order to expose the **esophagus**, retract the sternohyoid muscles in the neck laterally. The **trachea** will be exposed ventrally. Grasp the trachea and observe its dorsal surface. Separate the collapsed esophagus from the trachea with your probe. Expose the length of the esophagus in the neck. Follow it inferiorly on the left side of the thoracic cavity behind the heart and lungs. Observe the esophagus as it descends through the diaphragm into the abdominal cavity and connects with the stomach at the **cardiac sphincter**.

Observe the greater and lesser curvatures of the **stomach**. Superior to inferior, the divisions of the stomach are **cardiac**, **fundus**, **body**, and **pylorus**. If your instructor so indicates, make an incision along the greater (convex) curvature of the stomach. Be sure to wear gloves if you do this. Examine the inside surface for folded walls, **rugae**, that allow for distension and more surface area. The constriction between the stomach and small intestine is the **pyloric sphincter**, a circular smooth muscle that prevents the backflow of food from the small intestine to the stomach. The sphincter can be felt as a hard mass (**Figure 12.10**).

The first portion of the small intestine is the **duodenum**, which begins at the pyloric sphincter and extends along the head of the **pancreas**. Where the

CAUTION

Wear gloves and use caution when making the incision.

Question 12A.1

Name the sphincters at the two ends of the stomach.

Figure 12.8

Abdominal organs with greater omentum

1. parietal peritoneum
2. diaphragm
3. gallbladder
4. greater omentum
5. liver, left lateral lobe

6. falciform ligament
7. liver, right lobe
8. small intestine
9. spleen

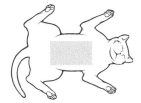

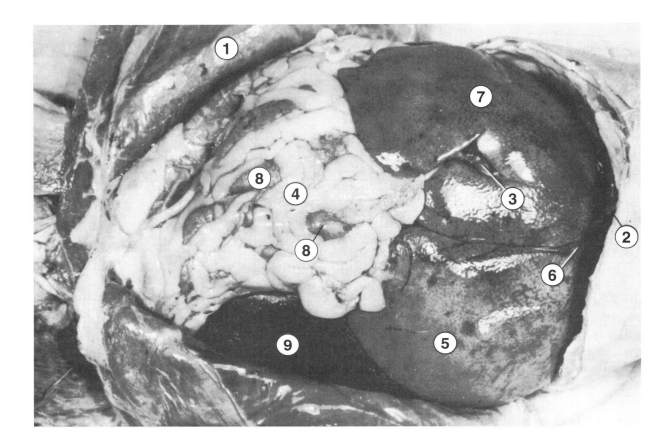

Figure 12.9

Abdominal organs exposed after portion of greater omentum removed

1. parietal peritoneum
2. cardiac sphincter
3. gallbladder
4. greater omentum
5. liver, left lateral lobe
6. liver, left medial lobe
7. liver, right lobe
8. small intestine
9. spleen
10. body of stomach
11. cardiac stomach
12. descending colon

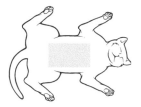

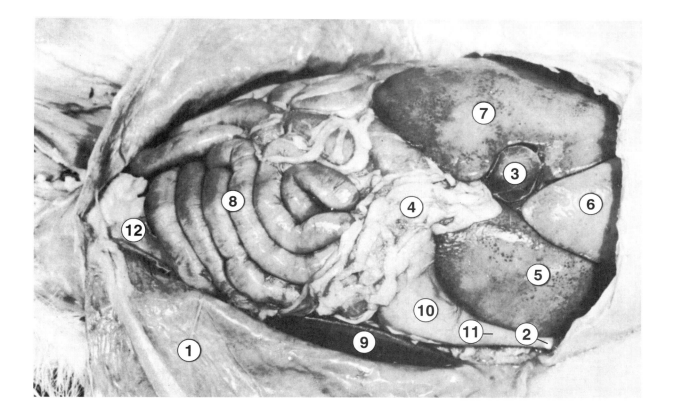

duodenum turns toward the left side of the body, it becomes the second portion of the small intestine, the **jejunum**. The third portion of the small intestine is the **ileum** (**Figure 12.10**). With your scapel make a 2″ longitudinal, ventral incision in the ileum. Flush out any contents with water and blot dry. Observe the carpetlike lining formed of **villi**, microscopic fingerlike projections that function in the process of absorption of digested foods. The jejunum and ileum are approximately equal in length in the cat. The ileum terminates on the right side of the body and joins the large intestine, or **colon**. The ileum and **cecum**, which is the first portion of the large intestine, are separated by

Figure 12.10

Ventral aspect of deep abdominal organs after reflection of small intestine

1. duodenum
2. pyloric portion of stomach
3. spleen
4. liver, right medial lobe
5. urinary bladder

6. horn of uterus
7. pyloric sphincter
8. pancreas
9. jejunum
10. ileum

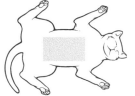

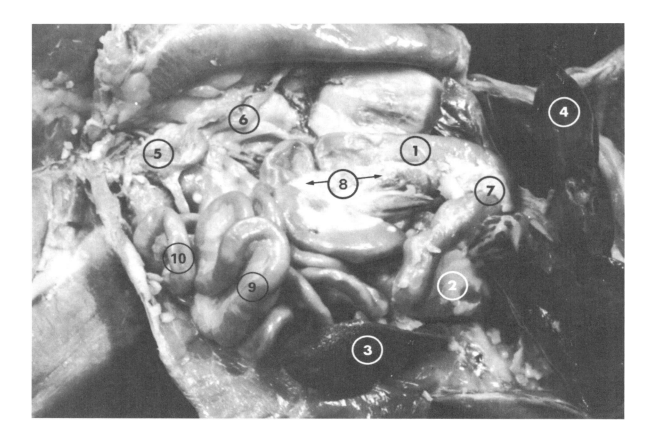

Question 12A.2

How do the rugae and villi differ in function?

the **ileocecal valve**, which you can feel in the cat. The large intestine is considerably larger in diameter than the small intestine and can also be recognized by the presence of **haustra**, or outpocketings, along its length. The **cecum** is a blind pouch found posterior to the junction of the small and large intestines. In the human being, the appendix is inferior to the cecum. Anterior to the cecum in the cat are the **ascending colon**, **hepatic flexure**, **transverse colon**, **splenic flexure**, **descending colon**, and **rectum**. The rectum opens to the outside through the anus. In the human being, the **anal canal** is the terminal inch of the rectum.

The accessory glands of the digestive tract include the pancreas, liver, gallbladder, and salivary glands. In the cat the **pancreas** is composed of two parts: (1) the anterior **head** of the pancreas, which lies between the pylorus and the duodenum; and (2) the **body**, which extends from the anterior head of the pancreas to the left side of the body dorsal to the stomach. It may be necessary to lift the stomach up toward the diaphragm to see this organ. The pancreas secretes digestive enzymes and also secretes the hormones insulin and glucagon, which maintain normal blood sugar levels.

The **liver** is the most prominent organ in the abdominal cavity and overlies other organs. In addition to its many other functions, it secretes bile, which is important for emulsifying and digesting fat. The cat liver has the following five lobes:

Right lobe, consisting of:
right lateral lobe
right medial lobe
right caudate lobe

Left lobe, consisting of:
left lateral lobe
left medial lobe

The **falciform ligament** is part of the ventral mesentery between the liver and ventral body wall. It is also found between the left and right lobes of the liver. On the ventral surface of the right medial lobe is the **gallbladder**, which usually appears as a greenish sac tapering to form the **cystic duct**. This combines with the **hepatic ducts** to form the **common bile duct**, which together with the main pancreatic duct, penetrates the duodenum at the **hepatopancreatic ampulla**. The opening of the ampulla is the **hepatopancreatic sphincter**.

In the human, the **spleen** is an elongated brown organ located on the left side of the body. It is part of the lymphatic system. Supporting the visceral organs are **mesenteries**, which are composed of simple squamous epithelium and loose connective tissue. The mesentery that suspends the small intestine from the dorsal body wall and stretches between the parts of the small intestine contains many blood vessels and mesenteric lymph nodes. **Figure 12.3** illustrates the relationship between the liver, gallbladder, and pancreas to the duodenum in the human being.

On the human torso or cadaver, note that the pancreas is composed of three parts: (1) the anterior **head**, which lies between the pylorus and the duodenum; (2) the **body**, which extends from the head to the left dorsal aspect of the stomach; and (3) the **tail**. As in the cat, the liver is the most prominent organ in the abdominal cavity and overlies other organs. It has two major lobes, **right** and **left**, and two minor lobes, the **quadrate**, which is adjacent to the gallbladder, and the **caudate**, which is next to the inferior vena cava.

Question 12A.3

Which ducts empty into the hepatopancreatic papilla?

EXERCISE 3   Microscopic Examination of Digestive Tissue

Observe the following slides and make a sketch of each.

Oral Cavity

Developing Tooth

In this slide, the developing deciduous tooth is embedded in the alveolar process of the jaw. Notice the bone of the **alveolus**, the central area of dental **pulp** that forms the core of the developing tooth, the broad layer of **dentin**, (which may stain pink) surrounding the pulp, and the **enamel** immediately overlying the dentin. An area known as the **dental sac** surrounds the developing tooth. Also, notice the darkly staining **odontoblasts** around the outer margin of the dental pulp. Sketch and label.

Dried Tooth

The broad portion of this section represents the **crown** of the tooth, which is covered with enamel. The narrower portion represents the part of the tooth beneath the gum line and is referred to as the **root**. Notice that beneath the enamel and extending into the root is a thick layer containing wavy, closely spaced parallel tubules. This layer is dentin. The junction of the root and crown of the tooth at the gum is the **neck**. In the region of the root, the dentin is covered by a thin layer of **cementum** and a **periodontal membrane** that is adjacent to the alveolar bone. In the center of your section, you will see a clear area within the dentin. The broader end of this space represents the **pulp cavity**. The more constricted extension of the pulp cavity is the **root canal**. The pulp cavity and root canal contain fibroblasts, odontoblasts, blood vessels, and nerves embedded in connective tissue. Sketch and label.

Tongue

The mucosa of the tongue is stratified squamous epithelium. In some sections, you will be able to see various types of **papillae**, which contain taste buds (**Figures 12.11–12.13**). Notice the longitudinal, transverse, and oblique planes of skeletal muscle fibers that occupy the interior of the tongue. Embedded in the muscle are **lingual glands**, **blood vessels**, and **nerves**. Sketch and label.

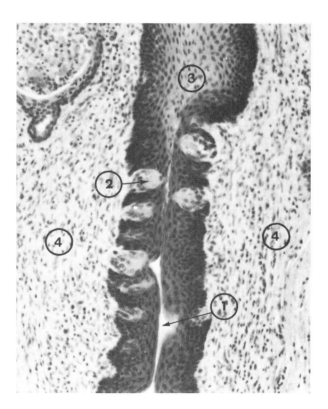

Figure 12.11

Tongue showing taste buds, (430×)

1. trench of vallate papilla
2. taste bud
3. stratified squamous epithelium
4. connective tissue

Figure 12.12

Tongue, filiform papillae (430×)

1. filiform papillae

Figure 12.13

Tongue, fungiform papilla (430×)

1. fungiform papilla

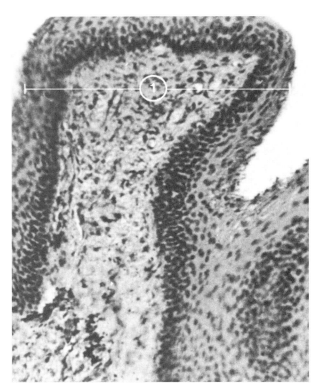

Parotid Gland

The **parotid glands** are one of the three types of salivary glands (the others being the **sublingual** and the **submaxillary**, or **submandibular**, glands) and are located inferior and anterior to the ear in the human being. The predominant structures on this slide are **serous alveoli**, which secrete a watery saliva. These glands, which are composed of pyramid-shaped cells arranged in a circular manner, contain darkly staining nuclei. In this section, there are also blood vessels, adipose tissue, connective tissue, and ducts that drain the glands (**Figures 12.14** and **12.15**). Sketch and label.

Question 12A.4

Which digestive enzyme do the parotid, submaxillary, and sublingual glands produce?

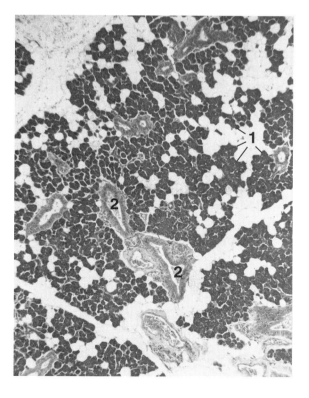

Figure 12.14

Parotid gland, overview (40×)

1. serous alveoli
2. excretory ducts

Figure 12.15

Parotid gland (100×)

1. excretory duct containing saliva
2. serous alveoli
3. arteriole
4. venule

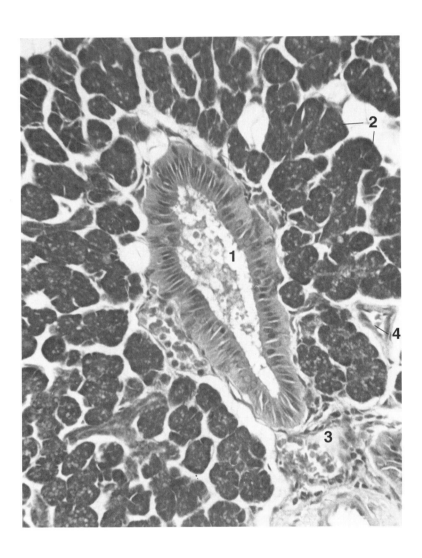

ALIMENTARY CANAL

The **alimentary canal** is essentially a long tube consisting of the esophagus, stomach, small intestine, and large intestine. The wall of the alimentary canal has four basic layers. The inner lining, the **mucosa**, most often consists of a surface **epithelium**, an underlying connective tissue layer of **lamina propria**, and a thin layer of smooth muscle, the **muscularis mucosae**. The second basic layer is a connective tissue layer beneath the mucosa known as the **submucosa**. The third layer is the **muscularis externa**, usually composed of two layers of smooth muscle—an inner **circular** layer and an outer **longitudinal** layer. The outer layer or coat is the **serosa**, which is a reflection of the peritoneum lining the walls of the body cavity. Some parts of the digestive tract have an adventitia rather than a serosa. The **adventitia** is a fibrous outer coat that blends in with surrounding structures. Modifications of the four basic layers occur in different regions of the digestive tract (**Figures 12.16** through **12.18**).

> **Question** **12A.5**
>
> Name the four layers that make up the alimentary canal wall.

> **Figure 12.16**
>
> Ileum showing general structure of layers of alimentary canal wall (100×)

1. serosa	3. submucosa
2. muscularis externa	4. mucosa
a. inner, circular muscle layer	5. plicae circulares
b. outer, longitudinal muscle layer	6. villi

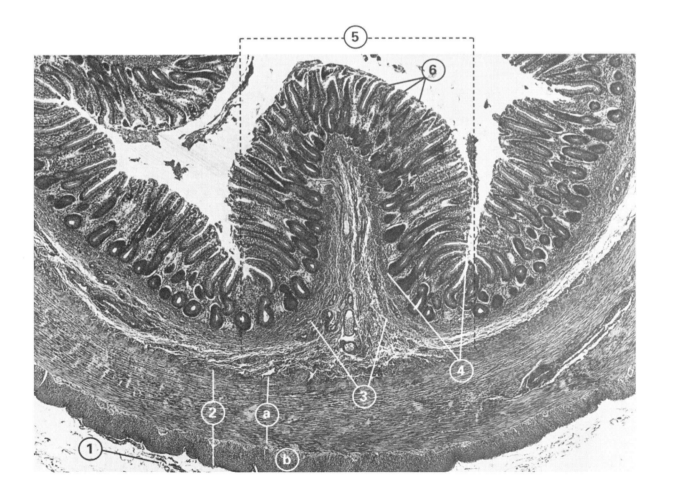

Esophagus

The mucosa is composed of **stratified squamous epithelium**, lamina propria, and muscularis mucosae layers. The submucosa contains **esophageal glands**, which appear similar to the serous alveoli of the parotid gland. The remainder of this layer is connective tissue containing adipose cells and blood vessels (**Figure 12.17**). The muscularis externa consists of two layers—an inner one of circular fibers and an outer longitudinal muscle layer. The muscularis externa of the superior third of the esophagus contains striated muscle, the middle third both striated and smooth muscle fibers, and the inferior third smooth muscle only. The outer layer of the esophagus is primarily adventitia, except at the inferior end, where it is serosa. Sketch and label.

Figure 12.17
Esophagus (100×)

1. stratified squamous epithelium
2. lamina propria
3. muscularis mucosae
4. esophageal glands
5. arteriole
6. venule

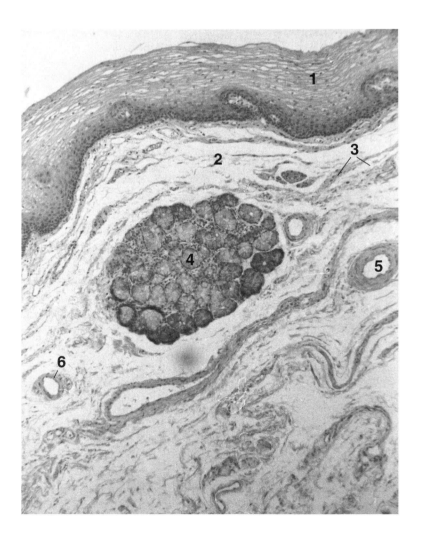

Stomach

The stomach has several regions that differ slightly in their histology. The four basic layers are present in all regions and will be described in general. Observe your section first under low power and then under high power to examine the layers in detail.

The mucosal layer of the stomach consists of a surface of simple columnar epithelium supported by lamina propria (**Figure 12.18**). If you are observing the fundic, or superior, portion of the stomach, notice that the epithelial layer dips into the mucosa, forming **gastric pits.** Farther down in the mucosal layer beneath the gastric pits are blue-staining cells, known as **chief** (or **zymogenic**) **cells** that contain pepsinogen, the precursor to the enzyme pepsin. There are also red-staining parietal cells, which secrete hydrochloric acid. Chief and parietal cells are arranged to form narrow, vertical **fundic (gastric) glands** (**Figures 12.18** and **12.19**). Toward the muscularis mucosae you may see some gastric glands in cross section.

Figure 12.18

Stomach, fundic region, overview (40×)

1. lumen
2. rugae
3. gastric pits
4. mucosa

5. muscularis mucosae
6. submucosa
7. muscularis externa

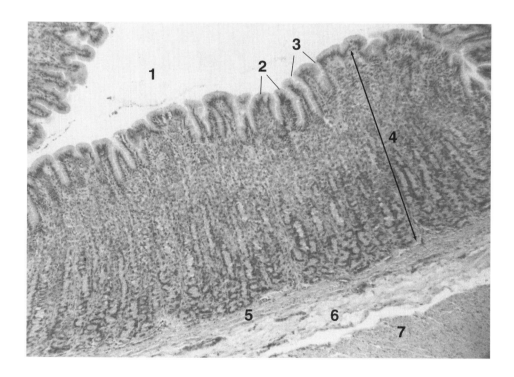

Figure 12.19

Stomach, fundic region, showing gastric pits and gastric glands (100×)

1. gastric pits
2. gastric glands
3. parietal cells
4. chief (zymogenic) cells

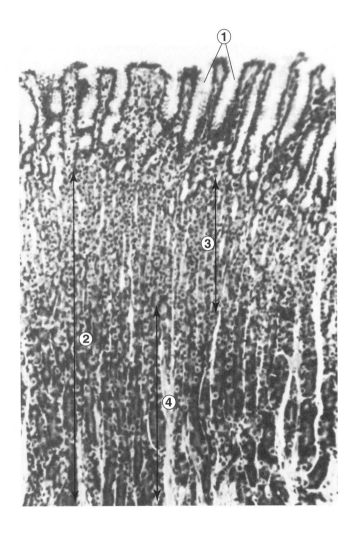

In all areas of the stomach, there are darkly staining **lymph nodules** at intervals immediately superior to the muscularis mucosae. Lamina propria serves as the connective tissue, both in the region of the gastric pits and in the region of the fundic and pyloric glands. The submucosa of the stomach contains blood vessels and adipose and connective tissue.

The muscularis externa region of the stomach consists of three layers. An inner layer of **oblique fibers** is present in addition to the middle and outer layers of circular and longitudinal fibers, respectively. The serosa is composed of connective tissue. Sketch and label your slide of the stomach.

The Small and Large Intestines

There are several features common to both the small and large intestines. The mucosa forms deep pits known as **intestinal glands** (or **crypts of Lieberkühn**). **Simple columnar epithelial cells** line the inner surface of the intestinal mucosa; **goblet cells** secrete mucus and are thought to be formed through the transformation of simple columnar cells (**Color Plates 5–6**). Other common features of the intestinal tract mucosa are the lamina propria, which is found underlying the mucosal epithelium and between the intestinal glands; the lymph nodules, which are more prominent in certain parts of the intestine than others; and a thin muscularis mucosae, which has an inner layer of circular and an outer layer of longitudinal smooth muscle fibers.

The submucosa of the intestinal tract consists of loose areolar tissue, adipose tissue, blood vessels, and nerves. A network of nerve fibers, the **submucosal plexus**, is in the submucosal layer. There may be other structures present in the submucosa, depending on the particular intestinal region. The muscularis externa consists of an inner, circular, smooth muscle layer and an outer, longitudinal layer. The outer layer of the intestinal tract is serosa. Another nerve plexus, the **myenteric plexus**, is located between the circular and longitudinal layers of the muscularis externa. The submucosal plexus and myenteric plexus aid in peristalsis (**Figure 12.26**).

Small Intestine   (Color Plate 5)

In the human, the small intestine is divided into three regions: the **duodenum**, which is approximately 10–12″ long; the **jejunum**, which is about 8′ long; and the **ileum**, which is about 12′ long. The entire small intestine is characterized by the presence of **plicae circulares** (or **valves of Kerckring**), which are folds in the mucosa and submucosa, and by **villi** (**Figure 12.16**), which are microscopic, fingerlike projections of the mucosa overlying the plicae. Villi further increase the absorptive surface area of the small intestine (**Figure 12.20**). **Crypts of Lieberkühn** (**intestinal glands**) can be seen inferior to the villi and superior to the muscularis mucosae.

Observe a section of duodenum. The primary histological characteristic of this region is the presence of **duodenal glands**, or **Brunner's glands**, that are found primarily in the submucosa (**Figures 12.21** and **12.22**) and, to some

Question 12A.6

Give the function of each of the following cells of the intestine:

goblet cells

simple columnar epithelium

Figure 12.20

Longitudinal section showing general structure of villi with lacteal detail visible (430×)

1. lacteal
2. goblet cells
3. striated border of microvilli
4. crypt of Lieberkühn (intestinal gland)

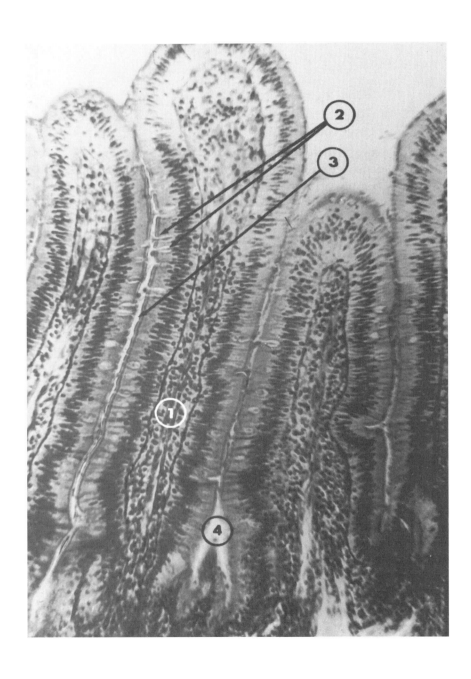

extent, in the **mucosa**. The secretions of these glands are of a mucoid nature. Also, notice the four basic layers in this section: the **mucosa**, with mucosal goblet cells and simple columnar cells, as well as the crypts of Lieberkühn and the muscularis mucosae; the submucosa; the muscularis externa; and the serosa (**Figures 12.21–12.24**). Sketch and label.

Figure 12.21

Duodenum (x.s.), overview (40×)

1. mucosal layer containing cross sections of intestinal glands
2. submucosa
3. circular layer of muscularis externa
4. longitudinal layer of muscularis externa

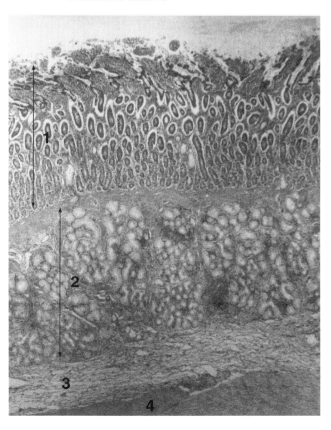

Figure 12.22

Duodenum, showing glands (100×)

1. intestinal glands (x.s.)
2. duodenal (Brunner's) glands containing mucous

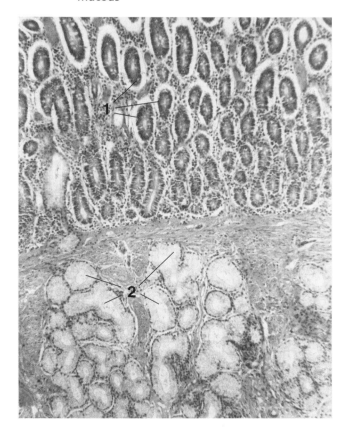

Figure 12.23

Cross section of jejunum (40×)

1. villi
2. plicae circulares
3. intestinal gland
4. submucosa
5. muscularis mucosae
6. inner circular layer of muscularis externa
7. outer longitudinal layer of muscularis externa

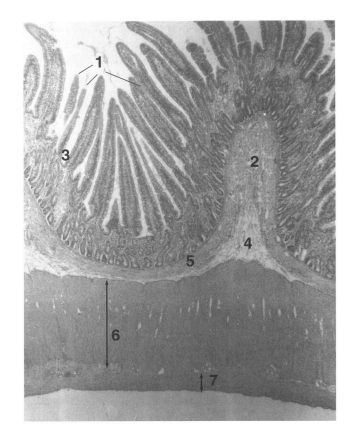

Figure 12.24

Ileum, showing prominent areas of lymphoid follicles (40×)

1. villi
2. submucosa
3. lymphoid follicle (nodule)
4. circular layer of muscularis externa

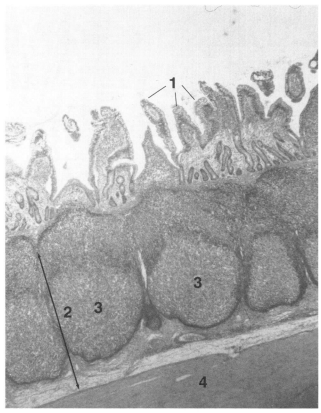

The sections of the jejunum and ileum are similar. Notice the presence of darkly staining lymphatic nodules in the submucosa. In the ileum, aggregations of these nodules are known as **Peyer's patches**. Also, notice that the villi become more elongated as one progresses from the duodenum to the jejunum to the ileum. Goblet and simple columnar cells should also be visible on these sections (**Figures 12.22–12.24**). Sketch and label.

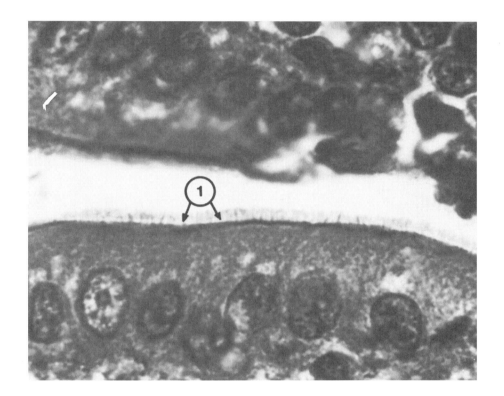

Figure 12.25

Microvilli, detail of small intestine (430×).

1. microvilli

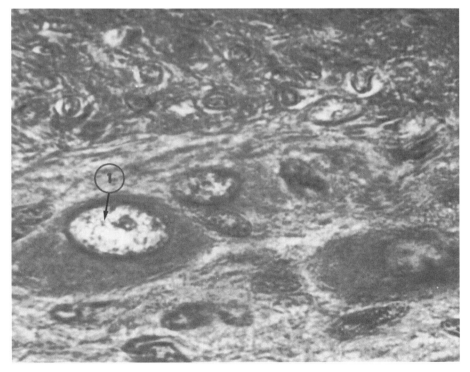

Figure 12.26

Detail of ganglion cells of myenteric plexus (430×).

1. ganglion cell

Large Intestine

The large intestine or colon lacks villi (**Color Plate 6**), but intestinal glands, or crypts, in the mucosa are very prominent. Notice the numerous goblet cells (**Figures 12.27–12.29**) in the lining of the intestinal glands in this section. **Solitary lymph nodes** may be seen in the lamina propria of the mucosa immediately adjacent to the muscularis mucosae. Also, observe the submucosa, muscularis externa, and serosa in this section. Sketch and label.

Figure 12.27

General structure of colon (100×)

1. mucosa
2. submucosa
3. muscularis externa
   a. inner circular layer
   b. outer longitudinal layer

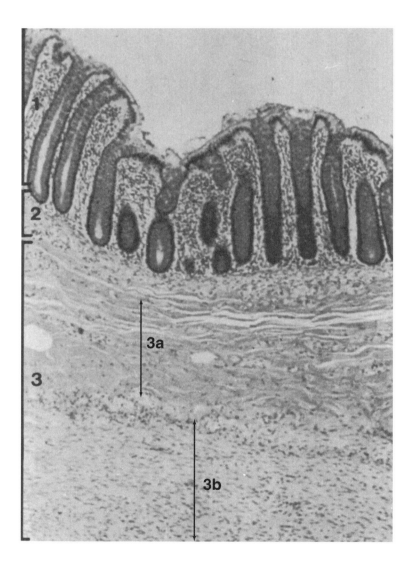

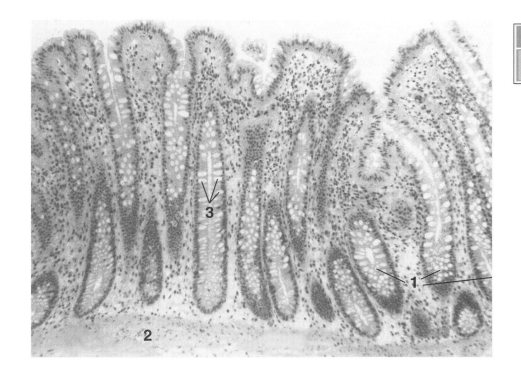

Figure 12.28

Colon mucosa (100×)

1. intestinal glands
2. muscularis mucosae
3. goblet cells

Figure 12.29

Colon mucosa (400×)

1. simple columnar epithelium
2. goblet cells
3. lamina propria
4. intestinal gland

Question 12A.7

Name the divisions of the large intestine.

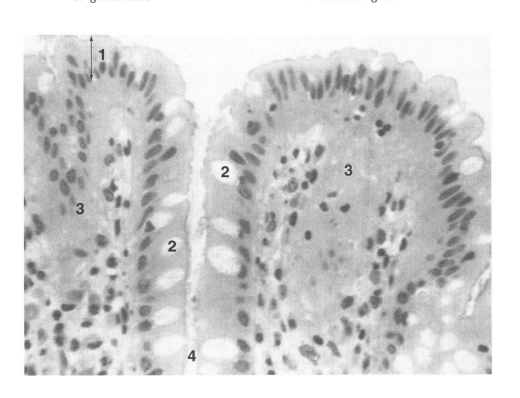

What is the function of the appendix?

Appendix

Notice the poorly developed intestinal glands and numerous lymph nodules in this section (**Figure 12.30**). Sketch and label.

Rectum

The rectum is characterized by the presence of large longitudinal folds of mucosa and submucosa known as **rectal columns**. These columns are seen in transverse section on your slide. Notice the four basic layers here, including the crypts of Lieberkühn in the mucosa. Sketch and label.

Anus (Anal Canal)

The mucosa of the anal canal is stratified squamous epithelium. The lamina propria contains a plexus of large veins that, when dilated, are known as hemorrhoids. Sketch and label.

Figure 12.30

Appendix, showing inflammation of lymph tissue (430×)

1. lymph nodes
2. submucosa
3. muscularis externa
4. occluded lumen

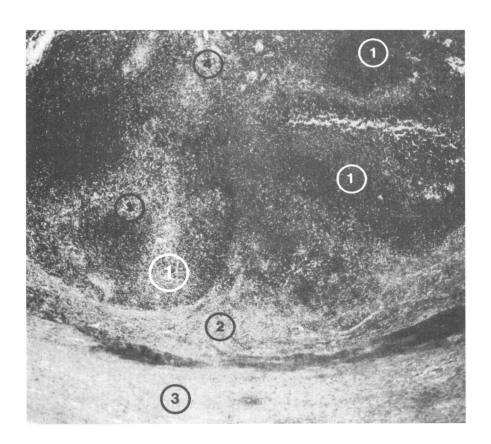

Question 12A.9

What structures make up a hepatic triad?

Liver

Under low power, observe the **lobules** of the **liver**, which are roughly hexagonal. In the center of each lobule is a clear area, the **central vein**, which may contain a few red blood cells. The liver cells, or **hepatocytes**, appear to radiate toward the periphery of each lobule from the central vein. The clear channels between the rows of hepatocytes are **hepatic sinusoids.** Focus on an area where the borders of three or more lobules come in contact with each other. First under low power and then under high, identify a thin-walled branch of the **hepatic portal vein**, a smaller in diameter but thicker-walled branch of the **hepatic artery**, and a **bile duct**. These three structures make up the **hepatic triad.** The bile duct can be identified by its intermediate size and walls of cuboidal and simple columnar epithelium (**Figures 12.31** and **12.32** and **Color Plate 41**). Sketch and label.

Figure 12.31

Liver, panoramic view, showing general structure of lobules (100×)

1. central vein
2. lobule

3. hepatic triad
4. branch of bile duct

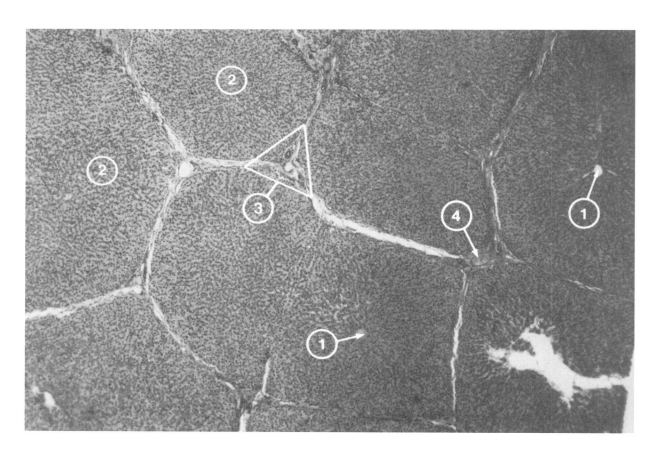

Figure 12.32

Liver, showing hepatic triad (430×)

1. branch of the hepatic portal vein
2. branch of the hepatic artery
3. bile duct
4. hepatocytes (liver cells)
5. hepatic sinusoid

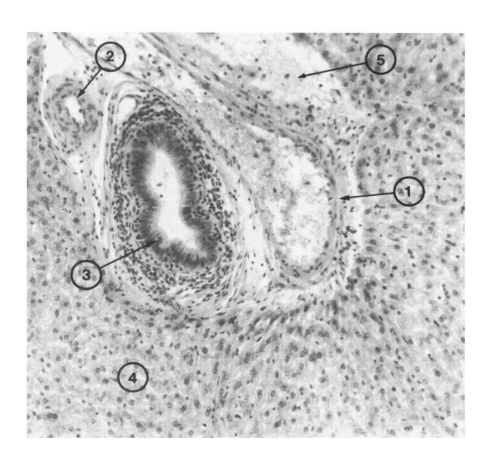

Gallbladder

The **gallbladder**, which is suspended from the right lobe proper of the liver, serves as a reservoir that stores the bile produced in the liver. It is also composed of the same four histological layers that are found in other organs of the digestive tract (**Figures 12.33** and **12.34**). Bile drains from the gallbladder through the **cystic duct**, which merges with the **common hepatic duct** to form the **common bile duct**. The common bile duct enters the duodenum at the ampulla of Vater. The opening of the **ampulla** is at the **hepatopancreatic ampulla** (Figure 12.3).

Figure 12.33

Gallbladder (100×)

1. mucosa
2. smooth muscle
3. arteriole containing blood cells
4. rugae

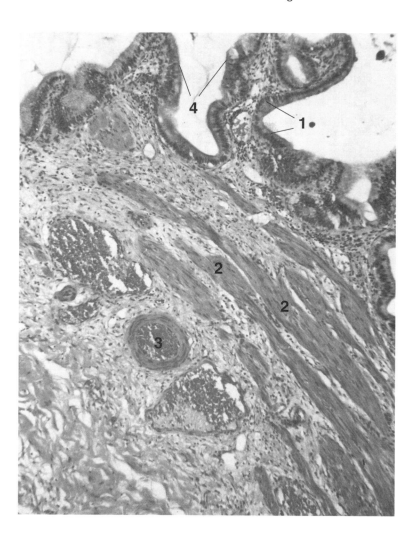

Figure 12.34

Gallbladder (400×)

1. simple columnar epithelium
2. lamina propria
3. smooth muscle

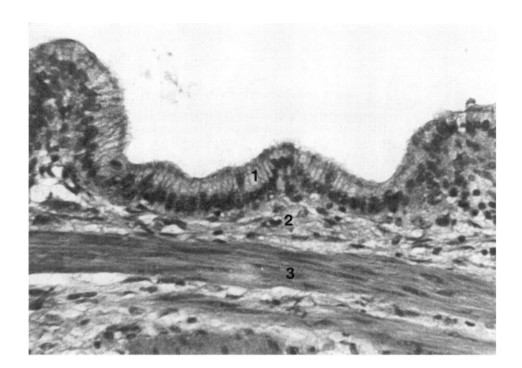

B. Digestive Chemistry

PURPOSE

Unit 12B will enable you to understand the basic chemical principles of digestion.

OBJECTIVES

After completing Unit 12B, you will be able to
- summarize the digestion of proteins, carbohydrates, and fats.
- perform chemical tests for the identification of various amino acids, proteins, carbohydrates, and fats.

EXERCISE 1   **Tests for Properties of Proteins**

Proteins are compounds composed of amino acids joined together with peptide bonds:

$$H-\underset{\underset{NH_2}{|}}{\overset{\overset{H}{|}}{C}}-COOH + NH_2-\underset{\underset{CH_3}{|}}{\overset{\overset{H}{|}}{C}}-COOH \overset{-H_2O}{\rightleftharpoons} H-\underset{\underset{NH_2}{|}}{\overset{\overset{H \quad O}{|\ \ ||}}{C}}-\overset{\,}{C}-\underset{\underset{H}{|}}{\overset{}{N}}-\underset{\underset{CH_3}{|}}{\overset{\overset{H}{|}}{C}}-COOH + H_2O$$

glycine   alanine   dipeptide with peptide bond (—)

All amino acids and proteins contain carbon, hydrogen, oxygen, and nitrogen. Some proteins also contain other elements.

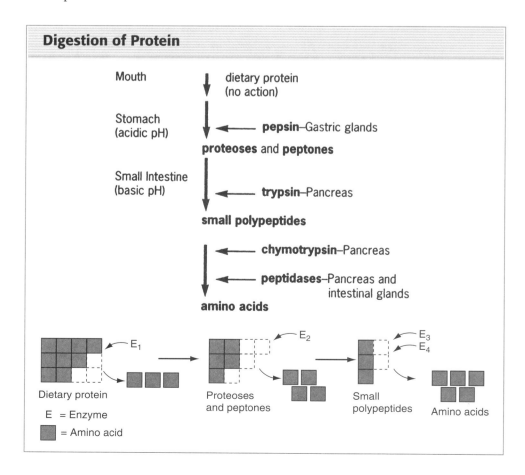

Digestion of Protein

Mouth — dietary protein (no action)

Stomach (acidic pH) ← **pepsin**–Gastric glands
proteoses and **peptones**

Small Intestine (basic pH) ← **trypsin**–Pancreas
small polypeptides

← **chymotrypsin**–Pancreas

← **peptidases**–Pancreas and intestinal glands
amino acids

E_1 E_2 E_3 E_4

Dietary protein   Proteoses and peptones   Small polypeptides   Amino acids

E = Enzyme

◼ = Amino acid

MATERIALS

1% albumin solution or egg white	asbestos pad
10% NaOH	ring stands
1% $CuSO_4$	5% trypsin solution
Millon's reagent	Bunsen burners or hot plates
0.1% alcohol solution of glycine or alanine	test tube holder
0.1% freshly prepared ninhydrin	test tube rack
undiluted egg white	medicine droppers
deionized water	either 1- and 10-mL pipettes
test tubes	with bulbs or graduated
beakers	cylinders
buffer solution (pH 8)	37°C water bath
	safety glasses

PROCEDURE

BIURET REACTION

The Biuret reaction is specific for compounds containing two or more peptide bonds. To a test tube containing 3 cc of 1% albumin solution or egg white, add 3 cc of 10% NaOH and one drop of 1% $CuSO_4$. Mix. Add additional drops of $CuSO_4$ until a violet color is obtained, indicating a positive reaction.

MILLON REACTION

The Millon reaction is specific for the amino acid tyrosine, which contains a benzene ring to which is attached a hydroxyl (—OH) group. Add five drops of Millon's reagent to a test tube containing 5 cc of 1% albumin solution. Heat in a beaker of boiling water. A brick-red color indicates the presence of tyrosine, an amino acid common in most proteins.

NINHYDRIN REACTION

The ninhydrin reaction is a test for alpha amino acids. To a test tube containing 5 cc of 0.1% alcohol solution of glycine or alanine, add 0.5 cc of 0.1% *freshly prepared* ninhydrin. Heat in a 37°C water bath. Note color formation. Repeat, using 3 cc of 1% albumin solution instead of an amino acid. Heat the albumin to which ninhydrin was added in a beaker of boiling water, and then cool. What is the color?

HEAT COAGULATION

Place 5 cc of undiluted egg white in a test tube and place in a beaker of water. Allow the water in the beaker to boil. Describe your results. Is this reaction reversible upon cooling and agitation?

DIGESTION OF PROTEIN

Place 1–2 pieces of chopped cooked egg white from the heat coagulation exercise into each of two test tubes. To Tube 1, add 10 mL distilled water (buffered to pH 8). To Tube 2, add 10 mL of 5% trypsin or pancreatin solution. Allow to stand in a 37°C water bath for 2 hours. Observe. Describe your results.

Question **12B.1**

What basic test allows you to identify an amino acid chain?

Record the results of Exercise 1 in the following table.

Tests for Proteins	Result	Conclusion
Biuret reaction		
Millon reaction		
Ninhydrin reaction		
Heat coagulation of protein		
Digestion of protein		

EXERCISE 2 **Tests for Properties of Carbohydrates**

All **carbohydrates** contain the elements carbon, hydrogen, and oxygen, the last two of which are present in the same ratio as in water (2:1). Carbohydrates may be classified as monosaccharides, disaccharides, or polysaccharides. The mono- and disaccharides resemble each other considerably, but polysaccharides bear little resemblance to the other two classes.

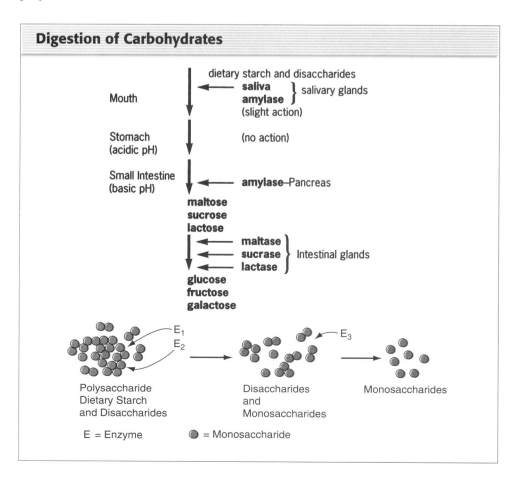

Digestion of Carbohydrates

MATERIALS

10% glucose solution	thin starch paste
5% alcoholic solution of alpha naphthol	Lugol's iodine
concentrated H_2SO_4	graduated cylinders
Benedict's solution	beakers
5% glucose solution	test tubes
5% sucrose solution	medicine droppers
10-mL pipettes with bulbs	ring stand
Barfoed's reagent	test tube holder
5% fructose solution	safety glasses
Seliwanoff's reagent	Bunsen burners or hot plates
1 M HCl	40°C water bath
1 M NaOH	

Because of the possibility of infections being transmitted from one student to another during exercises involving body fluids, take great care in collecting and handling saliva. *The saliva used in the following exercise should be your own. Follow proper procedures for collecting, handling, and disposing of saliva as directed by your instructor.*

CAUTION
BLOODBORNE PATHOGENS
Use care when working with body fluids.

PROCEDURE

TEST FOR STARCH AND STARCH HYDROLYSIS

Place 5 mL of thin starch paste into a test tube. Add one drop of Lugol's iodine solution. Is there a color change?

Put 10 mL of the same starch solution into a test tube and add four or five drops of saliva. Set in a 37°–40°C water bath for 1/2 hour or more. Is there a color change when iodine is added? Explain.

THE MOLISCH REACTION

The Molisch reaction is a general test for carbohydrates. Place 5 cc of 5% glucose solution in a test tube. Add two drops of 5% alcoholic solution of alpha naphthol. Mix. Using a test tube holder and a medicine dropper (or bulb pipette), **CAREFULLY** place 5 cc concentrated sulfuric acid in a second tube. With the tube containing glucose in your left hand, pour the sulfuric acid (H_2SO_4) slowly and gently down the side of the tube containing glucose. The object is to get a layer of H_2SO_4 under the glucose solution. If a purple or pink color is formed at the boundary, the substance is a carbohydrate. What is the result?

CAUTION
Use care when working with sulfuric acid.

Question **12B.2**

What test differentiates
fructose from glucose?

BENEDICT'S TEST

Benedict's test is a test for reducing sugars, which include glucose, fructose, galactose, mannose, maltose, and lactose. To a test tube containing 5 cc of Benedict's solution, add four to five drops of 5% glucose solution. Place in a beaker of boiling water for 2 minutes. Cool slowly. A red, green, or yellow precipitate indicates a positive reaction. Repeat using sucrose. Compare the results.

BARFOED'S TEST

Barfoed's test is a test for monosaccharides. To a test tube containing 5 cc of Barfoed's reagent, add five to six drops of 10% glucose solution. Place in a beaker of boiling water bath for 1 minute. Set aside for 15–20 minutes and watch for a red precipitate, which indicates a positive reaction. What is the result?

SELIWANOFF'S TEST

Run this test in duplicate, using one tube with glucose and another tube with fructose. Add five drops of 5% glucose solution to one tube containing 5 cc Seliwanoff's reagent and five drops of 5% fructose to another. Place in a beaker of boiling water for 30 seconds. What does Seliwanoff's test enable you to do?

INVERSION OF SUCROSE

Pour 5 cc of 5% sucrose solution into each of two test tubes. To the first tube, add 5 cc of Benedict's solution and place in a beaker of boiling water. Remove from heat after five seconds. To the second tube, add two drops of dilute (1 M) HCl and boil for 1 minute. Neutralize with an equal amount and strength of NaOH. (Why?) Add 5 cc of Benedict's solution to the second tube and bring to a boil. How do you explain the different results for these two procedures?

Record the results of Exercise 2 in the following table.

Tests for Carbohydrates	Result	Conclusion
Starch test		
Starch hydrolysis test		
Molisch reaction		
Benedict's test		
Barfoed's test		
Seliwanoff's test		
Inversion of sucrose		

EXERCISE 3   Test for Properties of Fats

Fats are the esters (organic salts) formed by the union of a fatty acid and an alcohol. The three common fats in our foods and body are oleic, palmitic, and stearic, which are formed, respectively, from oleic, palmitic, and stearic acids united chemically with the alcohol glycerol (glycerin). One molecule of glycerol can unite with three fatty acids as follows:

$$H_3C-(CH_2)_{16}-\overset{O}{\overset{\|}{C}}-OH$$

$$H_3C-(CH_2)_{16}-\overset{O}{\overset{\|}{C}}-OH \ + \ \begin{matrix} HO-CH_2 \\ | \\ HO-CH \\ | \\ HO-CH_2 \end{matrix} \ \underset{\xleftarrow{\hspace{1cm}}}{\overset{-3H_2O}{\rightleftharpoons}}$$

$$H_3C-(CH_2)_{16}-\overset{O}{\overset{\|}{C}}-OH$$

3 molecules of stearic acid   1 molecule
(fatty acid)   of glycerol

$$\begin{matrix} H_3C-(CH_2)_{16}-\overset{O}{\overset{\|}{C}}-O-CH_2 \\ | \\ H_3C-(CH_2)_{16}-\overset{O}{\overset{\|}{C}}-O-CH \\ | \\ H_3C-(CH_2)_{16}-\overset{O}{\overset{\|}{C}}-O-CH_2 \end{matrix} + 3H_2O$$

tristearin (a triglyceride)

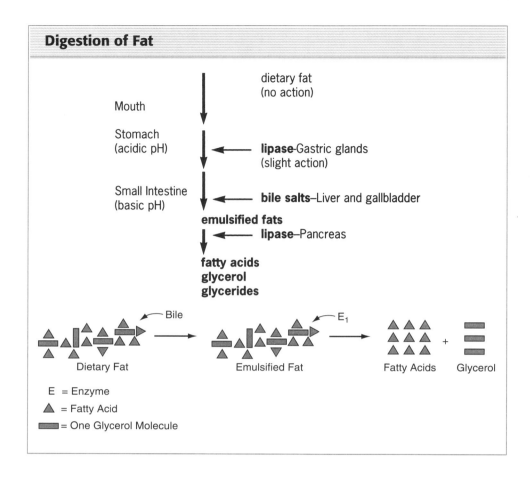

Digestion of Fat

MATERIALS

fresh vegetable oil
deionized water
ethyl alcohol
ether
5% trypsin
beakers
benzene
carbon tetrachloride
bile salts
graduated cylinders

sweet cream (fresh)
0.5% NaHCO$_3$ solution
test tubes
test tube rack
thermometer
blue litmus solution
10 mL pipettes with bulbs
40°C oven or electric water
     bath
safety glasses

PROCEDURE

SOLUBILITY OF FAT
Set up five test tubes as follows. **Make sure that there are no Bunsen flames burning in the area in which you are working.**

CAUTION

Organic solvents are flammable.

Tube No.	Solvent (5 mL)
1	water
2	ethyl alcohol
3	ether
4	benzene
5	carbon tetrachloride

Add five drops of fresh vegetable oil to each tube and shake well. Observe the tubes immediately and again after 15 minutes. What is the result?

EMULSIFICATION OF FAT

Emulsification is the mechanical dispersion of large fat molecules into small, water-soluble particles. Dietary fats need to be emulsified before pancreatic lipase can further digest them to yield fatty acids, which are then absorbed into the blood.

Set up two test tubes and label one "bile salts," the other "control."

Add 3 mL vegetable oil and 3 mL water to each tube.

Using a spatula, add the equivalent of a 0.5 cm sphere of bile salts to the test tube so labeled.

Place a rubber stopper, a 2-inch square of parafilm, or your thumb over the top of each test tube, and shake vigorously for 15 seconds.

Place the test tubes in a rack and immediately observe the droplet size in each. Compare the test tubes again after 30 minutes.

DIGESTION OF EMULSIFIED FAT

Pour 10 mL sweet cream into a clean test tube.

Add 8 to 10 drops of 1% blue litmus solution (an indicator) to the cream to impart a blue color. Mix.

Pour 5 mL of this mixture into another test tube. Add 2 mL of 5% trypsin to one tube and 2 mL 0.5% $NaHCO_3$ to the second tube.

Place both tubes in a 40°C water bath. Observe after 1 hour. Did you observe a color change? Explain.

Question 12B.3

What is the action of bile salts?

Question 12B.4

Why did you add $NaHCO_3$ to the second tube?

Record the results of Exercise 3 in the following table.

Tests for Fats	Result		Conclusion
Solubility of fat			
Emulsification of fat	Immediately	After 30 minutes	
Digestion of emulsified fat			

Question 12B.5

How would you identify a fat?

Complete this chart.

Digestive Enzymes

Enzyme	Organ of Production	Organ of Action	Substrate	End Product(s)
1	2	Mouth	3	4
Pancreatic amylase	5	6	7	8
9	10	11	Dietary protein	12
13	14	15	16	Fatty acids and glycerol
Maltase	Small intestine	17	18	19

EXERCISE 4 **Nutrition**

Normal metabolism, cellular growth and repair, and the production of energy require adequate amounts of carbohydrates, proteins, and fats. Additionally, metabolic reactions require small amounts of vitamins, organic substances found in food, which often function in enzymatic reactions. Minerals, which are inorganic substances that form chemical bonds with organic substances, function in numerous body activities such as bone formation, enzymatic reactions, and muscle and nerve function.

The United States Department of Agriculture (USDA) Food Guide Pyramid is considered a major guide to healthful eating. Currently, in addition to the USDA needs, a number of other guides and dietary plans have gained popularity.

Another method used to maintain a healthful diet is to consume the following:

Food Component	Amount	% of Total Caloric Intake*
Carbohydrate	300 g	60
Protein	50–60 g	10–15
Fats	60–65 g	25–30

*Based on a 2000 calorie/day diet.

Serving size is important. To correctly follow the USDA Pyramid, you need to know the amount of food that constitutes a serving. See the chart below.

Serving Sizes						
Food group	Bread, cereal, rice, pasta	Vegetable	Fruit	Milk, yogurt, cheese	Meat, poultry, fish, dry beans, eggs, and nuts	Fats, oils, and sweets
Number of servings per day	6–11 servings	3–5 servings	2–4 servings	2–3 servings	2–3 servings	Sparingly
Examples of a serving	1 slice of bread 1/2 C. cooked rice, pasta or cereal 1 oz. ready-to-eat cereal	1/2 C. cooked or raw vegetables 1 C. leafy raw vegetables	1 piece medium size or 1 melon wedge 3/4 C. juice 1/2 C. canned fruit 1/4 C. dried fruit	1 C. milk or yogurt 1.5 oz. natural cheese 2 oz. processed cheese	2–3 oz. cooked lean meat, poultry, or fish Count 1/2 C. cooked beans or 1 egg or 2 T peanut butter as 1 oz. of lean meat	Limit these

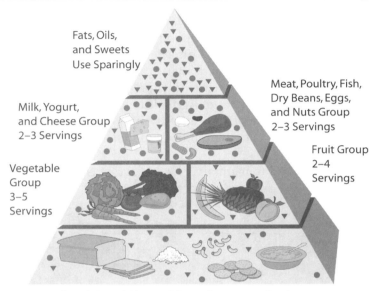

Key
● Fat (naturally occurring and added)   ▼ Sugars (added)
These symbols show that fat and added sugars come mostly from fats, oils, and sweets, but can be part of or added to foods from the other food groups as well.

Fats, Oils, and Sweets Use Sparingly

Milk, Yogurt, and Cheese Group 2–3 Servings

Meat, Poultry, Fish, Dry Beans, Eggs, and Nuts Group 2–3 Servings

Fruit Group 2–4 Servings

Vegetable Group 3–5 Servings

Bread, Cereal, Rice, and Pasta Group 6–11 Servings

In the table below, record all that you ate yesterday. Enter the names of foods <u>and</u> serving sizes.

One Day Food Intake

Meals	Carbohydrate Servings	Protein Servings	Fat Servings
Breakfast			
Snack			
Lunch			
Snack			
Dinner			
Snack			

Analyze your food intake and compare your intake with the USDA recommended number of servings. Enter your intake in the following chart:

Comparison of Your One Day Food Intake with the USDA Food Guide Pyramid

Your Intake	Recommended	Comparison (=, ate required amount; –, ate less than required amount; +, ate more than required amount)
Carbohydrate servings		
Protein servings		
Fat servings		

Question 12B.6

What recommendation do you have for yourself with regard to your nutrition?

The Digestive System

WORD RELATIONSHIPS

Indicate which of the following terms in each group is *unrelated* to the others:

Name _____

Section _____

Date _____

_____ 1. a. mucosa   c. muscularis
       b. submucosa   d. rugae

_____ 2. a. insulin and glucagon   c. usually is 6–9 inches
       b. is roughly fish   in lengh
         shaped   d. accumulates bile

_____ 3. a. the largest body gland   c. falciform ligament
       b. has two main lobes   d. contains a main and an
                 accessory duct

_____ 4. a. mastication   c. peristalsis
       b. deglutition   d. hydrolysis

_____ 5. a. completes digestion of foods   c. 20 feet long
       b. duodenum   d. serves as a reservoir

_____ 6. a. dentin   c. pulp
       b. periodontal membrane   d. haustra

_____ 7. a. ninhydrin reaction   c. Molisch reaction
       b. Biuret test   d. Millon reaction

_____ 8. a. transverse colon   c. sinusoids
       b. haustra   d. cecum

MULTIPLE CHOICE

_____ 9. The organ that produces bile is the
       a. gallbladder.   c. stomach.
       b. pancreas.   d. liver.

_____ 10. The stomach
       a. secretes gastric juice.   c. provides for limited
       b. mechanically churns food   absorption of alcohol
         into chyme.   and water.
                 d. all of the above.

_____ 11. The structure that connects the pharynx to the stomach is the
       a. trachea.   c. esophagus.
       b. larynx.   d. epiglottis.

12. The hard covering over the crown of a tooth is known as
    a. enamel.
    c. dentin.
    b. cementum.
    d. pulp cavity.

13. The waves of muscular contraction that push food down through the esophagus, stomach, and small intestine are known as
    a. periosteum.
    c. peristalsis.
    b. pericardium.
    d. peritoneum.

14. The accessory digestive gland which secretes digestive enzymes to digest carbohydrates, proteins, and lipids is
    a. the spleen.
    c. the small intestine.
    b. the pancreas.
    d. the liver.

15. An example of a monosaccharide is
    a. glucose.
    c. maltose.
    b. glycogen.
    d. cellulose.

16. Which of these is the correct pathway for the secretion of bile?
    a. common bile duct, left and right hepatic ducts, common hepatic duct, hepatopancreatic papilla
    c. common hepatic duct, left and right hepatic ducts, hepatopancreatic papilla, common bile duct
    b. hepatopancreatic papilla, common bile duct, common hepatic duct, left and right hepatic ducts
    d. left and right hepatic ducts, common hepatic duct, common bile duct, hepatopancreatic papilla

17. Benedict's reagent is a test for
    a. reducing sugars.
    c. sucrose.
    b. starch.
    d. amino acids.

18. Which of these lines the wall of the abdominal cavity?
    a. visceral peritoneum
    c. parietal peritoneum
    b. mesentery
    d. adventitia

19. In which portion of the colon would feces have the most liquid consistency?
    a. ascending
    c. descending
    b. transverse
    d. rectum

20. Maltase is to maltose as peptidase is to
    a. polypeptide.
    c. sucrose.
    b. trypsin.
    d. pepsin.

21. In which of these solvents would a fat be insoluble?
    a. water
    c. benzene
    b. ethyl alcohol
    d. carbon tetrachloride

22. Which of these chemical elements is found in proteins but not in lipids or carbohydrates?
    a. carbon
    c. nitrogen
    b. hydrogen
    d. oxygen

_____ **23.** The final breakdown products of carbohydrate digestion are
primarily
    a. amino acids.   **c.** enzymes.
    b. monosaccharides.   **d.** glycerol and fatty acids.

_____ **24.** Chemical digestion of a piece of bread begins in the
    a. stomach.   **c.** duodenum.
    b. ileum.   **d.** mouth.

_____ **25.** In which portion of the alimentary canal is digested food
absorbed to the greatest extent?
    a. esophagus   **c.** small intestine
    b. stomach   **d.** large intestine

DISCUSSION QUESTIONS

26. Why is it important to have increased surface area in the small intestine? _____

   _____

   _____

   _____

   _____

27. Compare the anatomy of the colon in the cat and human being with respect to the presence or absence of:

   a. appendix _____

   b. cecum _____

   c. hepatic flexure _____

   d. sigmoid colon _____

28. Which digestive glands secrete the following?

   a. amylase _____

   b. pepsin _____

   c. trypsin _____

   d. peptidase _____

   e. lactase _____

   f. lipase _____

CASE STUDY

DIGESTIVE SYSTEM-1

Michelle, a respiratory therapist, experienced abdominal pain every evening after supper. She described her pain as feeling like severe gas build up. She recently suffered bouts of bloating followed by alternating constipation and diarrhea. Upon seeking medical advice, she had an abdominal MRI, which showed an abnormality in the right hypochondrium.

Why?

What is a possible cause of Michelle's symptoms?

Why did her discomfort occur after supper? Explain.

What organ(s) are most likely involved?

CASE STUDY

DIGESTIVE SYSTEM-2

George is a 68-year-old retired fireman who drank alcohol for 50 years. During the past month he has felt very fatigued. Additionally, although he ate and exercised as usual, he noted a weight loss. He reported a change in bowel movements and felt a dull pain in his back. The doctor ordered tests for pancreatic enzyme function.

What are these enzymes?

Why are they being investigated?

Why does George have pain in his back?

The Urinary System

A. Urinary System Anatomy

PURPOSE

Unit 13A will familiarize you with the anatomy of the urinary system.

OBJECTIVES

After completing Unit 13A, you will be able to
- identify the parts of a nephron.
- identify gross anatomical features of the kidney.
- identify the anatomical features of the human urinary system.
- identify the anatomical structures of the urinary system in a male and a female cat.

MATERIALS

sheep or beef kidneys	kidney models
preserved male and female cats	human torso
microscope slides of kidney	disposable gloves
microscope	model of a nephron
dissecting instruments	model of human urinary system
dissecting pans	cadaver (optional)

PROCEDURE

Homeostasis is partially maintained by the urinary system, which is a major regulator of the composition of the blood. Substances essential to life, such as water, electrolytes, monosaccharides, and amino acids, are reabsorbed into the blood. Toxins, nitrogenous waste products such as urea, uric acid, and creatinine, other metabolic wastes such as ketone bodies, and excess electrolytes are excreted in the urine. The urinary system also functions in maintaining the acid-base balance of the blood.

 **Microscopic Examination of Renal Tissue (Color Plates 3, 42, and 43)**

Focus the slide of renal tissue or kidney under high power. Observe these boldfaced structures: a **renal corpuscle**, which includes a tight network of capillaries, a **glomerulus**, and a **Bowman's capsule**, or **glomerular capsule**, which is a single epithelial cell layer surrounding the glomerulus. Follow a

Question 13A.1

Is the collecting duct part
of a nephron?

Bowman's capsule as it narrows and extends from the renal corpuscle. This twisted tubule is the **proximal convoluted tubule**. Note that the tubule continues in an inferior direction where the epithelium becomes thinner and then turns superiorly, forming a loop. This section is referred to as the **loop of Henle**. As the loop of Henle continues and extends from the renal corpuscle, it is referred to as the **distal convoluted tubule**. As the distal convoluted tubule continues, note that the walls become slightly thicker. This thicker-walled, straight tubular extension is called a **collecting duct**. A **nephron**, the structural and functional unit of the kidney, is composed of a glomerulus, a Bowman's capsule, a proximal convoluted tubule, a loop of Henle, and a distal convoluted tubule. If you have a model of a nephron, identify the same structures that you have observed microscopically. Distal convoluted tubules from several nephrons merge into a common collecting duct (**Figures 13.1–13.4**).

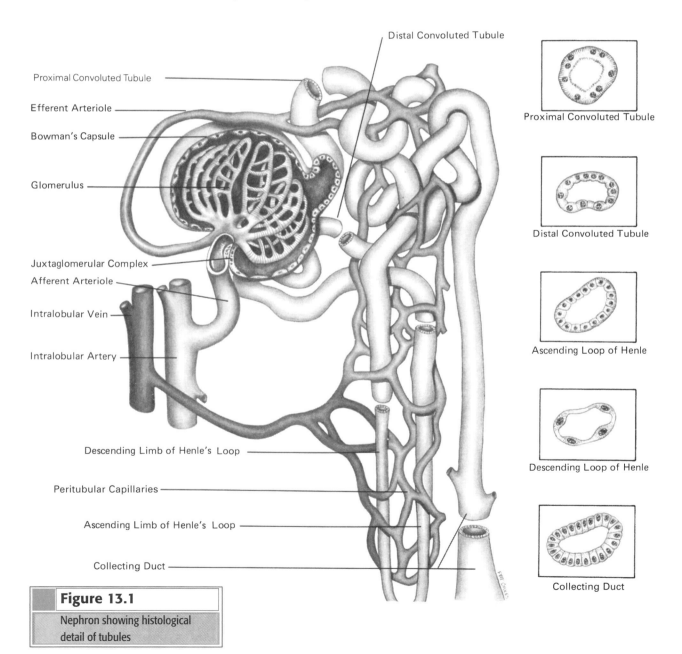

Proximal Convoluted Tubule

Efferent Arteriole

Bowman's Capsule

Glomerulus

Juxtaglomerular Complex

Afferent Arteriole

Intralobular Vein

Intralobular Artery

Distal Convoluted Tubule

Descending Limb of Henle's Loop

Peritubular Capillaries

Ascending Limb of Henle's Loop

Collecting Duct

Proximal Convoluted Tubule

Distal Convoluted Tubule

Ascending Loop of Henle

Descending Loop of Henle

Collecting Duct

Figure 13.1

Nephron showing histological detail of tubules

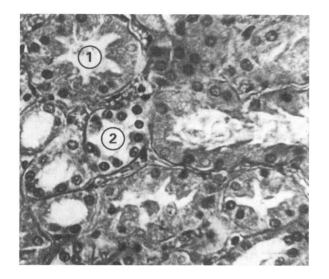

Figure 13.2

Kidney tubules, cortical region (430×)

1. proximal convoluted tubule
2. distal convoluted tubule

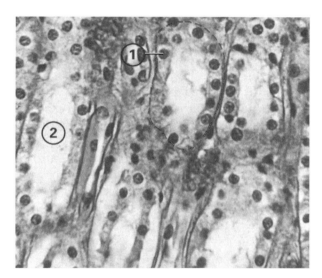

Figure 13.3

Kidney tubules, medullary region showing collecting ducts (430×)

1. collecting duct
2. lumen of collecting duct

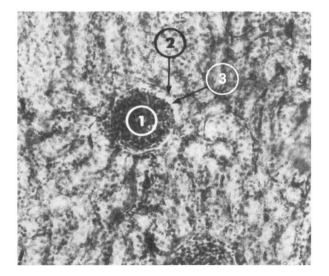

Figure 13.4

Glomerulus of kidney (430×)

1. glomerulus
2. Bowman's capsule
3. capsular space

Question 13A.2

What is the difference between a renal corpuscle and a renal capsule?

EXERCISE **2**   Gross Anatomy of a Kidney

On a model of a human kidney, find the structures printed in boldface. The **renal cortex** is the peripheral region of the kidney. The **renal medulla** is the central region containing the fine, tubular **medullary rays** that are grouped into triangular **renal pyramids**. The **papillae** are openings into the calyces, which receive urine from the medullary rays. From the calyces, urine drains into the **renal pelvis**. A **renal column**, which is the region between adjacent renal pyramids, extends medially (as part of the cortex) to the pelvis. The renal columns therefore resemble the cortex in appearance (**Figure 13.5** and **Color Plate 48**).

On the same model, identify the **renal artery**, which branches off the descending abdominal aorta. It divides into **interlobar arteries** that project peripherally between the renal pyramids toward the cortex. These arteries branch and connect with one another, forming the **arcuate artery**, which arches over each pyramid. **Interlobular arteries** branch into the cortex from the arcuate artery. An **afferent arteriole** from an interlobular artery extends into each glomerulus. The glomerulus contains a network of **glomerular capillaries**; blood exits the glomerulus through an **efferent arteriole**. From the efferent arteriole, blood flows into the **peritubular capillary network** surrounding the proximal and distal convoluted tubules. The **vasa recta** is the section of the peritubular capillary that intertwines around the loop of Henle in the medulla. Blood returns to the **renal vein** through **interlobular, arcuate**, and **interlobar veins**, respectively.

Using a preserved beef or sheep kidney, carefully remove the excessive adipose tissue. Be careful not to cut the ureter or any attached renal arteries or veins. The adipose tissue holds the kidney in position in a living animal. Observe the fibrous connective tissue covering that surrounds and protects the kidney. It is called the **renal capsule**. Also note the **hilum**, which is the indentation where the ureter and blood vessels enter and exit the kidneys. Make a midsagittal section through the kidney starting at the outer convex region. Observe the following structures: **renal cortex**; **renal medulla**, including the **pyramids, medullary rays, papillae, calyces**, and **renal columns**; and the **renal pelvis** (**Figure 13.6**). Note that the renal pelvis is formed by a wall of thick, white, fibrous tissue and forms the expanded proximal end of the ureter.

Question 13A.3

Using words connected by arrows, diagram the pathway of urine from the collecting duct to the urethral orifice.

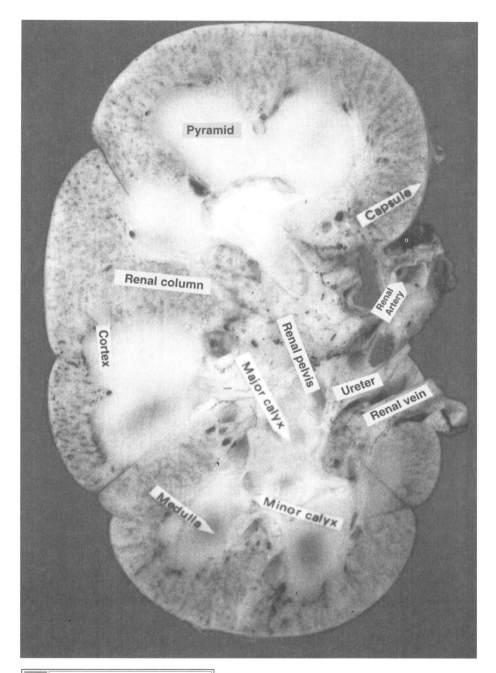

Figure 13.5

Human kidney, midsagittal section

Figure 13.6

Sheep kidney, midsagittal section, showing internal detail

1. medullary region
2. calyx
3. renal pelvis
4. ureter

5. renal pyramids with medullary rays
6. renal column
7. cortex

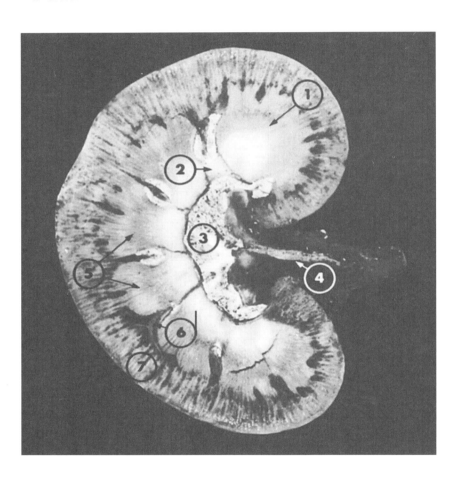

EXERCISE 3 **Anatomical Features of the Human Urinary System**

The human urinary system consists of two kidneys, two ureters, a urinary bladder, and a urethra. On a torso, a model of the human urinary system, or a human cadaver, if available, locate the structures printed in boldface (**Figure 13.7**). The **kidneys** are found retroperitoneally in the lumbar region. A pad of adipose tissue also secures each kidney to the abdominal wall. Observe the **hilum** where the renal arteries, renal vein, and ureter enter and exit the kidney. The ureters, bladder, and urethra are located more anteriorly, in the abdominopelvic cavity. Review and locate the **renal capsule**, **renal cortex**, **medulla**, and **pelvis** of the kidney. In the medullary region, find the **medullary rays**, **pyramids**, **papillae**, **calyces**, and **columns**.

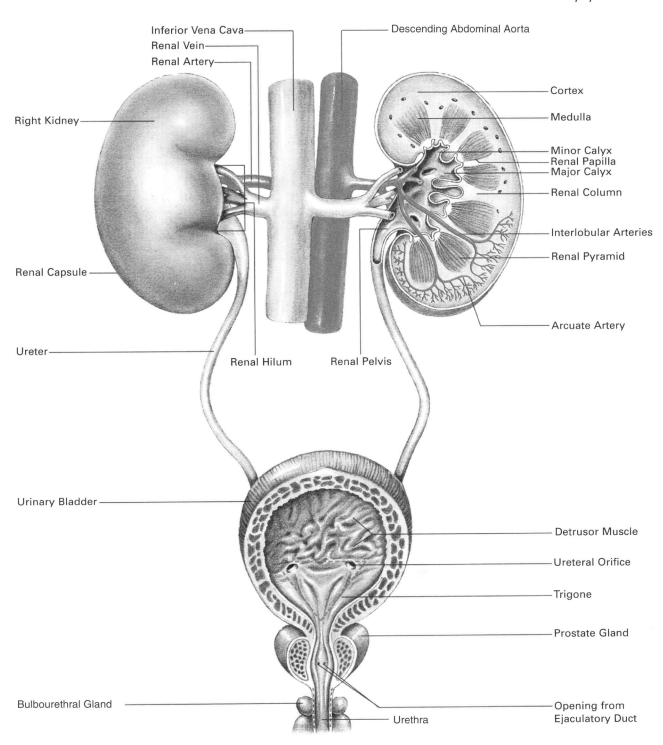

Figure 13.7

Frontal view of human urinary system (male)

Question 13A.4

Identify the following:

cuplike structures in the medulla that receive urine

structures made up of medullary rays

The **ureters**, slender tubes that transport urine to the **urinary bladder**, are extensions of the renal pelvis. The bladder is located posterior to the pubic symphysis. The walls of the bladder are primarily composed of the **detrusor muscle**. The **trigone** is the central area of the bladder in which urine is collected. It is named for its triangular shape and three openings. The two superior openings receive urine from the ureters. The inferior opening is surrounded by sphincter muscles, which allow for the controlled release of urine into the **urethra**. The urethra in the female is about 1½″ long, whereas in the male it extends through the penis and is approximately 8″ long. In the male, both sperm and urine exit the body through the urethra. The **urinary meatus** is the terminal opening of the urethra in the female. In the male it is called the **urogenital orifice**.

Identification 13.1

 13.1

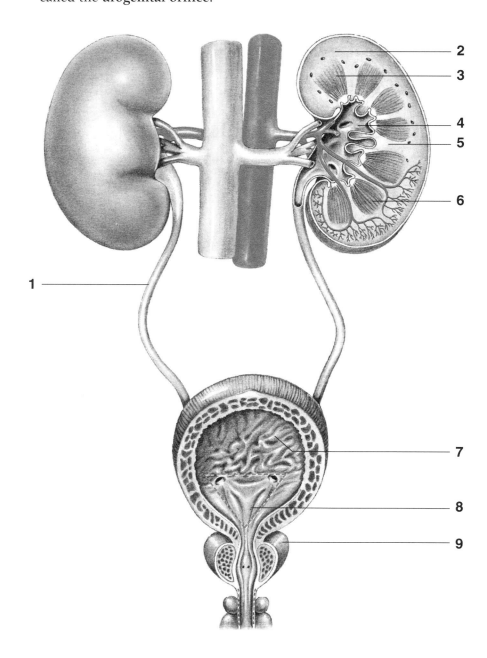

EXERCISE 4 | **Anatomical Features of the Male and Female Cat Urinary Systems**

This exercise may be completed on your own specimen, but be sure to observe a specimen of the opposite sex as well. Find the structures written in boldface.

The urinary organs are located in the pelvic region. As in the human, the **kidneys** are found retroperitoneally. They are surrounded by adipose tissue and are anchored to the dorsal body wall by a tough serous membrane known as the **peritoneum.** Reflect the small intestine, colon, and pancreas to one side in order to provide an unobstructed view of the urinary structures. Carefully remove the fat, connective tissue, and peritoneum that cover the ventral aspect of the kidneys. Recall that the **renal artery** branches from the **descending abdominal aorta,** and the **renal vein** empties into the **posterior vena cava** (Figure 13.8).

Identify the creamy-orange **adrenal gland** that lies on the superior border of each kidney. This endocrine gland secretes various hormones that influence metabolism, cardiac output, and fluid and electrolyte balance. The hormones secreted by this gland have physiological influences on the kidneys, but the adrenal glands are not an anatomical component of the urinary system. Identify the **renal hilum,** where the renal blood vessels, nerves, and ureter enter and exit from the kidney. Identify the **ureters** and trace them to the muscular saclike **urinary bladder.** Posteriorly, the bladder narrows to form a short canal, the **urethra** (Figure 13.8–13.10).

Carefully lift up the right kidney. Using your scalpel, slit the kidney from the lateral border to the medial border. Identify the **cortex, medulla,** and **pelvis** inside. Also identify the **hilum,** where the ureter exits from the kidney.

Question 13A.5

Draw and label the following structures of the medulla: pyramids, medullary rays, and calyces.

Figure 13.8

Cat urinary system

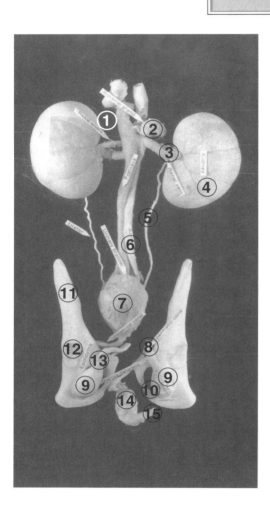

1. right renal artery
2. left adrenal gland
3. left renal vein
4. left kidney
5. left ureter
6. descending abdominal aorta
7. urinary bladder
8. vas deferens
9. testes
10. obturator foramen
11. ilium
12. acetabulum
13. epididymis
14. penis
15. urogenital orifice

Figure 13.9

Urogenital system, pregnant cat

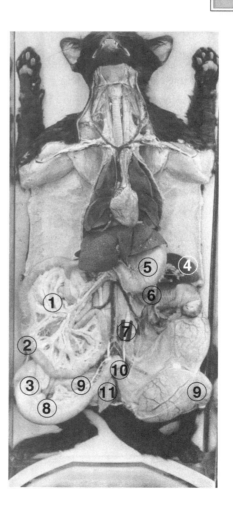

1. mesentery
2. small intestine
3. cecum
4. spleen
5. stomach
6. left kidney
7. ureter
8. colon
9. pregnant uterus
10. vagina
11. urinary bladder

Figure 13.10

Urogenital organs of the cat, with colon reflected

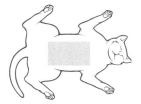

1. kidney, left
2. urinary bladder
3. horn of uterus
4. body of uterus
5. descending colon
6. descending abdominal aorta
7. inferior vena cava
8. renal vein
9. renal artery
10. adrenal gland

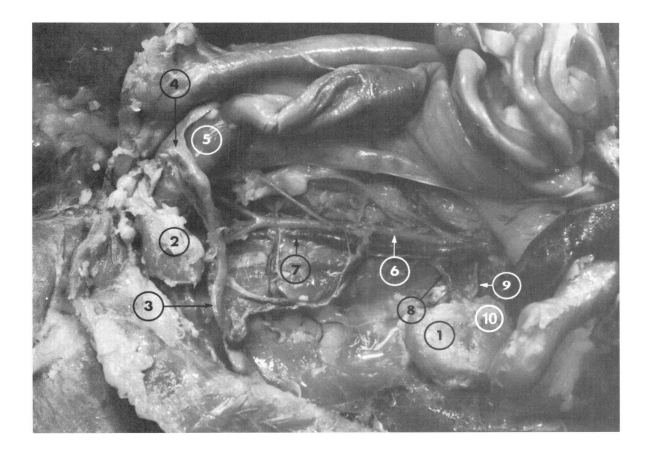

B. Urinary System Physiology

PURPOSE

Unit 13B will familiarize you with the general physiology of the urinary system and enable you to understand the basic components of a urinalysis.

OBJECTIVES

After completing Unit 13B, you will be able to

- correlate parts of a nephron with the process of urine formation.
- perform basic tests included in a urinalysis.

EXERCISE 1   Formation of Urine

The formation of urine in the nephrons involves three physiological processes. They are glomerular filtration, tubular reabsorption, and tubular secretion. **Glomerular filtrate** is the fluid that is filtered out of the blood as it circulates through the glomeruli of the nephrons. Its composition is similar to blood plasma, except that it normally does not contain plasma proteins or formed elements of the blood because of their large size and high molecular weights.

Glomerular filtration is a passive process and is maintained by blood pressure in the glomerulus itself. Under pressure, water, glucose, amino acids, and electrolytes are filtered out of the glomerular capillaries into Bowman's (glomerular) capsule.

Tubular reabsorption is an important means of conserving water, glucose, amino acids, and electrolytes in the body. These molecules are reabsorbed by active transport, diffusion, and osmosis. Of the 125 mL of glomerular filtrate formed per minute by the kidneys, 124 mL are reabsorbed through the tubules into the peritubular capillaries, which are part of the circulatory system. Therefore, less than 1% of glomerular filtrate is excreted as urine.

Tubular secretion is the opposite of tubular reabsorption. Certain molecules are selectively removed from the peritubular capillaries and are added to the filtrate in the kidney tubules by the processes of active transport and diffusion. Tubular secretion is a major mechanism for eliminating toxic or excessive substances from the blood and for controlling blood pH.

The table entitled Summary of Events Occurring in Urine Formation summarizes the events occurring in glomerular filtration, tubular reabsorption, and tubular secretion. After studying the table, fill in the blanks in **Figure 13.11** corresponding to each of these events.

Summary of Events Occurring in Urine Formation

Portion of Nephron	Substance Moved	Process
Glomerulus (glo-MER-yoo-lus)	Water Electrolytes Amino acids Glucose	Filtration
Proximal convoluted tubule	Sodium ions (Na^+) Glucose Amino acids	Reabsorption by active transport
	Chloride ions (Cl^-)	Reabsorption by diffusion
	Water	Reabsorption by osmosis
Descending limb of loop of Henle	Na^+ Cl^-	Reabsorption by diffusion
	Water	Reabsorption by osmosis
Ascending limb of loop of Henle	Na^+	Reabsorption by active transport; not followed by osmosis of water
	Cl^-	
Distal convoluted tubules and collecting ducts	Na^+	Reabsorption by active transport; influenced by aldosterone
	Potassium ions (K^+)	Secretion by active transport; influenced by aldosterone
	Water	Reabsorption by osmosis, controlled by antidiuretic hormone (ADH)
	Ammonia (NH_3)	Secretion by diffusion
	Hydrogen ions (H^+)	Secretion by active transport
	Certain drugs	Secretion by active transport

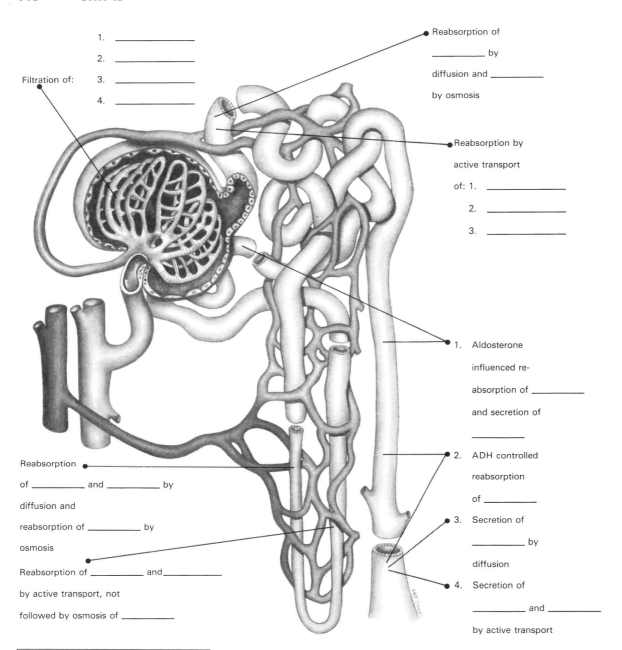

1. _____
2. _____

Filtration of:
3. _____
4. _____

Reabsorption of
_____ by
diffusion and _____
by osmosis

Reabsorption by
active transport
of: 1. _____
     2. _____
     3. _____

1. Aldosterone
   influenced re-
   absorption of _____
   and secretion of
   _____

2. ADH controlled
   reabsorption
   of _____

3. Secretion of
   _____ by
   diffusion

4. Secretion of
   _____ and _____
   by active transport

Reabsorption
of _____ and _____ by
diffusion and
reabsorption of _____ by
osmosis
Reabsorption of _____ and _____
by active transport, not
followed by osmosis of _____

Figure 13.11

Urine formation

EXERCISE 2 Routine Urinalysis

MATERIALS

students' urine samples or stock solutions
    of synthetic urine, each containing
    glucose, albumin, acetoacetic acid,
    blood, or hyrdochloric acid
graduated cylinders
Pasteur pipettes
blank microscope slides

coverslips
Multistix® 10SG Reagent Strips
    for Urinalysis*
disposable gloves
centrifuge
conical-tipped centrifuge tubes

CAUTION
BLOODBORNE PATHOGENS
Use care when working with
body fluids.

Precautions need to be taken when handling body fluids. **Therefore, any urine used in this exercise should be your own.** Collect your sample in an unbreakable, leak-proof container. Place the container in a sealable plastic bag for transport to the laboratory. Wear disposable gloves, and follow proper procedures for handling and disposing of urine samples as directed by your instructor.

PROCEDURE

Urine is approximately 95% water and includes the following dissolved substances: pigments, electrolytes (for example, sodium and potassium), metabolites, hormones, and nonprotein nitrogen waste products urea, uric acid, creatinine, and ammonia. A few epithelial cells, white blood cells, and crystals may also be seen in a normal urine sample. Bacteria (**Figures 13.12** and **13.13**), blood, glucose, and protein may be found in the urine of individuals with certain medical conditions. Casts, which are often associated with pathological conditions, are collections of mucous and inorganic materials that have hardened over time and have assumed the shape of the tubule in which they were formed. An adult excretes an average volume of 1200–1500 mL of urine during a 24-hour period.

If you will be using your own urine, collect a 30–50 mL sample after cleansing the area around the urinary meatus. The first urination of the day is best but is often difficult to collect in a school situation. Observe the physical characteristics of your urine or of an "unknown" synthetic sample provided by your instructor, and record in the table titled Physical Characteristics of Urine.

To observe microscopic structures such as cells, bacteria, and crystals that may be found in urine, you will need to first centrifuge a urine sample, which will separate it into particulate matter (sediment) at the bottom and a clear liquid supernate at the top. (Microscopic structures, of course, will not be present if using synthetic urine.)

Carefully rotate your covered container of urine to suspend any particles, then measure 5–8 mL into a centrifuge tube. Place the tube in the centrifuge, balancing each tube with another containing a urine specimen or an equal amount of water.

Centrifuge the sample for 5 minutes at 1500 rpm.

Slowly pour off the liquid supernate. The sediment should remain at the bottom of the centrifuge tube.

Using a Pasteur pipette, remove a small amount of sediment and place it on a microscope slide, then cover with a coverslip.

*Courtesy of Bayer Healthcare Diagnostics Division, Terrytown, NY.

Figure 13.12

Yeast cells and bacteria as seen in sediment of urine from urinary tract infection (100×)

1. yeast cells
2. bacteria

Ames Division, Miles Laboratory, Inc.

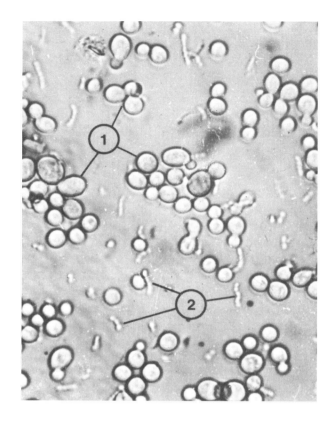

Figure 13.13

Bacteria (430×) as seen in urinary tract infection. The bacteria are shown as small rod-shaped structures.

1. rod-shaped bacteria

Ames Division, Miles Laboratory, Inc.

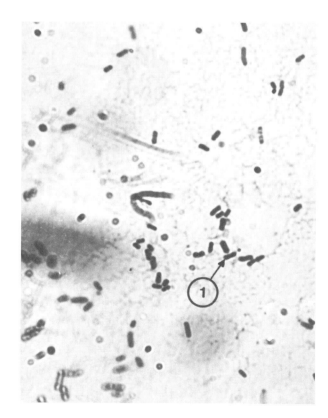

Observe the sediment, first under low power, then high power. You may find it helpful to reduce the light penetrating the slide by closing the diaphragm slightly.

Draw any cells, bacteria, or crystals you see in the Physical Characteristics of Urine table under "Your Urine or Synthetic Unknown." Some examples of crystals are illustrated in Figure 13.14.

Uric Acid   Calcium Oxalate   Calcium Phosphate   Cholesterol   Ammonium Magnesium Phosphate

Figure 13.4

Examples of Urine Crystals

Physical Characteristics of Urine		
Characteristic	**Normal**	**Your Urine or Synthetic Unknown**
Color	Amber to straw-colored; may vary with diet, medications, or state of hydration	
Odor	Fresh, no odor; ammonia odor develops in standing urine because of bacterial breakdown and formation of ammonia-substances	
Turbidity	Clear; becomes cloudy upon standing; heavier particles settle to bottom	
Specific gravity (record from Multi-stix® 10SG strips)	1.003–1.030 in normal adult; may be higher if an early-morning specimen. Best results at 59–86°F.	
Microscopic examination (**also see Color Plates 2, 20, and 23**)	May include epithelial cells, leukocytes, or crystals	

Also included in a routine urinalysis are tests to determine the absence or presence of substances that are normally not found in urine. These tests may be performed by using various dipsticks such as the Multistix 10SG. Directions on the package insert must be followed exactly for each of the tests, and the results must be carefully compared with the color chips on the container of test strips. Avoid touching the test strips to the color chart to prevent soiling or contamination. Also replace the top of the jar securely after removing the strip(s), and use before the expiration date on the container for more accurate results. To complete the following common tests, dip the treated strip of paper into your sample or a synthetic unknown stock solution, wait the exact required times, then record the results in the table titled Test Results for Urinalysis.

Test Results for Urinalysis

Indicator	Substance Indicated	Presence (+) or Absence (−) in Your Urine or Synthetic Unknown
Multistix® 10SG Reagent Strips for Urinalysis (recommended)	Glucose	
	Bilirubin	
	Ketone (acetoacetic acid)	
	Specific gravity (record on Physical Characteristics table)	
	Blood	
	pH	
	Protein	
	Urobilinogen	
	Nitrite	
	Leukocytes	

Question 13B.1

What is specific gravity?

How did the specific gravity of the urine sample you tested compare to that of water, which has a specific gravity of 1.000?

EXERCISE 3   Urine Screening Tests

MATERIALS

Test tubes

400-mL beakers

Hot plate

Benedict's reagent

10% acetic acid

No. 2 filter paper

Glass Petri dishes

Nitric acid (concentrated)

Plastic or glass specimen
    containers with covers

Compound microscopes

Urine sample

Microscope slides

Coverslips

PROCEDURE

TEST FOR URINE PROTEIN

Normal adult urine may contain a trace of plasma proteins (2–8 mg/100 mL). Plasma proteins are normally prevented from entering the glomerular filtrate by the glomerulus itself and due to their large molecular weights. An increase in protein of 0.5 g to 4 g per day may be evidence of glomerular disease.

Pour 10 mL of urine into a test tube. Put the test tube into a beaker of tap water and onto a hot plate. Bring the contents of the tube to a boil. As soon as the urine boils, add three drops of 10% acetic acid to the test tube. If a precipitate results, then albumin and globulins are present; this indicates a positive test for proteins. What are the results?

CAUTION

BLOODBORNE PATHOGENS

Use care when working with body fluids.

Question 13B.2

List and describe three disease states that may be indicated by proteinuria.

1.

2.

3.

BENEDICT'S QUALITATIVE METHOD FOR DETECTING URINARY GLUCOSE

Benedict's is a copper reduction test that is used for the qualitative screening of glucose in urine. If glucose is present, it reduces Benedict's reagent from blue alkaline copper sulfate to a precipitate of red cuprous oxide. If a yellow, green, orange, or brick red precipitate forms, a reducing sugar (e.g., glucose) is present in the specimen. The particular color produced depends upon the concentration of reducing sugar present. The brick red color indicates a maximum concentration. A disadvantage of this test is that any reducing sugar will give a positive test, not only glucose.

Place 0.5 mL, or eight drops, of urine into a clean, dry test tube. Add 5 mL of Benedict's reagent. Mix by agitating the tube. Place the test tube in a beaker of tap water and onto a hot plate. Let boil for 5 minutes. Remove from the boiling water and observe immediately for any color change.

Color Interpretation	Concentration of Reducing Sugar
Clear blue or green with no precipitate	0–100 mg/100 mL
Green with a yellow precipitate	100–500 mg/100 mL
Yellowish green with a yellow precipitate	500–1400 mg/100 mL
Brownish orange with a yellow precipitate	1400–2000 mg/100 mL
Orange to brick red precipitate	2000 or more mg/100 mL

What are the results?

Question 13B.3

List and describe two diseases that may be indicated by glycosuria.

1.

2.

....

TEST FOR URINARY BILIRUBIN

Bilirubin results from the breakdown of hemoglobin within the liver. Free bilirubin within the plasma normally does not pass through the glomerulus and is not found in the glomerular filtrate. A positive test for urinary bilirubin may indicate hepatic damage.

Place a small strip of no. 2 filter paper in a Petri dish or watch glass. Add two drops of urine and one drop of concentrated nitric acid (HNO_3). *Please use caution.* A green color is positive for biliverdin. Bilirubin is not stable in urine and is oxidized to biliverdin. Other colors are negative. What are the results?

Use care when working with nitric acid.

TEST FOR UREA

Normal adult urine contains a large amount of urea. The major portion of dietary nitrogen is excreted in the form of urea.

Place two drops of urine on a clean, dry microscope slide. *Carefully* add two drops of concentrated nitric acid (HNO_3). Slowly warm the slide on a slide warming tray or hot plate. Do not allow it to boil. Cool. When crystals begin to form, examine under the low power of your microscope. The crystals that form are urea nitrate. Draw your results.

Question 13B.4

List and describe two disorders that may be indicated by high levels of urinary bilirubin and/or biliverdin.

1.

2.

The Urinary System

MULTIPLE CHOICE

Name _____

Section _____

Date _____

_____ 1. Which of these is the correct sequence of vessels through which blood normally flows going to and from a nephron?

   a. intralobular vein, capillary plexus, efferent arteriole, afferent arteriole, glomerulus

   b. glomerus, afferent arteriole, capillary plexus, efferent arteriole, intralobular vein

   c. afferent arteriole, glomerulus efferent arteriole, capillary plexus, intralobular vein

   d. glomerulus, Bowman's capsule, proximal convoluted tubule, loop of Henle, distal convoluted tubule, collecting tubule

_____ 2. Which of these structures is located in the cortex of the kidney?

   a. glomerulus

   b. loop of Henle

   c. calyx

   d. collecting tubule

_____ 3. Considering the length of the urethra, who would be more prone to urinary tract infections?

   a. men

   b. women

_____ 4. Why is it advantageous that kidney tubules be only one cell layer thick?

   a. They occupy less space.

   b. Molecules and ions may be transported more readily.

   c. Diffusion of gases takes place more quickly.

   d. They provide more surface area for water absorption.

_____ 5. If a person neither eats nor drinks anything for eight hours, what would you expect the specific gravity of his or her urine to be?

   a. high

   b. low

   c. state of hydration has no effect on specific gravity

   d. zero

_____ 6. Which of these pituitary hormones influences the volume of urine excreted?

   a. oxytocin

   b. parathyroid hormone

   c. antidiuretic hormone

   d. aldosterone

_____ 7. A diagnostic test for diabetes mellitus is the presence of _____ in the urine.

   a. protein

   b. sodium

   c. bilirubin

   d. glucose

_____ 8. Which of these may be found in the urine if fats are not completely metabolized?

   a. ketones

   b. glucose

   c. bilirubin

   d. urea

_____ 9. In most people, urine pH
   a. is acidic.   c. varies.
   b. is basic.   d. is neutral.

_____ 10. The volume of glomerular filtrate produced per day is normally
   a. about the same as urine.   c. about 100 times less than
   b. about 100 times greater than   urine.
      urine.   d. not related to urine volume.

_____ 11. Which of the following represents the correct sequence of
           structures through which glomerular filtrate and urine must
           pass?
   a. Bowman's capsule, proximal   c. Bowman's capsule,
      tubule, glomerulus, distal   glomerulus, proximal
      tubule, loop of Henle, ureter   tubule, loop of Henle, distal
   b. glomerulus, Bowman's   tubule, ureter
      capsule, proximal tubule,   d. ureter, loop of Henle,
      loop of Henle, distal tubule,   proximal tubule, Bowman's
      ureter   capsule, distal tubule,
                                       glomerulus

_____ 12. Which of the following would be considered abnormal
           constituents of urine?
   a. protein   c. hemoglobin
   b. glucose   d. all of the above

_____ 13. The structural and functional unit of the kidney is the
   a. ureter.   c. nephron.
   b. glomerulus.   d. calyx.

_____ 14. Which of the following is *not* a function of the kidney?
   a. maintains acid-base balance   c. eliminates by-products of
   b. maintains H_2O balance   catabolism
                                    d. all of the above are functions

_____ 15. The average daily amount of urine excreted is
   a. 700 mL.   c. 1000 mL.
   b. 100 mL.   d. 1500 mL.

_____ 16. The muscle of the bladder is the
   a. myometrium.   c. detrusor.
   b. rectus abdominis.   d. levator ani.

_____ 17. From the renal pelvis, urine flows to the
   a. glomerulus.   c. ureter.
   b. calyces.   d. bladder.

_____ 18. In which part of the nephron are water, glucose, and salt
           primarily reabsorbed into the blood?
   a. proximal convoluted tubule   c. distal convoluted tubule
   b. loop of Henle   d. collection tubule

_____ 19. The body's blood supply initially comes into contact with the
           nephron at the
   a. glomerulus.   c. papillae.
   b. pyramids.   d. calyces.

Photo Identification

Refer to the following photograph of the kidney for Questions 20–23.

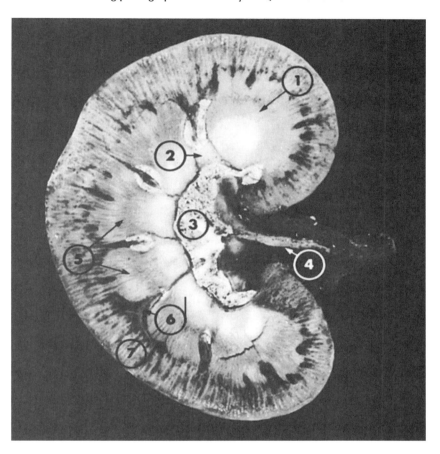

_____ 20. Regions 5 and 6 are parts of the _____ of the kidney.
   a. cortex   c. capsule
   b. medulla   d. ureter

_____ 21. The nonvascular tubular structure that exits from the hilum as
            indicated by 4 on the photograph is the
   a. renal artery.   c. ureter.
   b. renal vein.   d. urethra.

_____ 22. The region identified by number 7 contains
   a. mostly collecting tubules.   c. neither a nor b.
   b. mostly glomeruli.   d. both a and b.

_____ 23. The region identified by number 1 contains
   a. mostly collecting tubules.   c. mostly glomeruli.
   b. mostly loops of Henle.   d. both a and b.

_____ 24. Which of the following tests was *not* performed using a dipstick?
   a. pH   c. glucose
   b. urea   d. bilirubin

_____ 25. Sodium reabsorption in the distal convoluted tubules is
            influenced by which of the following hormones?
   a. antidiuretic   c. oxytocin
   b. aldosterone   d. insulin

DISCUSSION QUESTIONS

26. Why is it advantageous that Bowman's capsule is one cell layer in thickness? _____

_____

_____

_____

27. How does the urinary system of the male differ from that of the female? _____

_____

_____

_____

28. Turbidity of urine and specific gravity are related. Why? _____

_____

_____

_____

29. Why is urinalysis considered to be a valuable diagnostic test? _____

_____

_____

_____

30. Which nerves and hormones influence urine formation? _____

_____

_____

_____

CASE STUDY

URINARY SYSTEM

Interpret These Results

Patient's
Name: *David*

Multistix
Expected Values

Glucose	Norm = Neg	Negative	100	250	500	1,000	2,000 mg/dL
Bilirubin	Norm = Neg	Negative	1+	2+	3+		
Ketones	Norm = Neg	Negative	Trace	Small	Moderate	Large	
Specific Gravity	Norm=1.0–1.030	1.000	1.005	1.010	1.015	1.020	1.025	1.030

			Nonhemolyzed		Hemolyzed		
Blood	Norm = Neg	Negative	Trace	Moderate	Trace	Small	Moderate	Lg
pH	Norm (5.0–8.5)	5.0	6.0	6.5	7.0	7.5	8.0	8.5
Protein	Norm = Neg	Negative	Trace	1+	2+	3+	4+	
Urobilinogen	Norm = 0.1–1.0 mg/dL	Norm	1	2	4	8 mg/dL		
Nitrite	Norm = Neg	Negative	Positive				
Leukocytes	Norm = Neg	Negative	Trace	1+	2+	3+	

What are possible problems?

Discuss possible additional avenues for investigation based on this urinalysis.

Acid-Base Balance

A. The Measurement of pH

PURPOSE

Unit 14A will enable you to understand the concept of pH.

OBJECTIVES

After completing Unit 14A, you will be able to
- define acid, base, pH, and buffer system.
- use litmus paper, pH paper, and a pH meter to determine the acidity or alkalinity of several common solutions and biological fluids.

In order for the body to maintain a state of **homeostasis**, or a stable internal environment, mechanisms exist for controlling concentrations of acids and bases within body fluids.

Before beginning the exercises, you will need to learn the following terms:

acid A hydrogen ion (H^+) donor.
base A hydroxyl ion (OH^-) donor. In biological systems, bases are often called **alkalies** (AL-ka-līz) and are defined as negatively charged ions that combine with hydrogen ions in solution.
pH A measure of the acidity (amount of acid, or H^+) or alkalinity (amount of base, or OH^-) of a solution. Mathematically, pH is equal to the negative logarithm of the hydrogen-ion concentration, or $-\log[H^+]$. It can also be represented as $\log 1/[H^+]$.
buffer system A combination of two or more chemicals (usually a weak acid and its salt) that minimize the pH change of a solution when a strong acid or base is added to it.

EXERCISE 1   Measuring the pH of Common Solutions

The pH scale ranges from 0 through 14. The lower the pH, the greater the hydrogen ion concentration, and therefore the greater the acidity of a solution. The higher the pH, the lower the hydrogen ion concentration, indicating an alkaline, or basic, solution. A pH of 7 is midway between 0 and 14 and represents **neutrality** (where the number of hydrogen ions equals the number of hydroxyl ions) and therefore the solution is neither acidic nor basic, but neutral. **Figure 14.1** illustrates the pH scale.

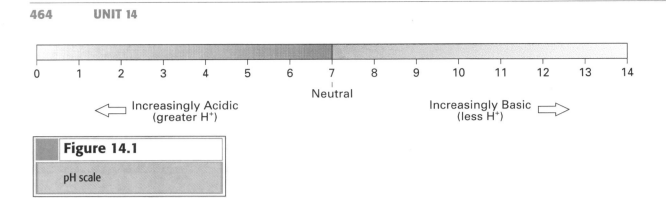

Figure 14.1

pH scale

Question 14A.1

What would you expect the pH of water to be?

It is important to understand that a shift of one pH unit represents a tenfold increase or decrease in hydrogen ion concentration. For example, a solution of pH 4 has 10 times the concentration of H^+ as a solution of pH 5. A solution of pH 4 has 100 times the concentration of H^+ as a solution of pH 6. Conversely, a solution of pH 10 has 1/10 the hydrogen ion concentration of a solution of pH 9 and 1/100 the hydrogen ion concentration of a solution of pH 8.

MATERIALS

red litmus paper	coffee
blue litmus paper	cola
pH meter	lemon juice
wash bottles with distilled water	baking soda solution
pH 7 buffer solution	distilled or deionized water
pH paper (or nitrazine paper)	beakers
milk	

PROCEDURE

DETERMINATION OF pH USING LITMUS PAPER

Dip a 2″ strip of red, then blue, litmus paper into each of the following solutions: milk, coffee, cola, lemon juice, baking soda solution, and distilled water. Watch for a color change, using fresh strips for each solution. Acidic solutions will turn blue litmus paper red. Basic, or alkaline, solutions will turn red litmus paper blue. Record your results in the table entitled Determination of pH of Common Solutions Using Litmus Paper.

Determination of pH Common Solutions Using Litmus Paper			
Solution	Red Litmus	Blue Litmus	Conclusion
Milk			
Coffee			
Cola			
Lemon juice			
Baking soda			
Distilled Water			

DETERMINATION OF pH USING pH PAPER

Dip a 2″ strip of pH paper into each of the following solutions: milk, coffee, cola, lemon juice, baking soda solution, and distilled water. Compare the colors that appear on the chart that accompanies the pH paper container. Record the pH of each solution and Record in the table entitled Determination of pH of Common Solutions Using pH Paper.

Determination of pH of Common Solutions Using pH Paper		
Solution	**pH**	**Conclusion**
Milk		
Coffee		
Cola		
Lemon juice		
Baking soda		
Distilled water		

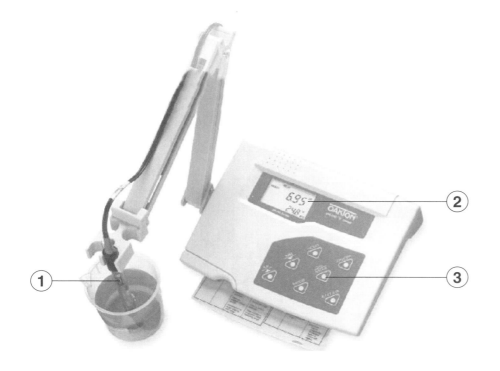

Figure 14.2

Oakton pH/mV/TempMeter

1. electrode
2. digital readout
3. calibrate/measure

Photo Courtesy of Oakton Instruments, Vernon Hills, IL.

DETERMINATION OF pH USING A pH METER

Plug in and turn on the pH meter at this time.(Your instructor will give you any additional information about the particular instrument you will be using.)

There are several types of pH meters; however, they all operate on a similar principle. Observe the pH meter you will be using. In general, pH meters have the following components that you should be able to identify (Figure 14.2):

- **Electrode.** The electrode projects in a downward direction and is attached to the body of the pH meter by a cable. Within this combination electrode is a glass electrode that contains a pH sensitive bulb and a reference electrode. If there is a protective cover over the electrode, remove it before testing the pH of a solution.
- **Calibrate button.** This is used to adjust the pH of the buffer.
- **Digital readout.** This display indicates pH and other data.

Using the instructions for the particular pH meter you are using, you will now determine the pH of each of the solutions. Record your results for each solution tested in the table entitled Determination of pH of Common Solutions Using a pH Meter.

Determination of pH of Common Solutions Using a pH Meter

Solution	pH	Conclusion
Milk		
Coffee		
Cola		
Lemon juice		
Baking soda		
Distilled water		

EXERCISE 2   Determining the pH of Biological Solutions

MATERIALS

saliva
urine (freshly voided and stagnant)
blood plasma (sheep)

pH paper (or nitrazine paper)
safety glasses
disposable gloves

Because of the possibility of infections being transmitted from one student to another during exercises involving body fluids, take great care in collecting and handling such fluids. The body fluids used in the following exercise should be your own. Follow proper procedures for collecting, handling, and disposing of body fluids as directed by your instructor.

CAUTION
BLOODBORNE PATHOGENS
Use care when working with body fluids.

PROCEDURE

The pH range in arterial blood that is compatible with life is relatively narrow: from about 6.8 to 7.8. In the healthy adult, the pH is between 7.35 and 7.45. The pH range for arterial blood may be represented as follows:

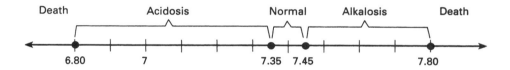

Using specimens of saliva, freshly voided urine, stagnant urine (urine that has been standing for several hours or overnight), and blood plasma, determine the pH of each solution using pH or nitrazine paper. Record the pH of each of these in the table entitled Determination of pH of Biological Solutions Using pH Paper.

Question 14A.2

Why is it possible for saliva and urine to have a pH below 6.8 or above 7.8 in a healthy individual?

Determination of pH of Biological Solutions Using pH Paper		
Solution	pH	Conclusion
Saliva		
Freshly voided urine		
Stagnant urine		
Plasma		

Question 14A.3

Was there a difference in pH between the freshly voided urine and the stagnant urine? Explain.

B. The Regulation of pH

PURPOSE

Unit 14B will enable you to understand the homeostatic role of chemical buffers and the respiratory system in the regulation of pH in the body.

OBJECTIVES

After completing Unit 14B, you will be able to
- demonstrate the effect of buffer systems on pH change.
- explain how the respiratory and renal systems function to maintain acid-base balance.
- define and identify the types of acidosis and alkalosis in certain disease states.

 EXERCISE 1   Buffer Systems

MATERIALS

pH meter
two beakers with 100 mL each of distilled water and pH 7 buffer solution
dropper bottles of 0.05 M HCl
dropper bottles of 0.05 M NaOH
250 mL and 60 mL beakers
wash bottles with distilled water
wax pencil

EFFECT OF BUFFERS ON ACIDIC AND ALKALINE SOLUTIONS

In this exercise, you will determine the effect of buffers on the extent of pH change in acidic and alkaline solutions.

PROCEDURE

Immerse the electrode in 100 mL of distilled water and determine the pH. Record.

Slowly add 0.05 M HCl dropwise to the distilled water, gently swirling the beaker after each drop. Record the number of drops of HCl added to change the pH one unit. In which direction did the pH change?

Remove the electrode from the solution and rinse with distilled water. Immerse in the beaker containing 100 mL of pH 7 buffer solution. Add 0.05 M HCl dropwise to the buffer, gently swirling after each drop. How many drops of HCl were added to change the pH of the buffer one unit?

After rinsing the electrode with distilled water, immerse them in the second beaker of distilled water. Add 0.05 M NaOH dropwise to the water, gently swirling the beaker after each drop. How many drops of NaOH were added to change the pH one unit? In which direction did the pH change?

Again remove the electrode from the solution and rinse with distilled water. Immerse the electrode in the second beaker of pH 7 buffer. Add 0.05 M NaOH dropwise to the buffer again, gently swirling after each drop. How many drops of NaOH were added to change the pH one unit?

Question 14B.1

What conclusion can you draw about the effect of buffers on pH change?

EFFECT OF BUFFERS IN BIOLOGICAL FLUIDS ON pH CHANGE

The body utilizes circulating chemical buffers, the lungs, and the kidneys to keep the pH of the blood within narrow limits. Three chemical acid-base buffer systems are found in the blood:

1. **Bicarbonate Buffer System.** This consists of carbonic acid (H_2CO_3) and sodium bicarbonate ($NaHCO_3$), which are present in extracellular and intracellular fluids. When minute amounts of a strong acid such as HCl are added to blood in the body, the following reaction takes place:

$$HCl + NaHCO_3 \rightarrow H_2CO_3 + NaCl$$

strong   buffer in   weak acid;   salt
acid   plasma   formed in
plasma

If a strong base such as NaOH is added to the bicarbonate buffer system, it will react with the carbonic acid component of this system:

$$NaOH + H_2CO_3 \rightarrow NaHCO_3, + H_2O$$

strong   carbonic acid;   weak base;   water
base   buffer in   formed in
plasma   plasma

2. **Phosphate Buffer System.** This system is also present in intracellular and extracellular body fluids. It primarily functions to regulate the H^+ concentration in the nephrons. The following reactions occur if a strong acid or strong base, respectively, is added to the system:

$$HCl + Na_2HPO_4 \rightarrow NaH_2PO_4 + NaCl$$

strong   buffer   weak acid   salt
acid

$$NaOH + NaH_2PO_4 \rightarrow Na_2HPO_4 + H_2O$$

strong   buffer   weak base   water
base

3. **Protein Buffer System.** This buffer system consists of plasma proteins and hemoglobin. Recall that amino acids contain carboxyl groups (—COOH) and amino groups (—NH_2).

Free carboxyl groups may function as acids by releasing a hydrogen ion:

$$—COOH \rightarrow —COO^- + H^+$$

If there is excess acid in the blood, (—COO^-) may accept a hydrogen ion and become a free carboxyl group (—COOH).

Likewise, free amino groups may accept hydrogen ions in the presence of excess acid:

$$—NH_2 + H^+ \rightarrow NH_3$$

Hydrogen ions may also be released in the presence of excess base, thus making the blood more acidic by the following reaction:

$$NH_3 \rightarrow —NH_2 + H^+$$

Question 14B.2

Why was a pH meter used to determine the pH of the distilled water and buffer solution, but pH paper used with the biological fluids?

EXERCISE 2   **Maintaining Acid-Base Balance in the Body**

MATERIALS

small paper bags	250-mL beakers
pH meter	straws
distilled or deionized water	clock or watch with second
0.1M NaOH	hand
dropper bottles of phenolphthalein solution	graduated cylinders

Two body systems assist in the regulation of blood pH: the urinary (renal) system and the respiratory system. The kidney tubules reabsorb or excrete bicarbonate ions, so that blood pH increases or decreases, respectively. The kidney tubules also reabsorb or secrete hydrogen ions, resulting in a decrease or increase in blood pH, respectively. The respiratory system functions to retain or release (through exhalation) carbon dioxide (CO_2), which results in a decrease or increase in blood pH, respectively. The term "pCO_2" (or "$paCO_2$") refers to the partial pressure of free carbon dioxide dissolved in arterial blood. An increased pCO_2 results in a low (acidic) blood pH; a decreased pCO_2 results in a high (basic) blood pH.

The general equation for maintaining acid-base balance in the body is:

$$CO_2 + H_2O \rightleftarrows H_2CO_3 \rightleftarrows H^+ + HCO_3^-$$

PROCEDURE

EFFECT OF pCO_2 LEVELS ON THE RESPIRATORY RATE

Observe your lab partner breathing normally. Record his respiratory rate (number of breaths per minute) _____.

Instruct your lab partner to take rapid, deep breaths, until he tires. After he stops hyperventilating, record his respiratory rate _____.

Instruct your lab partner to breathe in and out of a paper bag through his nose and mouth until breathing becomes difficult. Remove the bag and record his respiratory rate _____.

It is recommended that you perform part A *or* part B of this portion of the exercise.

EFFECT OF pCO_2 CHANGES ON pH

A. An estimate of the effect of CO_2 on pH may be demonstrated by using the indicator phenolphthalein, which appears pink in a basic solution and colorless in an acidic solution.

   Place 50 mL of distilled or deionized water in a beaker, then add 5 mL of 0.1M NaOH and 3 drops of phenolphthalein. Mix with a straw until the color is uniformly distributed.

   Pinch your nostrils together, and *exhale* through the straw. *Do not inhale or suck the phenolphthalein solution.* Continue exhaling through the straw, stopping briefly to take breaths of air if necessary, until the solution clears. Explain the reason for the color change.

B. Using the pH meter, determine the pH of a beaker of distilled water. Pinching your nostrils shut, exhale into the bottom of the beaker through a straw. Have your lab partner determine and record the pH of

---

CAUTION

Do not perform this exercise if you have a neurological or respiratory condition.

Question 14B.3

Explain the difference between respiratory rates when hyperventilating and breathing into a paper bag.

the water. Repeat, blowing into the water five additional times, taking a deep breath each time. Record the pH after each exhalation in the table entitled Effects of pCO_2 Changes on pH.

Rinse the electrodes and determine the pH of the water in a second beaker. Breathe into the paper bag until breathing becomes difficult. Remove the bag and, pinching your nostrils shut, exhale through the straw into the water. Have your lab partner record the pH after the first exhalation and after each of five additional normal breaths of room air in the table entitled Effects of pCO_2 Changes on pH.

Question 14B.4

Write the part of the acid-base balance equation that explains the results of the effect of pCO_2 changes on the pH exercise.

Effects of pCO_2 Changes on pH

Solution	pH After Breathing Room Air	pH After Breathing into Paper Bag
Original distilled H_2O		
H_2O after exhalation 1		
H_2O after exhalation 2		
H_2O after exhalation 3		
H_2O after exhalation 4		
H_2O after exhalation 5		
H_2O after exhalation 6		

EXERCISE 3 Disorders of Acid-Base Balance: Acidosis and Alkalosis

The pH of a healthy person's blood is maintained within narrow limits by chemical and physiological buffers. In certain pathological conditions, however, the pH may increase, resulting in alkalosis, or decrease, resulting in acidosis. Fortunately, the respiratory and renal systems are able to adjust for these changes to some extent by a process known as **compensation**, thus returning the pH to within normal or nearly normal limits.

ACIDOSIS

Acidosis occurs when the pH of arterial blood drops below 7.35. It may result from an accumulation of acids or a loss of bases. There are two types of acidosis:

1. **Respiratory acidosis.** In this imbalance, there is too much dissolved carbon dioxide in the plasma, thus elevating the pCO_2. Disease states that may cause this condition include injury to the respiratory center in the brainstem, which results in decreased rate and depth of breathing; emphysema, in which movement of air within the alveoli is impeded; and pneumonia. Respiratory acidosis is often compensated for by metabolic alkalosis, in which the kidneys conserve bicarbonate ions, thus raising the pH of the blood.

2. **Metabolic acidosis.** In this imbalance, the pH of arterial blood drops as a result of the accumulation of nonrespiratory acids or the loss of bases. Disease states that may lead to this condition include uremia, in which the kidneys fail to excrete acids formed through metabolic processes; prolonged diarrhea, in which there is an excessive loss of alkaline intestinal secretions; and diabetes mellitus, in which keto acids accumulate in the blood and bicarbonate ions are excessively excreted in the urine. Metabolic acidosis is often compensated for by respiratory alkalosis, in which the lungs eliminate excess CO_2 through rapid, deep breathing, thus raising the pH of the blood.

ALKALOSIS

Alkalosis occurs when the pH of arterial blood rises above 7.45. It may result from a loss of acids or an accumulation of bases. There are two types of alkalosis:

1. **Respiratory alkalosis.** In this imbalance, there is too little dissolved carbon dioxide in the plasma, thus decreasing the pCO_2 of arterial blood, usually as a result of hyperventilation. Conditions that may cause hyperventilation include anxiety, early stages of salicylate poisoning, or high altitudes. Lightheadedness, agitation, tingling sensations of the extremities, and vertigo may result. Respiratory alkalosis is often compensated for by metabolic acidosis, in which the kidneys excrete more bicarbonate ions, thus lowering the pH of the blood.

2. **Metabolic alkalosis.** In this imbalance, the pH of arterial blood rises as a result of an excessive loss of hydrogen ions or a gain in bases. Disease states that may lead to this condition include vomiting or gastric suctioning and excessive intake of alkaline drugs such as sodium bicarbonate (baking soda) to relieve heartburn or other gastric distress. Metabolic alkalosis is often compensated for by respiratory acidosis, in which there is decreased rate and depth of breathing; which will raise the arterial pCO_2 and thus lower the pH of the blood.

By analyzing arterial blood gases, it is possible to determine whether a person is in a state of acidosis or alkalosis and to determine the type (metabolic or respiratory) of acidosis or alkalosis that results in a pH shift. Normal values and the result of increased or decreased values are as follows:

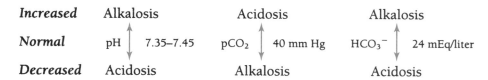

When the pH shift is primarily reflected by pCO_2 values, it is respiratory in nature. When the pH shift is primarily reflected by HCO_3^- values, it is metabolic in nature. This pH shift is usually resisted or compensated for by an increase or decrease in respiratory center activity or by chemical buffer activity. If compensation does not occur, death may result.

ACID-BASE PROBLEMS

Solve the acid-base balance problems presented in the accompanying table by filling in the columns.

	Acidosis or Alkalosis?	Why?	Respiratory or Metabolic?	Why?	Compensated or Uncompensated?	Why?
1. pH = 7.48 pCO$_2$ = 48 HCO$_3$ = 32						
2. pH = 7.31 pCO$_2$ = 50 HCO$_3$ = 32						
3. pH = 7.30 pCO$_2$ = 41 HCO$_3$ = 18						
4. pH = 7.50 pCO$_2$ = 30 HCO$_3$ = 24						
5. pH = 7.27 pCO$_2$ = 52 HCO$_3$ = 18						

Acid-Base Balance

MATCHING

Select the letter that best describes the acid-base imbalance for each condition listed.

Name _____

Section _____

Date _____

_____ **1.** Diarrhea

_____ **2.** Oral intake of excessive NaHCO$_3$

_____ **3.** Emphysema

_____ **4.** Early stage of overdose of aspirin

_____ **5.** Vomiting

_____ **6.** Diabetes mellitus

_____ **7.** Pneumonia

_____ **8.** Anxiety

a. respiratory acidosis

b. metabolic acidosis

c. respiratory alkalosis

d. metabolic alkalosis

MULTIPLE CHOICE

_____ **9.** The most accurate method of determining pH is to
  a. use litmus paper.
  b. use pH or nitrazine paper.
  c. use a pH meter.
  d. use a spectrophotometer.

_____ **10.** A solution that has the same number of hydrogen ions as hydroxyl ions is
  a. neutral.
  b. acidic.
  c. basic.
  d. salty.

_____ **11.** If one solution is 1,000 times as acidic as another, how many pH units would there be between the two solutions?
  a. one
  b. two
  c. three
  d. four

_____ **12.** A person with an arterial blood pH of 7.62 would be in a state of
  a. acidosis.
  b. alkalosis.
  c. neutrality.
  d. death.

_____ **13.** The two organs primarily responsible for maintaining acid-base balance in the body are
  a. liver and spleen.
  b. liver and lung.
  c. kidney and heart.
  d. lung and kidney.

_____ 14. Which of these would be formed if a strong acid were added to a weak base in a buffered solution?
  a. weak acid
  b. water
  c. strong base
  d. weak base and strong acid

_____ 15. Which of these acids is the strongest?
  a. 0.01 M HCl
  b. 1 M HCl
  c. 1 M NaOH
  d. 0.05 M HCl

_____ 16. What effect does breathing in and out of a paper bag have on a person's arterial pCO_2 level?
  a. The level decreases.
  b. The level increases.
  c. It has no effect.
  d. The level alternately increases and decreases.

_____ 17. The strength of an acid refers to
  a. the number of hydrogen ions in each molecule.
  b. the type of inorganic salt it forms.
  c. the degree to which its molecules ionize in water.
  d. the concentration of acid molecules.

_____ 18. The most common buffer in plasma and intercellular fluid is
  a. bicarbonate.
  b. phosphate.
  c. hemoglobin.
  d. sodium and potassium.

TRUE OR FALSE

_____ 19. The pH of blood is constant.

_____ 20. The pH of venous blood is usually lower than that of arterial blood.

_____ 21. Buffer solutions neutralize acidic or alkaline solutions.

_____ 22. Phosphate buffers may act as either acids or bases.

_____ 23. More drops of acid are required to change the pH of a buffered solution than a nonbuffered solution.

_____ 24. The respiratory system is the major compensatory mechanism for respiratory acidosis.

_____ 25. A base is a hydroxyl ion acceptor.

DISCUSSION QUESTIONS

26. Rank the body fluids you tested in this unit from acidic to basic.

_____

_____

_____

27. List and describe three general properties of acids and bases. _____

_____

_____

_____

28. Explain how the kidneys and respiratory center regulate acid-base balance in the body. _____

_____

_____

_____

CASE STUDY

ACID-BASE BALANCE

Sam Smith is 75 years old, has smoked two packs of cigarettes per day for 50 years, and has been diagnosed as having pulmonary emphysema.

Would Sam's condition represent acidosis or alkalosis? Which type?

Would his pCO_2 probably be elevated or decreased? Explain.

What compensatory mechanism would his body utilize to stabilize this condition?

The Reproductive System

A. Reproductive System Anatomy

PURPOSE

Unit 15A will enable you to understand the anatomy of the male and female reproductive systems.

OBJECTIVES

After completing Unit 15A, you will be able to
- identify the structures of human male and female reproductive systems.
- identify the structures of the male and female cat reproductive systems.
- observe specimens of an ovary, a testis, and a uterus containing a fetus.
- identify the stages in oogenesis and spermatogenesis.
- using a microscope, recognize sections of ovary, testis, penis, and uterus.

MATERIALS

human torso	preserved male and female cats
cadaver (optional)	model of female breast
dissecting microscope	disposable gloves
compound microscopes	dissecting instruments

gross specimens of prepared bull testis, pig uterus with fetus and attached ovary in situ

microscope slides of ovary, testis, penis, uterus, sperm smear, and cat or rat testis

model of midsagittal section of female pelvis

model of midsagittal section of male pelvis

PROCEDURE

 **EXERCISE 1**   **Human Reproductive Organs**

FEMALE REPRODUCTIVE SYSTEM

On the model of a midsagittal section of the female pelvis (**Figure 15.1**) or on the human torso or cadaver, if available, locate and identify the female reproductive structures printed in boldface.

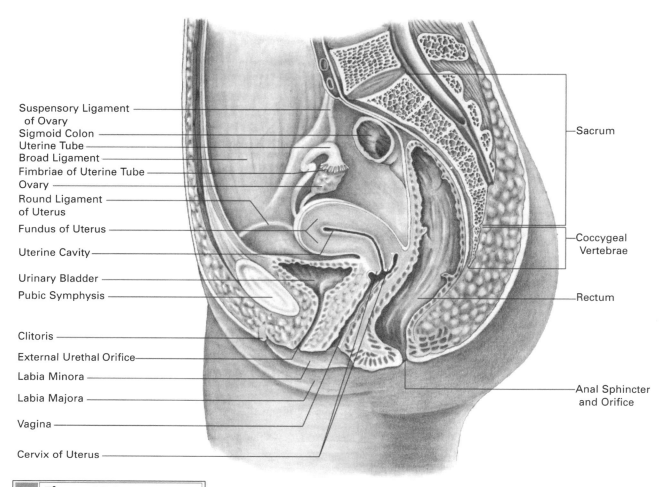

Suspensory Ligament of Ovary
Sigmoid Colon
Uterine Tube
Broad Ligament
Fimbriae of Uterine Tube
Ovary
Round Ligament of Uterus
Fundus of Uterus
Uterine Cavity
Urinary Bladder
Pubic Symphysis
Clitoris
External Urethal Orifice
Labia Minora
Labia Majora
Vagina
Cervix of Uterus

Sacrum
Coccygeal Vertebrae
Rectum
Anal Sphincter and Orifice

Figure 15.1

Midsagittal section of female pelvis showing internal organs and external genitalia

The female gonads, the paired **ovaries**, are held in place against the lateral walls of the pelvis by the **broad ligaments**. The **uterine**, or **fallopian tubes**, each with their **fimbriae** (fingerlike enlargements at the distal end) partially overlie but do not touch the ovaries. The fallopian tubes increase in diameter as they merge with the **uterus**. The uterus is positioned medially between the urinary bladder and the rectum. This muscular organ has great capacity for enlargement in pregnancy but in a nonpregnant female is about the size and shape of an upside-down pear. The **fundus**, or superior region, is the domed muscular area above the entrance of the two fallopian tubes. The **corpus**, or **body**, is the main portion of the uterus in which prenatal development takes place. The **cervix**, or neck region, extends inferiorly and interfaces with the **vagina**. The **vagina**, or birth canal, is a short muscular tube extending from the cervix of the uterus to the exterior of the female body (**Figures 15.1** and **15.2**). The **vulva**, or **external genitalia**, of the female include two pairs of lip-like structures, the **labia majora** and **labia minora**, that enclose the orifices. The **clitoris**, analogous to the male penis, is anteriorly positioned in the

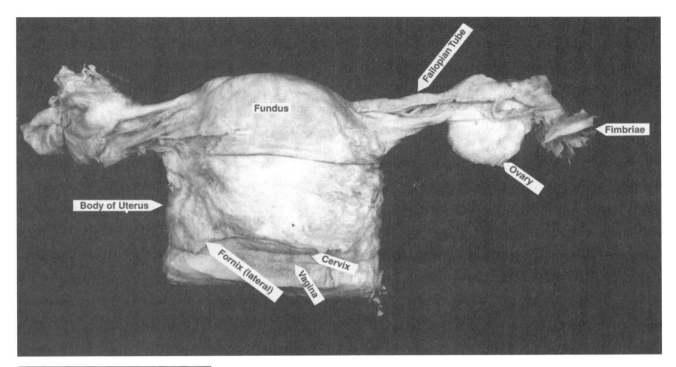

Figure 15.2

Female reproductive structures

vestibule, which is the region that extends from the clitoris to the vagina. On either side of the **vaginal orifice**, find the openings from the paired **Bartholin glands**. The **urinary meatus** is anterior to the vaginal orifice and the **anus** is posterior. The **perineum** extends from the anterior region of the labia majora to the anus. The **mons pubis** is the area overlying the **pubic symphysis**. It is covered with adipose tissue and pubic hair.

From a model of a human female breast. Notice that central to this skin-covered gland, which is normally attached to the dense fascia over the pectoralis major muscle, is the **nipple**, surrounded by its pigmented **areola**.

From a sagittal view you can identify approximately 20 lobes or divisions formed by extensions of the pectoralis fascia, which extend to the skin. Each lobe contains secretory **glandular lobules** embedded in adipose tissue. From each lobule is a narrow tubule, or **lactiferous duct**, which transports milk produced in the lobules to a **lactiferous sinus**, an enlargement of the duct. The sinuses continue to form a **lactiferous tubule**, which terminates at the base of the nipple and serves as an exit for the milk produced (**Figure 15.3**).

Hormones greatly influence mammary gland development and function. **Estrogens** contribute to the development of the tubular network and **progesterone** influences the development of the secretory lobules. Sucking causes the anterior pituitary gland to secrete **prolactin**, which stimulates milk production. The posterior pituitary gland secretes **oxytocin**, which causes lactation.

Figure 15.3

Female breast, sagittal section

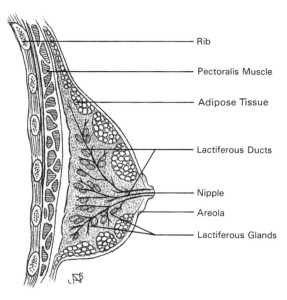

- Rib
- Pectoralis Muscle
- Adipose Tissue
- Lactiferous Ducts
- Nipple
- Areola
- Lactiferous Glands

Identification 15.1

 15.1

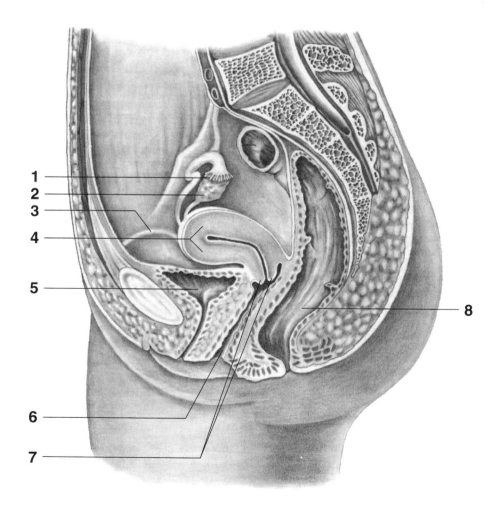

1
2
3
4
5
6
7
8

MALE REPRODUCTIVE SYSTEM

On the model of a midsagittal section of the human male pelvis (**Figure 15.4**) or on the torso or cadaver, identify the structures printed in boldface. The **scrotum**, the muscular sac that is suspended outside the male body, contains the paired gonads, or **testes**. Each testis (**Figure 15.16**) contains **seminiferous tubules**, where gametogenesis, or sperm formation, takes place. The **rete testis** is an extension of the seminiferous tubules through which sperm travel to the **epididymis**, where sperm are stored and mature. The **vas deferens**, or **ductus deferens** (**Figures 15.4 and 15.5**) an extension of the epididymis, is the duct that carries the sperm from outside the body to the inside. Where the vas deferens is encased in a connective tissue sheath also containing blood vessels and nerves, it is referred to as the **spermatic cord**. Sperm and seminal fluid continue through the **ejaculatory ducts**, which lead to the **urethra** (**Figure 15.4 and 15.16**). The urethra extends through the shaft of the penis and terminates at the **urogenital orifice**. **Semen**, the fluid that transports sperm, is produced in the **seminal vesicles, prostate gland** (**Figure 15.6**), and **Cowper's**, or **bulbourethral, glands**. The seminal vesicles produce about 60% of the semen, and they also produce fructose for nourishing the sperm. Seminal vesicles lie dorsal and lateral to the bladder. The prostate is a doughnut-shaped gland immediately inferior to the bladder. It surrounds the urethra and produces an alkaline fluid that contributes about 30% of the volume of semen (**Figures 15.4 and 15.16**). The very small Cowper's glands located inferior and lateral to the prostate produce about 5% of the semen. Like the scrotum, the **penis**, which is the male organ of copulation, is also external. The penis is composed of the **shaft** and contains the **corpus spongiosum**, which surrounds the urethra, and the paired **corpora cavernosa**, superior to the urethra. All three corpora are cylinders composed of porous tissue that become engorged, or filled with blood, allowing for penile erection. The enlarged tip of the penis is the **glans penis**. The **prepuce**, or **foreskin**, surrounds the proximal region of the glans penis (**Figures 15.4 and 15.16**).

Question **15A.1**

What structure is removed in a circumcision?

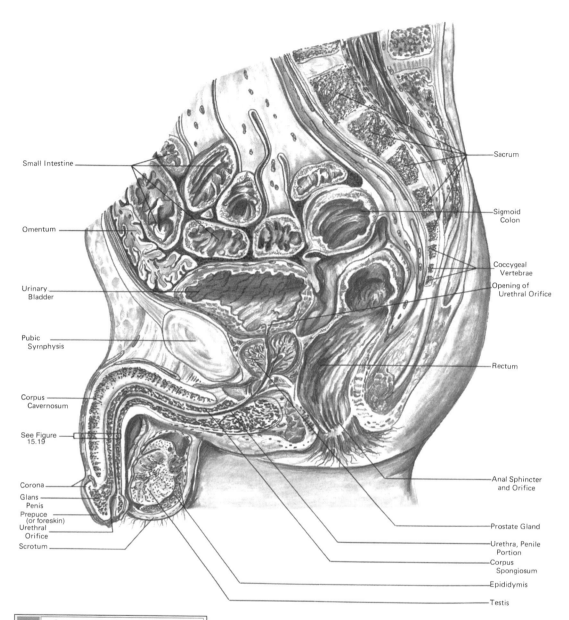

Small Intestine

Omentum

Urinary
Bladder

Pubic
Synphysis

Corpus
Cavernosum

See Figure
15.19

Corona

Glans
Penis

Prepuce
(or foreskin)

Urethral
Orifice

Scrotum

Sacrum

Sigmoid
Colon

Coccygeal
Vertebrae

Opening of
Urethral Orifice

Rectum

Anal Sphincter
and Orifice

Prostate Gland

Urethra, Penile
Portion

Corpus
Spongiosum

Epididymis

Testis

Figure 15.4

Midsagittal section of male pelvis
showing internal organs and
external genitalia

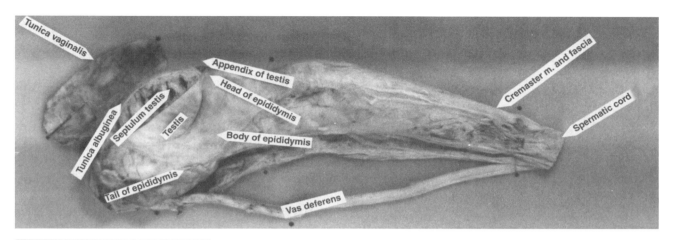

Figure 15.5

Human male reproductive structures, lateral view

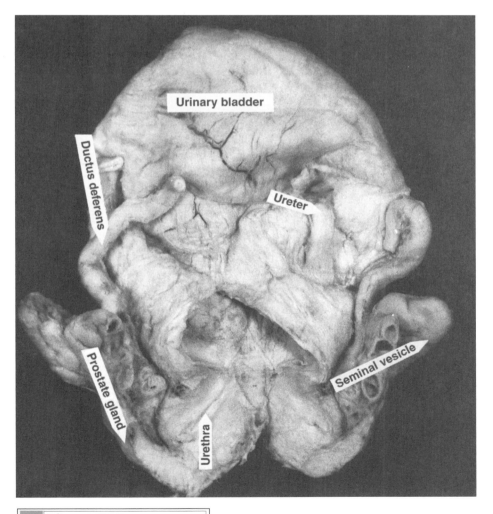

Figure 15.6

Human male urogenital structures, posterior view

EXERCISE 2 Cat Reproductive Organs

FEMALE REPRODUCTIVE SYSTEM

In order to observe the reproductive organs of a female cat, reflect the small and large intestines to one side and locate a small cream-colored, oval-shaped **ovary** lying just below each kidney. Each ovary is connected to a short coiled **oviduct (fallopian tube)** that conveys **ova**, or eggs, to the **uterus.** The uterus in the cat is bicornate, that is, made up of two **uterine horns** that are continuous with the oviducts. The uterine horns extend medially and caudally and then merge to form the **body of the uterus** dorsal to the base of the urinary bladder. The human uterus has no horns. Cut completely through the pubic symphysis and follow the body of the uterus posteriorly. The **vagina** is dorsal to the urethra that extends from the bladder to the **vestibule,** into which the urethra and vagina merge. Posterior to the vestibule is the **urogenital sinus,** which opens to the outside.

To observe the vagina and uterus internally, make a ventral incision into the opening of the urogenital sinus and extend it to the horn of the uterus. Locate the **cervix,** the inferiorly projecting neck portion separating the body of the uterus from the vagina (**Figures 15.7**).

If your cat is pregnant, observe the swollen bicornate uterus and feel for a developed fetus. Carefully, make an incision at that point and remove the fetus that is attached to its umbilical cord (**Figure 15.7**).

Figure 15.7

Female reproductive system, pregnant cat

1. small intestine
2. renal vein
3. gallbladder
4. liver
5. stomach
6. inferior vena cava
7. spleen
8. kidney
9. pregnant bicornate uterus
10. ureter
11. descending abdominal aorta
12. mesentery
13. bladder

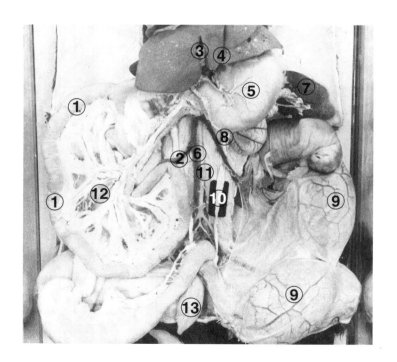

MALE REPRODUCTIVE SYSTEM

Observe the reproductive organs of a male cat (**Figure 15.8**). Immediately posterior to the pelvic region, locate the **scrotum**, a sac covered with skin and fur. Carefully cut through the skin and connective tissue at the ventral portion of the scrotum, and expose the paired **testes**. Remove the **tunica albuginea**, the tough covering of the testes and observe the **epididymis**, a large coiled duct on the anterior lateral surface of each testis. Extending from the epididymis is an ascending tube, the **spermatic cord**, which consists of the **vas deferens (ductus deferens)**, muscle, blood vessels, and nerves. Trace the spermatic cord and notice that it enters the body cavity through an opening, the **inguinal canal**. When the spermatic cord reaches the urinary bladder, the vas

---

Figure 15.8

Male reproductive system of cat

1. small intestine
2. heart
3. left liver lobe
4. kidney
5. ureter
6. urinary bladder
7. skin reflected
8. penis
9. testes
10. large intestine

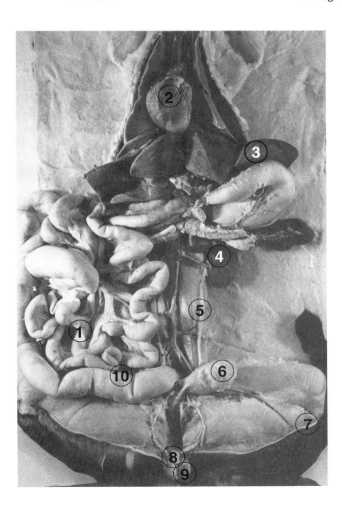

deferens leaves the spermatic cord and bends medially and posteriorly over the ureter. It continues posteriorly until it connects with the urethra at the **prostate gland.** Cut through the pubic symphysis and trace the urethra posteriorly until it enters the **penis,** an external structure. The opening at the anterior end of the penis, where urine and semen are released, is the **urogenital opening (Figures 13.8** and **15.8).**

EXERCISE 3 Observation of Pregnant Pig Uterus and Bull Testicle

On preserved specimens of a pregnant pig uterus and a bull testicle, observe the structures printed in boldface. On the pregnant uterus of the pig you can usually observe the paired almond-shaped **ovaries.** They are usually encased in the **fimbriae** of the **oviducts,** or **uterine tubes.** Each of the two oviducts is a short tube that is continuous with the muscular **uterine horns** of the **bicornate uterus.** Inferiorly, the merger of the horns forms the body of the uterus, which tapers into the **vagina** distally. Make a longitudinal incision into the swollen uterine region containing the fetus. Note, as you cut, the thick, outer layer of connective tissue, the **perimetrium,** and the inner muscular **myometrium.** The **placenta,** which is within the sac containing the developing pig, is adherent to the uterine lining, or **endometrium.** Reflect the tissues and observe the fetal pig attached by the **umbilical cord** to the placenta.

On the preserved specimen of the bull testicle observe the thick connective tissue covering, the **tunica albuginea.** Find the outer medial border of the testis and observe the highly coiled tube, the **epididymis.** This tube becomes wider as it forms the **vas deferens** which is encased in the **spermatic cord.** Make a longitudinal incision through the testicle and open both halves so that you can observe the internal structures. Medially, the **rete testis** is a tubular network formed by the **straight tubules.** These carry sperm from the **seminiferous tubules** where sperm are produced to the epididymis. The seminiferous tubules are tightly coiled tubules that are sometimes more visible through the tunica albuginea. Internally, they appear as a yellow mass.

EXERCISE 4 Microscopic Study of an Ovary and Uterus

Before birth, the ovaries contain millions of undifferentiated cells termed **oogonia,** which by birth have undergone many mitotic divisions in order to increase their number. As the oogonia mature, they develop into larger **primary oocytes.** These cells divide by meiosis into the **secondary oocyte** and the **first polar body.** The first polar body may or may not divide into two second polar bodies. The **secondary oocyte** undergoes further meiosis after fertilization, when it becomes a mature **ovum (egg).** The polar bodies reduce the excess number of chromosomes. Thus the ovum contains half the number (haploid) of chromosomes as the diploid oogonium (**Figure 15.9**). The process by which the mature ovum is formed is called **oogenesis.**

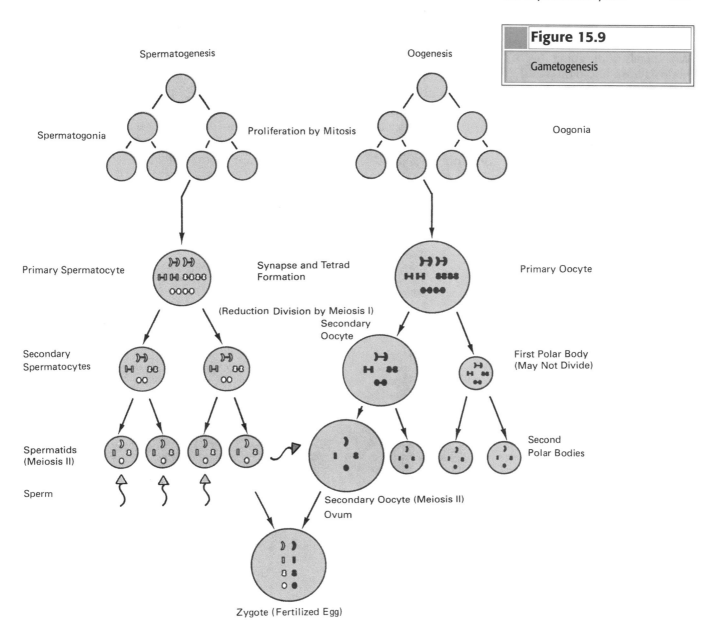

Spermatogenesis

Oogenesis

Figure 15.9

Gametogenesis

Spermatogonia

Proliferation by Mitosis

Oogonia

Primary Spermatocyte

Synapse and Tetrad Formation

Primary Oocyte

(Reduction Division by Meiosis I)

Secondary Spermatocytes

Secondary Oocyte

First Polar Body (May Not Divide)

Spermatids (Meiosis II)

Sperm

Second Polar Bodies

Secondary Oocyte (Meiosis II)
Ovum

Zygote (Fertilized Egg)

Question 15A.4

How many chromosomes are in a human oogonium?

How many chromosomes are in a human secondary spermatocyte?

Question 15A.5

Which structure in female gametogenesis is actually fertilized?

Question **15A.6**

Which cells undergoing meiosis form haploid cells?

Observe a slide of a sectioned ovary showing **Graafian follicles**, or **vesicular ovarian follicles** (**Figure 15.10** through **15.12** and **Color Plate 44**). Notice the **cuboidal (germinal) epithelium** around the periphery of the ovary. Medial to the cuboidal epithelium, observe **primary follicles**. After puberty, under hormonal influence, several follicles begin to mature prior to ovulation and become first primary oocytes and then **secondary oocytes** in Graafian follicles. Graafian follicles occur only in mammals and are hollow sacs containing a secondary oocyte surrounded by **follicular fluid**, or **liquor folliculi**. The cavity containing the follicular fluid is the **antrum**. Immediately underlying the secondary oocyte within the Graafian follicle is a mound of cells known as the **cumulus oophorus** (**Figure 15.12**). Draw what you observe and label the various structures.

Figure 15.10

Ovary, showing generalized structure (100×)

1. germinal epithelium
2. primary follicle
3. egg nest containing growing follicles
4. medulla containing blood vessels
5. mature Graafian follicle

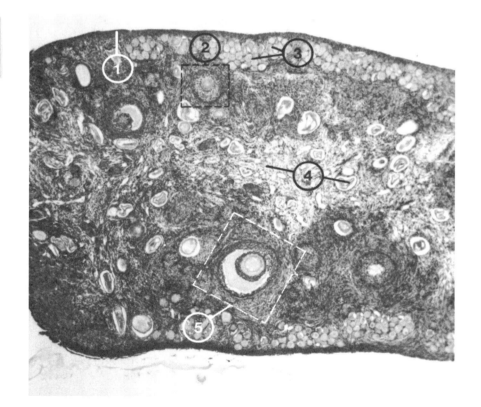

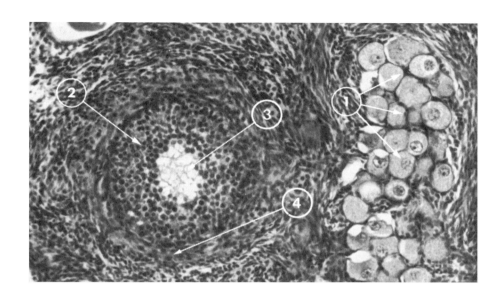

Figure 15.11

Egg nest with follicles in the cortex of the ovary (430×)

1. egg nest containing growing follicles
2. primary follicle
3. cytoplasm of primary oocyte
4. theca externa

Figure 15.12

Graafian follicle (100×)

1. theca externa
2. theca interna containing basement membrane of follicle
3. corpus luteum
4. antrum containing follicular fluid with follicle cells
5. corona radiata
6. secondary oocyte
7. nucleus of secondary oocyte
8. zona pellucida
9. cumulus oophorus

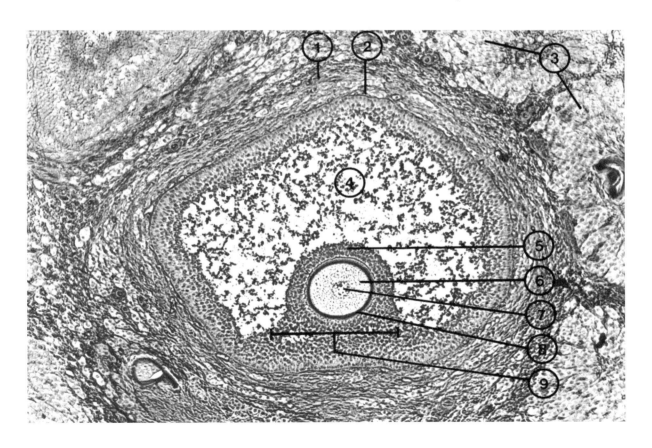

Observe a slide of an ovary showing corpora lutea (**Figre 15.12**). A **corpus luteum** is a collapsed follicle formed after the expulsion of the secondary oocyte. Corpora lutea first appear yellow and then regress into paler, whitish scar tissue known as the **corpus albicans**. The active corpus luteum produces progesterone, a hormone essential to pregnancy. Draw a corpus luteum and corpus albicans (**Figure 15.13**) in the space below and label these structures.

Figure 15.13

Corpus albicans (430×)

1. cortex of ovary
2. theca interna cells
3. fibrous connective tissue

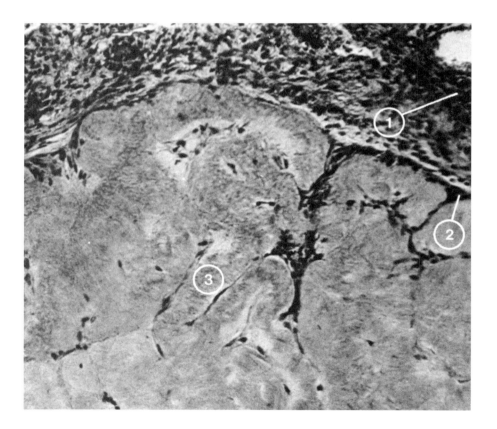

Obtain a section of uterus (any phase) and identify the smooth muscle area and endometrial lining. The **endometrium** is composed of two layers: the underlying **stratum basale**, which contains blood vessels, and the more superficial **stratum functionale**, which is composed of secretory glands and columnar epithelium. The stratum functionale is sloughed off during **menstruation** (**Figures 15.14** and **15.15**). Draw and label this slide.

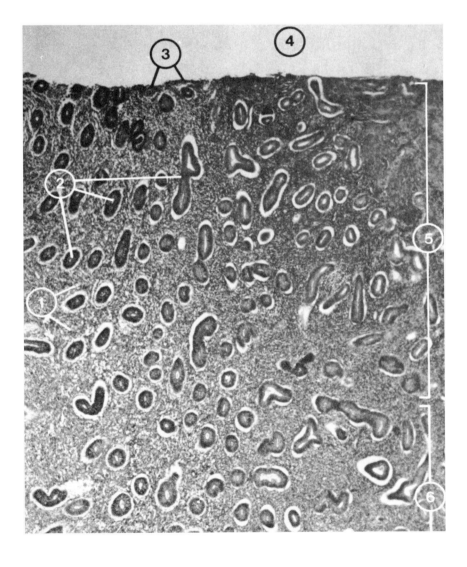

Figure 15.14

Human uterus, proliferative stage (100×)

1. lamina propria
2. uterine glands
3. columnar epithelium of endometrium
4. lumen of uterus
5. stratum functionale layer of endometrium
6. stratum basale layer of endometrium

Figure 15.15

Human uterus, menstrual phase
(100×)

1. lumen of uterus
2. endometrium
   undergoing erosion
3. uterine glands

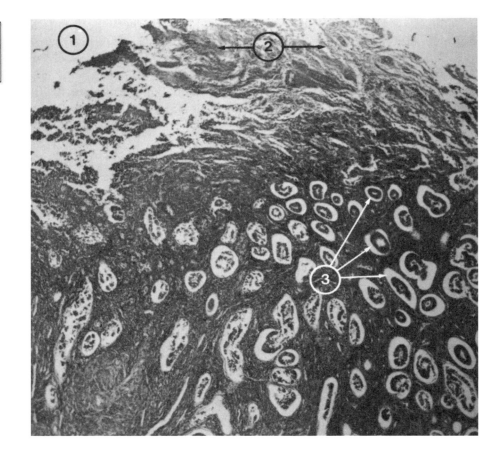

Question 15A.7

Compare the structure of
the human and cat uterus
and explain the reasons
for the difference.

EXERCISE 5 **Microscopic Study of a Testis and Penis**

The **testes**, which are contained in the **scrotum**, consist of thousands of coiled **seminiferous tubules** in which sperm develop. These tubes are lined with undifferentiated germ cells termed **spermatogonia**, which are formed before birth. Spermatogonia divide by the process of mitosis to increase their number. At puberty, the spermatogonia undergo **spermatogenesis**. The first stage of this process is the growth of these cells into much larger cells called **primary spermatocytes**. These cells in turn divide by meiosis into two cells of equal size known as **secondary spermatocytes**. Meiosis continues as the secondary spermatocytes immediately divide into four equal-sized cell types termed **spermatids**. Spermatids contain half the number of chromosomes as diploid primary spermatocytes and require further minor changes, which occur in a process known as **spermiogenesis** before becoming functional **spermatozoa (sperm)** (**Figure 15.9** and **Figures 15.16** through **15.18**).

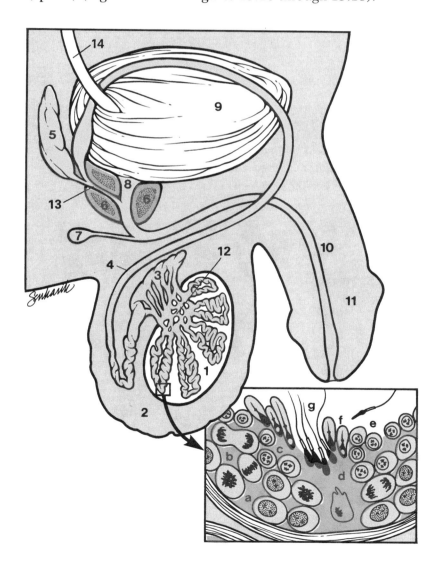

Figure 15.16

Sperm production in the adult male

1. testis
2. scrotum
3. epididymis
4. sperm duct (vas deferens)
5. seminal vesicle
6. prostate gland
7. Cowper's gland
8. urethra
9. bladder
10. penile urethra
11. penis
12. rete testis
13. ejaculatory duct
14. ureter

Inset: Enlarged segment of a seminiferous tubule

a. diploid spermatogonia
b. primary spermatocytes
c. primary spermatocytes undergoing reduction division
d. haploid secondary spermatocytes
e. spermatids
f. spermatids undergoing spermiogenesis
g. spermatozoa

Question **15A.8**

Trace the tubular structures through which sperm are transported from the place of their production to their ejaculation.

The sperm are transported from the testes to the **epididymis,** where they undergo modifications to enhance motility. They then travel through the **vas deferens** and receive contributions of seminal fluid from the **seminal vesicles, prostate gland,** and **Cowper's glands.** The distal end of the seminal vesicle joins with the vas deferens to form the **ejaculatory duct.** The ejaculatory duct is about 1″ long and is partially surrounded by the prostate gland. The ejaculatory duct then joins with the **urethra** as it exits from the **bladder.** The urethra continues through the **penis** (**Figure 15.16**).

Observe a slide containing a section of a testis. Within the testis, you will find numerous **seminiferous tubules** (**Figure 15.17** and **Color Plate 45**). Under high power, focus on a cross section near the periphery of a seminiferous tubule where you will find larger, elongated **Sertoli,** or **sustentacular cells,** which aid in sperm development. Starting at the periphery and moving toward

Figure 15.17

Seminiferous tubules of testes (430×)

1. interstitial cells (of Leydig)
2. spermatogonia undergoing mitosis
3. lumen of seminiferous tubules
4. Sertoli (sustentacular) cells
5. spermatozoa

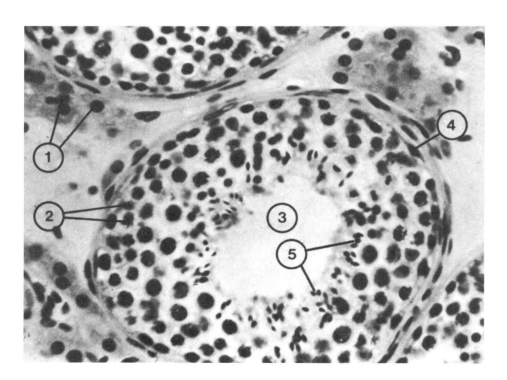

the center, you should also be able to see the maturation of sperm (spermato-genesis) from **spermatogonia** to **primary spermatocytes** to **secondary spermatocytes** to **spermatids** to mature sperm in the center of the tubules (**Figure 15.18**). A mature sperm can be recognized by the presence of a tail, or **flagellum**. Between the seminiferous tubules are the **interstitial cells of Leydig**, which produce and secrete the male hormone **testosterone**. Draw and label a cross section of a **seminiferous tubule**.

Figure 15.18

Detail of seminiferous tubule of rat testis showing spermatogenesis (400×)

1. spermatogonia
2. spermatogonia in mitosis
3. primary spermatocytes
4. secondary spermatocytes
5. spermatozoa

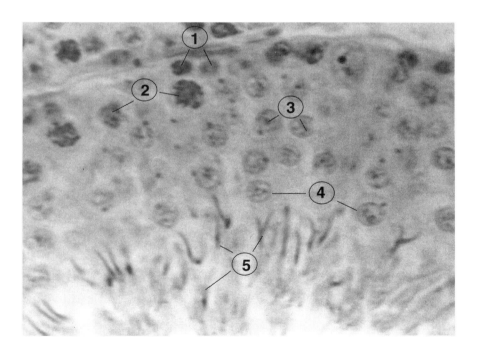

Observe and draw a slide of a sperm smear. Identify the head and tail of a spermatozoan.

Obtain a slide of a cross section of a penis. This slide is best observed under a dissecting microscope. Identify two **corpora cavernosa**, the **corpus cavernosum urethrae (corpus spongiosum)** and the **urethra**, which is found in the corpus spongiosum. Also, observe the connective tissue, muscle, sinuses, and blood vessels in this section (**Figure 15.19**). Draw and label:

Figure 15.19

Cross section of human penis (100×)

1. corpora cavernosa
2. urethra
3. corpus spongiosum (corpus cavernosum urethrae)

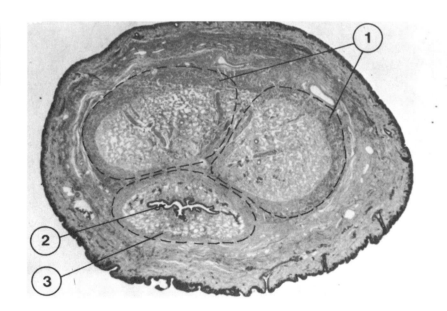

B. Reproductive System Physiology

PURPOSE

Unit 15B will enable you to understand selected physiological characteristics of the reproductive system and basic stages of human embryological development.

OBJECTIVES

After completing Unit 15B, you will be able to

- correlate stages of the female sexual cycle with events occurring in the anterior pituitary gland, ovaries, uterus, and blood.
- demonstrate use of a human pregnancy test kit.
- identify basic stages of embryological development.

MATERIALS

model of human female sexual cycle

PROCEDURE

EXERCISE 1 The Female Sexual Cycle

The female sexual cycle occurs every 21 to 45 days (28 days is average) during the active reproductive phase of a nonpregnant woman's life. This cycle, illustrated in **Figure 15.20**, is regulated by cyclical hormonal events occurring in the hypothalamus, anterior pituitary, ovaries, endometrium, and blood. These events can be correlated with days of the menstrual cycle, which are represented at the bottom of **Figure 15.20**. Refer to **Figure 15.20** or a model of the human female ovarian and uterine cycles. Note that the hypothalamus secretes **gonadotropin releasing hormone (GnRH)**, which is secreted into the anterior lobe of the pituitary gland and stimulates the release of **follicle stimulating hormone (FSH)** and **luteinizing hormone (LH)**. FSH stimulates development of the ovarian follicles and also secretion of the **estrogen** beta estradiol by the ovaries. FSH is primarily secreted during the menstrual and proliferative phases of the cycle. LH, which is secreted in greater amounts later in the proliferative phase, triggers ovulation from the mature follicle on day 14 of a 28-day menstrual cycle. Following ovulation, the edges of the ovarian follicle, from which the sex cell (secondary oocyte) was released, begin to thicken and turn inward by a process known as involution. The resulting structure is known as a **corpus luteum**, which secretes the hormone **progesterone**. In response to rising estrogen and progesterone levels in the blood, the endometrium thickens and becomes more glandular during the proliferative phase and through approximately the first half of the secretory phase. If fertilization of the ovum does not occur, estrogen and progesterone levels decline. This results in sloughing of the superficial layers of the endometrium, which is characterized by menstrual bleeding, on day 28 of an average cycle.

Figure 15.20

Female sexual cycle

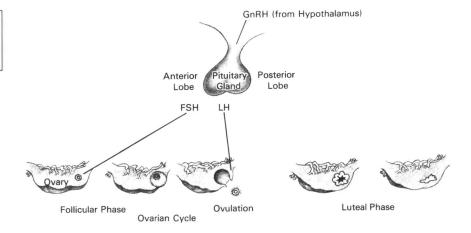

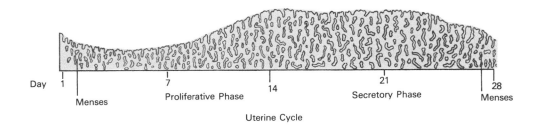

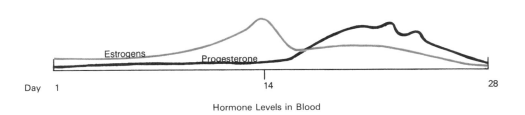

EXERCISE 2 Urinary hCG as an Indicator of Pregnancy

Pregnancy tests are based on the detection of the hormone **human chorionic gonadotropin** in the urine of pregnant women. This hormone is secreted within a few days after fertilization by the cells of an embryological structure, the trophoblast, and later by the developing placenta. hCG also stimulates the corpus luteum to secrete large quantities of estrogens and progesterone during the first few weeks of pregnancy. The level of hCG in the mother's blood and urine peaks at 8 weeks of pregnancy then declines after the placenta has formed.

Commercially available pregnancy test kits have an immunological basis, in which monoclonal antibodies are specific for and react with two different regions of the hCG molecule. Most tests have a built-in positive (contains hCG) control window or area, that is is compared with the urine sample being tested.

MATERIALS

nonpregnant urine (or stock solution synthetic urine)
urine from a woman who is 6–12 weeks pregnant
vials of hCG lyophilized powder diluted to 100 IU hCG/mL
   in stock solution synthetic urine
immunological pregnancy test kits
disposable gloves

PROCEDURE

Collect a urine sample from yourself or use 50-mL samples of stock solution synthetic urine with and without added hCG.

Following the instructions on the particular test kit you are using, dip a test strip or stick into the sample for the recommended time, using a separate strip for each sample.

Place each test strip on a flat surface with the indicator surfaces facing up for the recommended time.

Read your results by comparing the presence or absence of color or figure development on each strip within the designated time limit.

Question 15B.1

How does hCG that is produced by an embryo reach the mother's urine?

EXERCISE 3 **Prenatal Development**

CAUTION

Wear safety glasses when flushing reagents down sink.

MATERIALS

model or preserved specimen of human placenta
models of stages of human embryonic and fetal development

PROCEDURE

During the first two months after fertilization, the developing human being is called an **embryo**. At this time the primary germ layers, embryonic membranes, and placenta are formed. After the eighth week of embryonic development, the developing offspring is known as a **fetus**.

The three primary germ layers—ectoderm, mesoderm, and endoderm—give rise to all the tissues and organs that compose the adult body systems.

Outside the embryo itself, there are three membranes, known as extraembryonic membranes, that protect and nourish the developing embryo. The outermost membrane is the **chorion**, which overlies the other two membranes and, in its early stage of formation, surrounds the embryo, forming a vascular network known as **chorionic villi.** It later becomes part of the placenta. The **amniotic membrane**, or **amnion**, secretes amniotic fluid. This fluid absorbs shock and therefore protects the developing fetus. The amnion ruptures just prior to birth and expels its fluid. The **allantois** is the third extraembryonic membrane. It is vascularized and becomes part of the umbilical cord, which connects the fetus to the placenta.

The **placenta** develops from the chorion of the embryo and the endometrium of the mother. Its development is complete by the third month of gestation. This organ serves as an endocrine gland, providing necessary hormones that support pregnancy. The placenta also provides for the exchange of nutrients and waste products between mother and fetus. On a model or preserved specimen of a placenta, identify the smooth **fetal** (or **amniotic**) **surface**, the roughly textured **maternal surface**, the **umbilical cord**, the two **umbilical arteries**, and the single **umbilical vein** (**Figure 15.21**).

On the models illustrating various stages of embryonic and fetal development, observe the changes in appearance of the extra-embryonic membranes and the formation of different tissues and organs (**Figures 15.22–15.25**).

Figure 15.21

Human placenta, fetal surface

1. umbilical cord
2. umbilical artery
3. arteries
4. veins
5. amnion

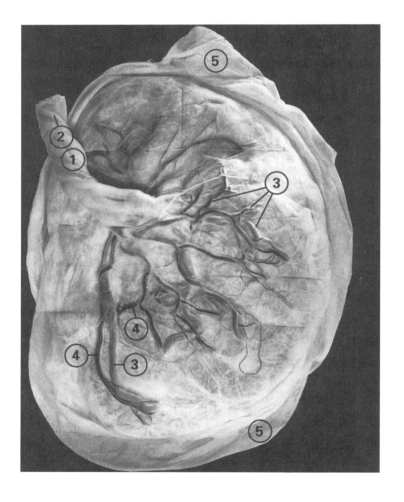

Figure 15.22

Embryonic development, eight weeks

1. chorionic villi
2. developing placenta
3. developing umbilical cord
4. embryo
5. amnion
6. amniotic cavity
7. uterine cavity
8. endometrium of uterus
9. myometrium of uterus
10. perimetrium of uterus
11. uterine (fallopian) tube
12. fimbriated end of uterine tube
13. ovarian ligament
14. round ligament
15. broad ligament
16. blood vessels

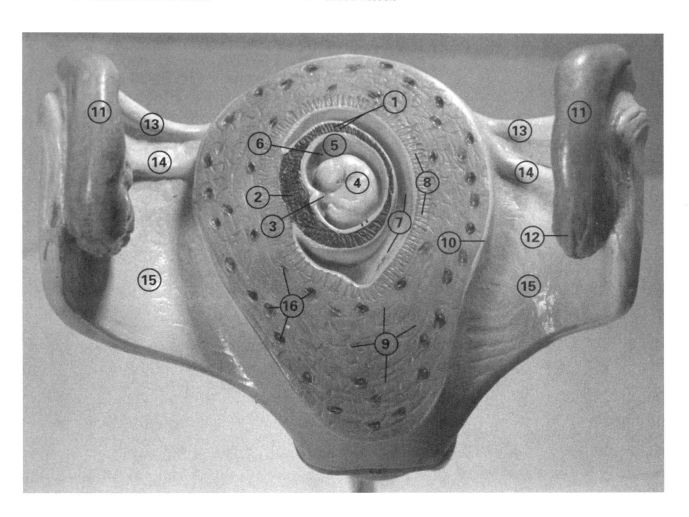

Figure 15.23

Fetal development, six months

1. fetus
2. amnion
3. umbilical cord
4. placenta
5. amniotic cavity
6. ovary
7. uterine tube
8. fimbriated end of uterine tube
9. cervix
10. internal os
11. external os
12. cervical canal

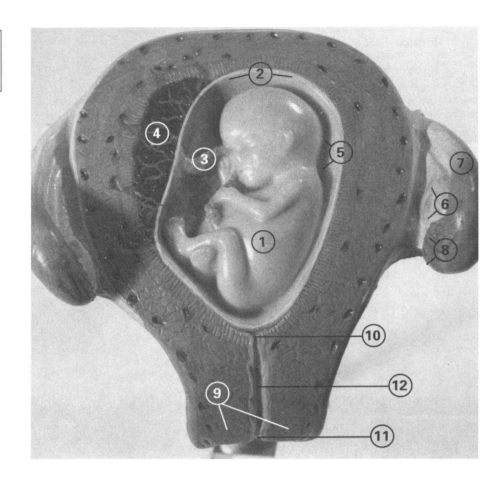

Figure 15.24

Fetal development, eight months, transverse position

1. umbilical cord
2. placenta
3. amnion
4. myometrium, stretched

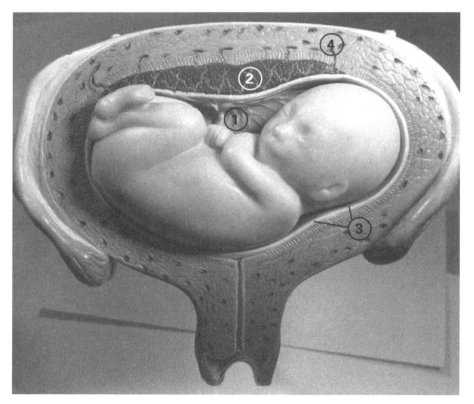

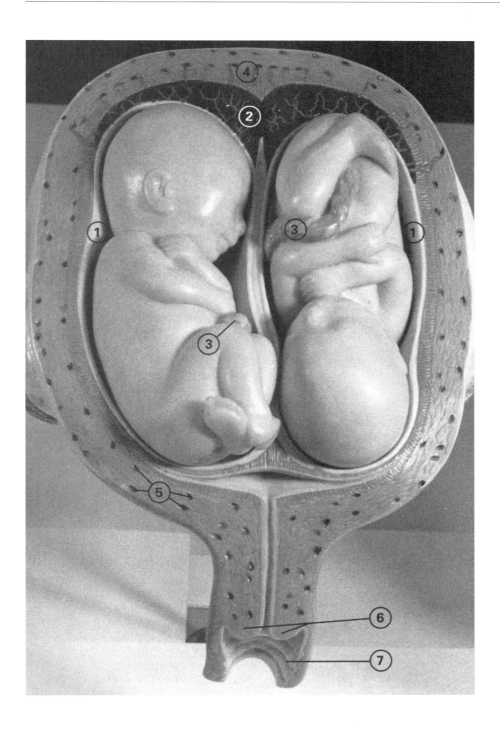

Figure 15.25

Fetal development, identical twins, full term

1. amniotic sacs
2. placenta, shared
3. umbilical cords
4. myometrium, stretched
5. blood vessels
6. cervix
7. vagina

The table, Fetal Development, lists major anatomical and physiological features of prenatal development.

Fetal Development	(Figures 15.22–15.25)
Month	**Selected Features of Development**
End of month 2	Cardiovascular system developed and functioning
	Heart pumping blood since end of week 4
	All brain regions present; initial waves are transmitted
	Ossification begins; limbs distinct, with digits formed
	Head with eyes, fused lids, and flattened nose present; head almost as large as rest of body
	All body systems present in rudimentary form
End of month 4	Head remains large, with basic facial features evident; eyes blink and lips move
	Skin, bones, and many joints formed
	All body systems continue developing
End of month 6	Body elongates so head less disproportionate in length
	Eyelids and eyelashes formed and functioning
	Rapid growth of all body systems
End of month 8	Testes in male descend to scrotum from abdominal cavity
	Subcutaneous fat in place
	All body systems well developed
End of month 9	Body has increased in length and weight
	Birth; average weight: 6–10 lbs (2.7–4.0 kg), average length: 16–20 inches (40–50 cm)

Question 15B.2

At what point in time in prenatal development does an embryo become a fetus?

Question 15B.3

Draw a two-month-old fetus showing the development of the placenta and limbs.

Question 15B.4

Draw an eight-month-old fetus showing the development of the placenta and limbs.

MULTIPLE CHOICE

Name _____

Section _____

Date _____

_____ 1. Which of these is the correct sequence for oogenesis?
- a. oogonium, primary oocyte, ovum, secondary oocyte
- b. primary oocyte, secondary oocyte, oogonium, ovum
- c. oogonium, primary oocyte, secondary oocyte, ovum
- d. ovum, primary oocyte, oogonium, secondary oocyte

_____ 2. How many spermatozoa are normally produced from one primary spermatocyte?
- a. one
- b. one plus three polar bodies
- c. two
- d. four

_____ 3. How many chromosomes does a human ovum normally contain?
- a. 23
- b. 23 pairs
- c. 46
- d. 46 plus 2 sex chromosomes

_____ 4. Hormones that promote ovulation are secreted by the
- a. pituitary gland.
- b. ovary.
- c. adrenal glands.
- d. uterus.

_____ 5. Which hormone levels are highest on day 14 of the female sexual cycle?
- a. FSH and LH
- b. FSH and progesterone
- c. LH and estrogens
- d. estrogens and progesterone

_____ 6. Spermatogenesis producing viable sperm begins
- a. immediately prior to ejaculation.
- b. at puberty.
- c. within the first year of life.
- d. during embryonic development.

_____ 7. What do human ova and spermatozoa have in common?
- a. They contain the same number of chromosomes.
- b. They secrete the same hormones in equal quantities.
- c. They are approximately the same size.
- d. They are produced in equal numbers.

_____ 8. Which of these could not occur within the body if the uterine tubes were cut and ligated?
- a. fertilization
- b. ovulation
- c. sexual intercourse
- d. menstruation

_____ 9. What is the purpose of having the testes suspended within the scrotum?

a. to protect the sperm from becoming contaminated with urine

b. to provide additional surface area for sperm maturation

c. to facilitate the passage of sperm through the vas deferens

d. to protect the sperm from higher temperatures within the abdominal cavity

_____ 10. Which of these hormones maintains male secondary sex characteristics?

a. corticosterone

b. FSH

c. testosterone

d. progesterone

_____ 11. Which of the following structures is not contained in the testes?

a. seminiferous tubules

b. rete testes

c. vas deferens

d. interstitial cells of Leydig

_____ 12. Which of the following glands manufacture(s) semen?

a. seminal vesicles

b. prostate

c. Cowper's

d. all of the above

_____ 13. The major anatomical feature contributing to penile erection is

a. bulbourethral contraction.

b. urethral distention.

c. engorgement of corporal cavities with blood.

d. scrotum distention.

Photo Identification

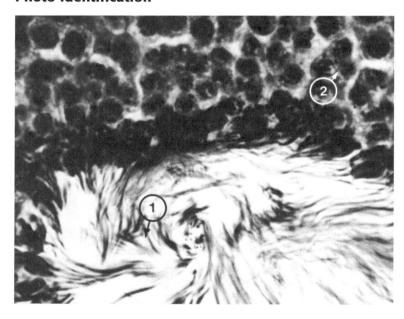

_____ 14. The cell type identified by 1 in the above photograph is a

a. spermatogonium.

b. sperm cell.

c. primary spermatocyte.

d. secondary spermatocyte.

_____ 15. The first meiotic division produces the cell type identified by 2 which is
   a. a spermatid.
   b. a sperm cell.
   c. a secondary spermatocyte.
   d. none of the above.

_____ 16. Of the following cell types, which is not found in the seminiferous tubule photograph?
   a. secondary spermatocyte
   b. primary spermatocyte
   c. interstitial cell of Leydig
   d. spermatid

_____ 17. In the female, the structure immediately anterior to the urethral opening is the
   a. hymen.
   b. clitoris.
   c. mons pubis.
   d. vaginal orifice.

_____ 18. In a vasectomy, the duct that is cut and ligated is the
   a. epididymis.
   b. vas deferens.
   c. ejaculatory duct.
   d. seminiferous tubule.

_____ 19. Human fertilization normally takes place in the
   a. ovary.
   b. fallopian tubes.
   c. uterus.
   d. vagina.

_____ 20. Menstruation is initiated by
   a. decreased progesterone.
   b. FSH production.
   c. corpus luteum formation.
   d. ovulation.

_____ 21. Which of the following does not secrete some substance that is ejaculated by the male?
   a. testis
   b. vas deferens
   c. seminal vesicle
   d. prostate gland

_____ 22. The spermatic cords enclose
   a. the seminal ducts.
   b. blood vessels.
   c. nerves.
   d. all of the above.

_____ 23. The hymen is structurally part of the
   a. cervix.
   b. clitoris.
   c. vagina.
   d. labia minora.

_____ 24. To which of the following structures does the term "fimbriated" refer?
   a. ovary
   b. uterine tube
   c. uterus
   d. vagina

_____ 25. Which of the following is exclusively found in the female?
   a. parotid gland
   b. Bartholin's gland
   c. prostate gland
   d. Cowper's gland

DISCUSSION QUESTIONS

26. Trace the passage of sperm from the seminiferous tubules to the point of ejaculation. _____

_____

_____

_____

27. Trace the passage of an egg from its expulsion from a Graafian follicle through implantation. _____

_____

_____

_____

28. What happens to a Graafian follicle after ovulation in a nonpregnant woman? _____

_____

_____

_____

29. Indicate whether each of the following contains the diploid or haploid number of chromosomes:

   a. sperm _____

   b. primary oocyte _____

   c. secondary spermatocyte _____

   d. interstitial cell _____

   e. ovum _____

30. In the human, which hormones are responsible for inducing ovulation? _____

_____

_____

_____

CASE STUDY

REPRODUCTIVE SYSTEM

Tara, a 32-year-old caterer, underwent premature menopause when she was 26 years old. She was unable to conceive normally. When embryo adoption became available she, with hormone therapy, successfully carried a set of twins to birth.

Describe the function of Tara's ovaries before and after embryo implantation.

Describe the conditions of Tara's uterus before and after embryo implantation.

Describe Tara's pituitary function before and after hormone therapy.

What hormones have been synthetically replaced? Explain your answer.

Glossary/Index

A band 149 Anisotropic band; myosin and actin filament in a myofibril.

ADH (antidiuretic hormone) or **vasopressin** (va′-sō-PRES-in) 280 A hormone that is secreted by the neurohypophysis.

ATP (adenosine triphosphate) 211 A compound that stores and releases energy for metabolic reactions.

abduct 137 Move away from body midline or from body part.

Achilles (a-KIL-ēz) **tendon** 210 The tendon posterior to the heel that connects the gastrocnemius and soleus muscles with the tuberosity of the calcaneus.

acid 463 A hydrogen ion (H+) donor.

acidophil 299 *See* eosinophil.

acidosis (as′-i-DŌ-sis) 471 A condition that occurs when pH of arterial blood drops below 7.35.

adduct (AD-ukt) 137 Move toward body midline or toward another body part.

adenohypophysis (ad′-e-nō-hī-POF-i-sis) 278 *See* pars distalis.

adipose (AD-i-pōs) 47 Fat tissue.

adrenal (a-DRĒ-nul) **glands** 278 Suprarenal glands; one located on superior border of each kidney.

adrenalin (a-DREN-a-lin) 281 Epinephrine.

adrenocorticotrophic (a-drē′-nō-kor′-ti-kō-TRŌ-fik) **hormone (ACTH)** 280 A hormone produced by the adenohypophysis, having its effect on the adrenal cortex.

adventitia (ad-ven-TISH-ē-uh) 399 The outermost fibrous coat of the digestive tract that blends in with surrounding tissues.

afferent (AF-er-ent) 227 Moving or conveying toward a central point, leading toward a central point.

afterimage 259 A visual sensation of complementary colors that occurs after the original stimulus ends.

agglutination (a-GLOO-ti-nā′-shun) or **coagulation** (koAG-yoo-la′-shun) 338 The clumping together of erythrocytes.

aldosterone (al-DOS-ter-ōn) 281 A mineralocorticoid produced by the adrenal cortex.

alkalosis (al-ka-LŌ-sis) 472 A condition that occurs when pH of arterial blood rises above 7.45.

alpha cells 281 Cells in the islets of Langerhans that produce the hormone glucagon.

alveolus (al-VĒ-o-lus) 361 Saclike structure found in lung and bone.

amino (uh-MĒ-nō) **acids** 416 The structural and functional unit of protein molecules.

amnion (AM-nē-on) 501 Thin extraembryonic membrane surrounding each fetus in utero.

amphiarthrodial (am′-fē-ar-THRŌ-dē-ul) 136 A type of joint in which bones are attached to each other by cartilage.

amplitude 264 The height of a sound wave; determines the volume of sound.

anaphase 32 Mitotic phase in which homologous chromosomes move apart to opposite poles.

anatomical position 1 The appearance of the human body as it stands erect, with palms facing forward.

androgen (AN-drō-jin) 281 A class of steroid hormones that produces masculinization.

antibody or **agglutinin** (a-GLOO-ti-nin) 339 A protein produced in response to an invasion by a foreign substance.

antigen or **agglutinogen** (ag-′looTIN-ō-jen) 338 Any foreign substance that, when introduced into the body, will cause the production of an antibody that reacts specifically with the foreign substance.

apex (Ā-pex) 313 The inferior pointed aspect of the heart; the superior portion of the lungs.

aponeurosis (ap´-ō-noo-RŌ-sis) 188 A sheath of connective tissue.

appendicular (ap´-en-DIK-yoo-ler) 92 Referring to the limbs. A division of the skeleton that includes the bones of the shoulder, hip, and limbs.

arbor vitae (AR-ber VĪ-tē) 230 White matter of the cerebellum assuming a treelike shape.

arch 77 A fingerprint pattern characterized by an elevation of dermal ridges in the middle of the pad of a finger.

areola (a´-rē-Ō-luh) 481 The pigmented circular area around the nipple.

arrhythmia (uh-RITH-mē-uh) 350 A variation of the heartbeat from normal sinus rhythm.

articulation (ar-tik´yoo-LĀ-shun) 134 Union of bones in the body; a joint.

ascending colon 392 Portion of large intestine between the cecum and transverse colon.

atria (Ā-trē-uh) 320 The upper chambers of the heart.

axial (AK-sē-ul) 91 A division of the skeleton that includes the vertebrae, sternum, ribs, and bones of the skull.

axon (AK-son) 223 A cytoplasmic extension from a neuron that transmits impulses away from the cell body.

Barfoed's test 420 Test for presence of monosaccharides.

base 313 Anatomically, the superior end of the heart; a hydroxyl ion (—OH) donor.

basement membrane 51 A supporting membrane underlying mucosal epithelium.

basophil (BĀ-sŌ-fil) 299 A granular leukocyte.

Benedict's test 420 A test for presence of reducing sugars.

beta (BĀ-tuh) **cells** 290 Cells in the islets of Langerhans that secrete the hormone insulin.

bicuspid (bĪ-KUS-pid) 383 A tooth posterior to the cuspid that has two points, or cusps; premolar.

bicuspid or **mitral valve** 313 Valve consisting of two flaps, or cusps, of tissue between the left atrium and left ventricle.

bile 414 An alkaline fluid secreted by the liver and concentrated and stored in the gallbladder that facilitates the emulsification of fats.

bile duct 412 A canal or passageway for transporting bile.

Biuret reaction 417 A reaction specific for compounds containing two peptide bonds united by a nitrogen or carbon atom.

body 112 The central portion of an organ.

Bowman's capsule 435 First collection unit of a nephron, shaped as a hemisphere with single-layered epithelial walls.

brainstem 232 Inferior section of the brain consisting of the midbrain, pons, and medulla oblongata.

broad ligament 479 One of the uterine ligaments.

Brunner's glands 403 Mucous glands in the submucosa of the duodenum.

buffer 468 A fluid substance that tends to reduce changes in hydrogen ion concentration.

buffer system 463 A combination of two or more chemicals that minimizes the pH change of a solution when an acid or base is added to the solution.

calcitonin (kal´-si-TŌ-nin) 281 Thyrocalcitonin; hormone of the thyroid gland that opposes parathyroid hormone.

calorie 425 The amount of heat required to raise 1 kg of water 1°C at atmospheric pressure.

calyx (KĀ-lix) 438 Cup-shaped structure for urine collection in the renal medulla.

canaliculus (kan´-uh-LIK-yoo-lus) 59 A small passageway found in osteon of bone.

cancellous (KAN-suh-lis) 88 Spongy; "soft" bone tissue.

canine (KĀ-nīne) 383 A conical pointed tooth located between the lateral incisor and first premolar; cuspid.

carbohydrate 418 A class of organic compounds that contain carbon and the elements of water in the ratio of 1:1.

carboxyl (kar-BOK-sil) **group** 421 COOH; represents an organic acid.

cardiac (KAR-dē-ak) 47 Pertaining to the heart; a muscle type occurring in the heart; portion of the stomach nearest the esophagus.

cardiac muscle 47 Striated, involuntary muscle comprising the heart wall; is characterized by intercalated disks.

cartilage (KAR-ti-lij) 47 A firm, flexible supporting connective tissue.

caudal (KAW-dul) 1 Inferior in human anatomy; tending toward the tail.

cecum (SĒ-kum) 391 A blind pouch; first section of the large intestine.

cell 29 The structural and functional unit of the body.

cementum (suh-MEN-tum) 394 A bony layer surrounding the dentin of the root of a tooth.

centriole (SEN-trē-ōl) 31 Structure within a centrosome that divides and functions in spindle formation during cell division.

centrosome (SEN-trō-sōm) 31 A cellular organelle, located outside the nucleus, that contains centrioles.

cerebellum (ser-e-BEL-um) 230 Second largest brain division; concerned mainly with coordination of movements.

cerebrum (se-RĒ-brum) 229 Largest brain division of man; consists of two hemispheres connected by the corpus callosum.

cervical (SER-vi-kul) **vertebrae** 113 Seven vertebrae found in the neck region; pertaining to the neck region.

cervix (SER-vix) 480 Neck, or necklike structure; the inferior end of the uterus.

chiasma (kī-AZ-ma) 230 X-shaped intersection, as of the fibers of the optic nerve on the ventral surface of the brain.

chief cells 284 Cells in the cardiac and fundic regions of stomach mucosa that secrete pepsinogen; cells in parathyroid tissue.

chordae tendineae (KOR-ay ten-DIN-ē-ay) 313 Slender fibers attached to papillary muscles and cusps of the bicuspid and tricuspid valves; serve to prevent the cusps of the valve from being forced open during systole.

chorion (KOR-ē-on) 501 An extraembryonic membrane surrounding the fetus in utero.

choroid (KOR-oyd) 251 Middle, highly vascular layer of the eyeball.

chromatin (KRŌ-ma-tin) 31 Strands of DNA and protein in the nucleus of the cell that condense into chromosomes during cell division.

chromophil (KRŌ-muh-fil) **cells** 282 Readily staining cells with granular cytoplasm in the adenohypophysis.

ciliary body 252 A thickened area connecting the choroid and iris at the anterior portion of the eye; supports and determines the thickness of the lens of the eye.

cisterna chyli (sis-TER-nuh KĪ-lī) 303 The saclike beginning of the thoracic duct originating in the region of the second lumbar vertebra.

clitoris (KLIT-or-is) 480 A small erectile body the size of a pea, located at anterior union of the vulva in the female; analogous to the penis in the male.

coccygeal (kok-SIJ-ē-ul) 113 Pertaining to the coccyx, or tailbone, at the base of the sacrum; three to five fused vertebrae forming the inferior end of the vertebral column.

cochlea (KOK-lē-uh) 261 A spiral shaped portion of the inner ear.

collecting duct 436 Thick-walled straight tubules extending from nephron into the renal medulla.

colloid (KOL-oyd) 35 A state of matter usually consisting of a liquid medium and a suspended solute.

colon (KŌ-lon) 384 The large intestine; bowel.

columnar (ko-LUM-ner) 47 Tall, having greater length than width; epithelial type.

column 436 Region between pyramids in the renal medulla.

common bile duct 393 A duct formed by the cystic and hepatic ducts, leading to the duodenum.

compensation 471 The process by which the respiratory and renal systems normally adjust to pH changes.

compound microscope 17 An instrument having two lens systems that is used to enlarge objects for viewing.

condyle (KON-dīl) 94 Bone marking; a rounded or knoblike projection.

contraction period 214 The time in which a muscle actually contracts or shortens.

corium (KŌ-rē-um) 69 Dermis of the skin.

cornea (KOR-nē-uh) 252 Anterior transparent area of the eyeball.

coronal (ko-RŌ-nul) 2 A transverse plane dividing the body or any of its parts into anterior and posterior portions.

coronal suture (SOO-cher) 134 An immovable or synarthrodial joint that joins the frontal bone with the parietal bones.

corpus albicans (AL-bi-kanz) 492 A regressed corpus luteum.

corpus callosum (ka-LŌ-sum) 229 A large band of white matter connecting the two cerebral hemispheres.

corpus cavernosum (ka′-ver-NŌ-sum) 483 One of the two lateral masses of erectile tissue found in the penis.

corpus luteum (LOO-tē-um) 492 A collapsed follicle that has expelled its ovum.

corpus spongiosum (spon-jē-Ō-sum) 483 A mass of erectile tissue in the penis through which the urethra passes (corpus cavernosum urethrae).

cortex 224 The outer portion of a structure.

cortisol (KOR-ti-sol) 281 A glucocorticoid; hormone of the adrenal cortex.

cranial (KRĀ-nē-ul) 1 Toward the head end of the body.

crenation (kre-NĀ-shun) 37 The shrinking or shriveling of a cell when placed in a hypotonic solution, as a result of water loss from the cell.

crest 94 Bone marking; an elevation or a ridge that serves as a point for muscle attachment.

crown 394 The exposed portion of a tooth.

crystalloid (KRIS-tuh-loyd) 36 A noncolloidal substance resembling a crystal that dissolves and forms a true solution; will pass through a dialyzing membrane.

cuboidal (kyoo-BOY-dul) 47 Cube shaped.

cumulus oophorus (KYOO-myoo-lus ō-OF-er-us) 490 A mound of cells immediately surrounding an oocyte within a Graafian follicle.

cuspid (KUS-pid) 383 A conical, pointed tooth situated between the lateral incisor and first premolar; canine.

cutaneous (kyooTĀ-nē-us) 266 Having to do with the skin.

cystic duct (SIS-tik) 393 A duct leading to the common bile duct connecting the gallbladder with the hepatic duct.

cytoplasm (SĪ-tō-plazm) 31 The portion of a cell outside the nucleus and within the plasma membrane.

cytoskeleton (SĪ-tō-SKEL-e-ton) 31 A framework-like structure of microfilaments and microtubules that contributes to a cell's shape.

D-antigen 339 The Rh antigen.

delta cells 281 Cells in the islets of Langerhans that produce the hormone somatostatin.

dendrite (DEN-drīt) 223 Short process extending from the nerve cell body to a presynaptic neuron.

dense, fibrous 47 A type of connective tissue consisting mainly of parallel rows of bundles of fibers.

dentin (DEN-tin) 394 A collagenous and calcified substance comprising the principal mass of a tooth.

depolarization (dē-pōl-ler-i-ZĀ-shun) 348 A process by which the inside of a muscle or nerve cell membrane becomes more positively charged with respect to the outside.

dermis (DER-mis) 69 The layer of dense, vascular connective tissue subjacent to the epidermis.

descending colon 392 A vertical portion of the large intestine in the left side of the abdomen, extending from the stomach to the level of the iliac crest.

diabetes mellitus (muh-LĪ-tus) 288 A chronic metabolic disease resulting from insufficient insulin production by the pancreas.

dialysis (dī-AL-uh-sis) 36 The separation of crystalloids from colloids utilizing a differentially permeable membrane.

diaphragm (DĪ-uh-fram) 308 A skeletal muscle separating the thoracic cavity from the abdominal cavity; important in breathing.

diaphysis (DĪ-AH-fi-sis) 88 The shaft of a long bone.

diarthrosis (dī′-ar-THRŌ-sis) 134 One of two major groups of joints in the body; movable joints.

diastole (dī-AS-tuh-lē) 345 The noncontractile phase of a heart cycle.

diencephalon (dī'-en-SEF-uh-lon) 229 Inferior medial brain section consisting of the thalamus and hypothalamus; primarily of embryological significance.

diffusion (di-FYOO-zhun) 34 Movement of molecules from a highly concentrated area to a less concentrated area until equilibrium is achieved.

digit (DIJ-it) 125 A finger or a toe.

dipeptide (dī-PEP-tīd) 384 Two amino acids linked together by a peptide bond.

disaccharide (dī-SAK-uh-rīd) 418 A sugar formed by the union of two monosaccharides.

distal 2 Farthest from the trunk or point of origin of a part.

distal convoluted tubule 436 Last tubular portion of a nephron; continuation of the loop of Henle.

dorsal 1 Pertaining to the back.

ductus arteriosus (ar-tē-rē-Ō-sus) 336 A fetal blood vessel that connects the aorta with the pulmonary artery, bypassing the lungs.

ductus deferens (DEF-uh-renz) 483 Vas deferens; sperm duct.

ductus venosus (vē-NŌ-sus) 336 A blood channel in the fetus that extends from the umbilical vein through the fetal liver to the inferior vena cava.

duodenum (doo'-Ō-DĒ-num) 389 The first portion of the small intestine, about 10 inches long.

effector (e-FEK-ter) 228 A muscle or gland stimulated by a motor neuron.

efferent (EF-er-unt) 227 Movement away from a central point; a component leading away from a central structure.

elastic 47 A connective tissue type characterized by fibers of elastin and few cells.

electrocardiogram (ē-lek'-trō-KAR-dē-ō-gram) 348 A tracing of the electrical impulse that initiates cardiac muscle contraction.

electrode 466 An internal sealed tube with a metallic tip surrounded by an external tube containing a standard solution and a pH-sensitive glass bulb; used to determine pH.

electrolyte (e-LEK-trō-līt) 435 A solution that is able to conduct electricity due to the presence of ions.

embryo (EM-brē-ō) 501 An animal in the early stages of development; a human being from implantation through the eighth week postconception.

emulsified fat (ē-MUL-si-fid) 422 Fat that has been converted into small fat droplets.

enamel (ē-NAM-ul) 394 A hard covering of the crown of a tooth composed primarily of calcium salts.

endocrine (EN-dō-krin) 277 Glandular organs that secrete hormones directly into blood.

endometrium (en'-dō-MĒ-trē-um) 493 The inner lining of the uterus.

endoplasmic reticulum (en'-dō-PLAZ-mik re-TIK-yoo-lum) 31 A protein and lipid membranous network in the cytoplasm through which biochemical components of the cell move.

endothelium (en'-dō-THĒ-lē-um) 309 Simple squamous epithelium that surrounds the lumen of blood vessels; innermost lining of blood vessels.

enzyme 418 A complex produced by living cells that catalyzes chemical reactions in organic matter.

eosinophil (ē'-ō-SIN-ō-fil) or acidophil (a-SID-o-fil) 299 A granular leukocyte somewhat larger than a neutrophil, with an irregularly shaped lobed nucleus.

epidermis (ep'-i-DER-mis) 69 The outermost layer of skin; consists of five layers.

epididymis (ep'-i-DID-i-mus) 487 A long coiled duct on the anterior lateral surface of each testis.

epigastric (ep'-i-GAS-trik) 5 An abdominal region lying superior to the stomach.

epiglottis (ep'-i-GLOT-is) 364 The most superior cartilage of the larynx; covers opening into larynx during swallowing.

epimysium (ep'-i-MIZ-ē-um) 150 Connective tissue surrounding a muscle.

epiphysis (e-PIF-i-sis) 88 End of a bone connected by cartilage to a long bone in the embryological stage of life; later becomes the end portion of a long bone.

epithelium (ep'-i-THĒ-le-um) 45 A sheet of cells that covers an external surface or lines an internal surface and functions in protection, absorption, secretion, and filtration.

erythrocytes (e-RITH-rō-sīts) 298 Red blood cells; biconcave-shaped anucleate disks.

esophagus (e-SOF-a-gus) 389 A muscular tube, about 10 inches long, extending from the pharynx to the stomach.

estrogen (ES-trō-jin) 281 A class of female hormones.

exocrine (EK-sō-krin) 46 A glandular organ that secretes outwardly through ducts.

expiratory reserve volume 373 The amount of air that an individual can forcibly exhale after a normal, quiet expiration.

external respiration 361 Exchange of gases between the blood and external environment; occurs in the alveoli of the lungs.

falciform (FAL-si-form) **ligament** 393 A ligament dividing the liver into right and left lobes.

Fallopian (fuh-LŌ-pē-an) **tubes** or **uterine tubes** 479 Paired tubes that serve as ducts from the ovaries to the uterus, through which ova pass.

fascia (FASH-ē-uh) 178 Dense, fibrous connective tissue covering muscles.

fascicle (FAS-i-kul) 150 A subdivision of an entire muscle; a bundle of muscle fibers.

fatty acid 420 A saturated or unsaturated organic monocarboxylic acid.

fertilization (fer′-ti-li-ZĀ-shun) 501 Union of sperm and egg; formation of a zygote.

fetus (FĒ-tus) 336 An unborn vertebrate; an unborn human being from the third month after conception until birth.

fibrous (FĪ-brus) 47 See **dense, fibrous**.

first polar body 488 A structure containing chromosomes formed by the meiotic division of a primary oocyte.

fissure (FISH-er) 94 A bone marking; narrow slit-like opening.

follicle (FOL-i-kul) 71 A small excretory sac or gland.

follicle stimulating hormone (FSH) 280 A gonadotropic hormone whose target organ is the ovary or testis; produced by the anterior pituitary gland.

follicular fluid (fol-IK-yoo-lr) 490 Liquid within a Graafian follicle surrounding an ovum.

fontanels (fon-tuh-NELZ) 134 Fibrous areas existing before the four cranial bones completely form and fuse together.

foramen (fo-RĀ-men) 94 An opening through a bone for transmission of nerves and/or blood vessels.

foramen ovale (o-VĀ-lē) 337 An opening in the atrial septum of the fetal heart, through which blood flows, bypassing the lungs.

formed elements 298 The cellular portion of blood; comprised of red blood cells, white blood cells, and platelets.

fossa 94 Bone marking; a cavity or hollow area in a bone.

fovea (FŌ-vē-uh) or **fovea centralis** (sen-TRĀ-lis) 255 The central area of the retina, where vision is most acute.

frenulum linguae (FREN-yoo-lum LING-wē) 388 A vertical fold of mucus membrane that attaches the tongue to the floor of the mouth in the midline.

frequency 264 The number of cycles per second of sound waves; determines the pitch of a tone.

frontal 2 A lengthwise plane dividing the body into anterior and posterior portions.

fructose (FRUK-tōs) 420 A monosaccharide found in honey and fruits; a component of sucrose.

fundus (FUN-dus) 389, 480 Enlarged portion of the stomach to the left of and superior to the cardioesophageal junction; the superior, rounded portion of the uterus.

galactose (ga-LAK-tōs) 420 A monosaccharide derived from lactose.

gallbladder 393 An accessory organ of the digestive tract that concentrates and stores bile.

ganglion (GANG-lē-on) 228 A collection of nerve cell bodies.

gastric pits 401 Narrow depressions containing glands that extend through the full thickness of the gastric (stomach) mucosa.

glomerular (glom-ER-yoo-ler) **filtration** 446 The process of filtering protein-free blood plasma through the glomeruli of nephrons into the glomerular capsule.

glomerulus (glom-ER-yoo-lus) 435 A cluster of blood vessels in close proximity to Bowman's capsule.

glucagon (GLOO-ka-gon) 281 A hormone produced by alpha cells of the islets of Langerhans; involved in carbohydrate metabolism.

glucocorticoids (gloo′-kō-KOR-ti-koydz) 281 A class of adrenal cortical steroid hormones that influence the metabolism of glucose, protein, and fat.

glucose (GLOO-kōs) 420 A simple sugar or monosaccharide; the chief source of energy in living organisms.

glyceride (GLIS-er-īd) 422 An ester (salt) of glycerol with fatty acids.

glycerol (GLIS-er-ol) 421 An alcohol found in neutral fats.

Golgi (GŌL-jē) **apparatus** 31 An organelle comprised of membranes stacked upon each other, functioning in glycoprotein synthesis and the movement of substances to the external cell environment.

gonadotropic (gō-nad′-ō-TROP-ik) **hormone** 281 A type of hormone secreted by the adenohypophysis (pars distalis) that stimulates the gonads to secrete hormones.

gonadotropin (gō-nad′-ō-TRŌ-pin) **releasing hormone** (GnRH) 499 A hormone secreted by the hypothalamus into the anterior lobe of the pituitary gland, stimulating the release of FSH and LH.

Graafian follicle (GRAF-ē-un FOL-i-kul) or **vesicular** (ves′-i-kyoo-lar) **ovarian follicle** 490 A hollow fluid-containing sac surrounding a primary or secondary oocyte contained within an ovary.

graded muscle contraction 212 An increased force of muscular contraction caused by the recruitment of additional motor units.

gray matter 224 Nerve tissue composed of cell bodies of neurons and nerve cell bodies; found in the brain and spinal cord.

greater omentum (ō-MEN-tum) 322 A large fold of peritoneum covering the small and large intestines.

growth hormone (GH) 280 Tropic hormone, secreted by the adenohypophysis, that influences metabolism of proteins, fats, and carbohydrates.

haustra (HOS-truh) 392 Sacculations of the wall of the large intestine.

head 94 Bone marking; a rounded surface of bone connected to the shaft by the neck.

helicotrema (hē-li-ko-TRĒ-muh) 263 The space between the scala vestibuli and scala tympani at the apex of the cochlea.

hemoglobinometer (hē′-mō-glob′-in-OM-e-ter) 341 A laboratory instrument that colorimetrically measures the hemoglobin content of blood.

hemolysis (hē-ma′-luh-sis) 37 The bursting of a cell when placed in a hypotonic solution as a result of water entering the cell.

hepatic (he-PAT-ik) **duct** 393 A duct from the liver that joins the cystic duct to form the common bile duct.

hepatic flexure (FLEK-sher) 392 The angle formed by the ascending and transverse colons.

hepatic sinusoid (SĪ-nuh-soyd) 412 Microscopic channels through which venous blood flows through the lobules of the liver.

hepatocyte (he-PAT-ō-sīt) 412 A liver cell.

hilum (HĪ-lum) 305 Medial indentation of kidney, lungs, lymph nodes.

homeostasis (hō-mē-ō-STĀ-sis) 436 A tendency toward maintenance of stability of the internal environment.

horizontal 2 A plane dividing the body or its parts into superior and inferior portions.

hormone 277 A chemical messenger secreted by an endocrine gland.

human chorionic gonadotropin (kō-rē-ON-ik gŌ-nad′-ō-TRŌ-pin) (hCG) 500 A hormone secreted by pre-embryonic trophoblast cells during pregnancy; maintains estrogen and progesterone secretion by the corpus luteum until placenta formation.

hyaline (HĪ-uh-lin) 47 The most common type of cartilage; glassy in appearance.

hyoid (HĪ-oyd) 110 A bone that supports the base of the tongue.

hyperglycemia (hī′-per-glī-SĒ-mē-uh) 290 Elevated blood glucose level.

hypersecretion (hī′-per-se-KRĒ-shun) 294 Secretion of greater than normal amounts of a substance.

hypertonic (hī′-per-TON-ik) **solution** 37 A solution that, in relation to another solution, contains a greater amount of dissolved material (solute).

kyphosis (kī-FŌ-sis) 112 An exaggerated thoracic curvature of the spine, resulting in a hunchback appearance.

lactase (LAK-tās) 418 An enzyme that hydrolyzes lactose into glucose and galactose.

lactiferous (lak-TIF-er-us) **duct** 481 The continuation of the mammary ducts beyond the ampullae, terminating at the nipple.

lactose (LAK-tos) 420 A disaccharide comprised of glucose and galactose.

lacuna (luh-KYOO-nuh) 58 A small depression or hollow cavity in an osteon; a depression containing osteocytes.

lambdoidal suture (lam-DOY-dul SOO-cher) 134 A joint that unites the parietal bones with the occipital bone.

lamella (luh-MEL-uh) 87 A thin bony plate, such as the Haversian lamella.

lamina propria (LAM-in-uh prō-PRĒ-uh) 399 A connective tissue layer in the mucosa underlying the mucosal epithelium and basement membrane.

laryngopharynx (la-ring′-gō-FAR-inks) 364 A portion of the pharynx (throat) posterior to the larynx.

latent period 214 The shortest period of muscle contraction; occurs between the stimulus and contraction period.

lateral 2 Toward the side.

leukemia (loo-KĒ-mē-uh) 302 A malignant disease of the spleen, bone marrow, and lymphoid tissue; characterized by uncontrolled proliferation of immature leukocytes.

leukocytes (LOO-kō-sīts) 298 White blood cells containing various shaped nuclei.

ligament 47 Dense fibrous band of connective tissue that attaches bone to bone.

ligamentum venosum (vē-NŌ-sum) 336 A remnant of the fetal ductus venosus found in the adult liver.

liminal (LIM-uh-nul) **stimulus** 214 Threshold stimulus; the amount of stimulus needed for a muscle to contract.

line 94 A bone marking; a slight ridge.

lingual tonsil (lin-gwal) 307 One of two paired masses of lymphatic tissue embedded in mucous membrane at the base of the tongue; functions to combat infections of the pharyngeal region.

lipase (LĪ-pās) 422 An enzyme that hydrolyzes emulsified fats to fatty acids and glycerol.

liver 412 The largest gland in the body, located immediately inferior to the diaphragm, occupying most of the right hypochondriac region and part of the epigastric region.

lobule (LOB-yool) 307 A small segment of a lobule; most commonly associated with the liver and lung.

loop 77 A fingerprint pattern characterized by elongated curves of dermal ridges that are skewed to the right or the left.

loop of Henle 436 A portion of the nephron between the proximal and distal convoluted tubules that extends from the renal cortex to the medulla and then proceeds superiorly toward the renal cortex.

loose, areolar (a′-rē-Ō-ler) 47 Type of connective tissue found in and around internal organs.

lordosis (lor-DŌ-sis) 112 An exaggerated lumbar curvature of the spine, resulting in a swayback appearance.

lumbar vertebrae 111 Five vertebrae inferior to the thoracic vertebrae in the spinal column; pertaining to the lower back.

luteinizing (LOO-tē-in-ī′-zing) **hormone** (LH) 280 A gonadotropic hormone whose target gland is the ovary in the female; called ICSH in the male.

lymph (LIMF) 303 The interstitial fluid that is contained and transported through the lymph vessels.

lymph nodes 304 Oval shaped glands associated with lymph vessels functioning to filter and remove bacteria and foreign matter from lymph and to manufacture lymphocytes.

lymphatic (lim-FAT-ik) **system** 303 A specialized system of vessels and tissues for the collection, purification and return of tissue fluid to the bloodstream.

lymphocyte (LIM-fō-sīt) 298 A type of leukocyte that has a large, round dark-staining nucleus surrounded by a thin rim of clear blue cytoplasm.

lymph vessel 303 A vessel of the lymphatic system.

lysosome (LĪ-sō-sōm) 31 A membrane-bound cytoplasmic organelle containing acid hydrolases.

macrophage 299 A large phagocytic cell that engulfs foreign particles outside the blood vessels; a derivative of a monocyte.

maltase 424 An enzyme that hydrolyzes maltose into two glucose molecules.

maltose 420 A disaccharide comprised of two glucose units.

matrix (MĀ-trix) 56 The basic background substance in which cells or tissues grow and develop, such as bone matrix.

meatus (mē-Ā-tus) 94 A canal running within a bone.

medial (MĒ-dē-ul) 2 Toward the midline of the body.

mediastinum (mē-dē-a-STĪ-num) 317 The center of the chest containing the heart, aortic arch, esophagus, and trachea.

medulla (me-DOO-luh) 305 The inner portion of a structure.

medulla oblongata (ob′-lon-GAH-tuh) 234 The most inferior section of the brain containing control centers for vital functions.

medullary (MED-yoo-lerr-ē) **rays** 436 Structures within renal pyramids leading toward papillae and calyces; involved in urine formation.

meiosis (mī-Ō-sis) 488 A process of nuclear division in which daughter cells are produced that have half the number of chromosomes as the parent cell; this process involves two cell divisions, the second yielding the haploid number of chromosomes; specific to gametes.

melanin (MEL-uh-nin) 69 A skin pigment formed in melanocytes in the stratum basale of the epidermis.

melanocyte (mel-LAN-ō-sīt) 69 A melanin-producing cell in the stratum basale of the epidermis.

melanocyte stimulating hormone (MSH) 280 A hormone secreted by the pituitary gland that stimulates melanin-producing cells in the epidermis.

meninges (men-IN-jez) 228 Covering of the brain and spinal cord.

menstrual (MEN-stroo-ul) **phase** 499 A stage of the menstrual cycle during which the stratum func-

tionalis of the endometrium is sloughed off, accompanied by bleeding in the uterine cavity; days 1 through 4 of the average menstrual cycle.

menstruation (men′-stroo-Ā-shun) 493 The periodic sloughing off of the stratum functionale of the endometrial lining, accompanied by bleeding.

mesentery (MEZ-en-terr-ē) 393 A fold of parietal peritoneum that anchors abdominal organs to the dorsal body wall.

metabolic (met-uh-BOL-ik) **acidosis** 472 The decrease in pH of arterial blood as a result of the accumulation of nonrespiratory acids or the loss of bases.

metabolic alkalosis 472 The increase in pH of arterial blood as a result of excessive loss of H+ ions or gain in bases.

metabolism (muh-TAB-uh-lizm) 425 The total of all the chemical processes and reactions taking place in a living organism; composed of two phases—anabolism, the building up phase, and catabolism, the breakdown phase.

metaphase 32 A stage in mitosis during which the chromosomes are aligned across the equator of the cell.

midbrain 232 That portion of the brain containing the corpora quadrigemina and cerebral peduncles; also called the mesencephelon.

midsagittal (mid-SAJ-i-tal) 2 A lengthwise plane from front to back dividing the body or any of its parts into equal right and left halves.

Million reaction 417 A reaction specific for the amino acid tyrosine.

mineralocorticoids 281 A class of steroid hormones secreted by the adrenal cortex that increases sodium reabsorption by the distal convoluted tubules of the kidney.

mitochondria (mī′-tō-KON-drē-uh) 31 Cytoplasmic, membrane-bound organelles; contain oxidative enzymes.

mitosis (mī-TŌ-sis) 32 A process of cell division in which daughter cells are produced that have the same number of chromosomes as the parent cell.

molar (MŌ-ler) 383 A tooth posterior to the bicuspid that contains three or more cusps.

Molisch reaction 419 A general test for carbohydrates.

monoclonal (mon′-ō-KLŌ-nul) **antibody** 500 A specific antibody, derived from a culture of identical cells, that reacts with a specific antigen.

monocyte (MON-ō-sīt) 298 A granular leukocyte.

monosaccharide (mon′-ō-SAK-er-īd) 418 A simple sugar.

mucosa (myoo-KO-suh) 399 The innermost lining of the digestive tract.

mucus (MYOO-kus) 403 The secretion of mucous cells that contains water, mucin, and various inorganic salts.

muscle fatigue 213 A refractory condition in which the contractile tissue of a muscle loses its response to stimulation as a result of overactivity.

muscle fiber 149 A muscle cell.

muscle tone 212 A normal state of balanced tension.

muscle twitch 214 A single muscle contraction.

muscularis externa (mus-kyoo-LĀ-ris EX-ter-nuh) 399 A layer of circular and longitudinal muscle beneath the submucosal layer of the digestive tract.

muscularis mucosae (myoo-KŌ-sē) 399 A smooth muscle layer between the mucosa and submucosa.

myelin sheath 223 A lipid covering that surrounds the larger axons of cranial and spinal nerves.

myeloid (MĪ-uh-loyd) 46 Relating to or derived from bone marrow.

myocardium (mī′-ō-KAR-dē-um) 316 The thick middle layer of the heart wall, composed of cardiac muscle.

myofibril (mī′-O-FĪ-bril) 149 A group of myofilaments.

myofilament (mī′-ō-FIL-uh-ment) 149 A molecular threadlike contractile element within a myofibril; a contractile component of a myofibril.

nasopharynx (nā′-zō-FAR-inx) 364 Portion of the throat located posterior and inferior to the nasal cavity.

neck 394 The portion of a tooth at the gum line where the crown and root meet: constricted portion below the head of a bone.

nephron (NEF-ron) 436 The structural and functional unit of the kidney.

neurilemma (noo-ri-LEM-a) 223 A thin outer sheath surrounding the axons of myelinated and unmyelinated peripheral nerves.

neuroglia (noo-ROG-lē-uh) 222 Supporting cells of the nervous system.

neurohypophysis (noo′-rō-hī-POF-i-sis) 278 The posterior lobe of the pituitary gland.

neuron (NOO-ron) 222 Nerve cell responsible for generation, conduction, and transmission of a nerve impulse.

neurovascular bundle 310 An anatomical unit of an artery, vein, and nerve.

neutrophils (NOO-truh-filz) 299 Granular leukocytes that usually contain three to five irregular oval-shaped lobes connected by thin strands of nuclear material.

Ninhydrin reaction 417 A test for alpha amino acids.

nipple 481 The external termination of lactiferous ducts of the mammary glands; point from which milk can be ejected.

nuclear membrane 31 A double protein-lipid membrane surrounding the nucleus.

nucleolus (noo-KLĒ-ō-lus) 31 A small dense structure in a nucleus consisting of RNA and protein.

nucleus (NOO-klē-us) 31 A structure containing DNA within a cell; control center of the cell.

nystagmus (nis-TAG-mus) 264 Continuous involuntary movement of the eyeball.

olfactory (ol-FAK-ter-ē) 230 Relating to the sense of smell. The olfactory nerve (the first cranial nerve) concerned with the sense of smell.

oogenesis (ō′-ō-JEN-e-sis) 488 The process of egg formation by meiosis.

oogonium (ō′-ō-GŌ-nē-um) 488 An immature diploid ovum.

ophthalmic (of-THAL-mik) 241 Relating to the eye; a branch of cranial nerve V.

organ of Corti 263 The organ of hearing, located within the scala media of the cochlea of the inner ear.

ora serrata (or-a ser-RAH-ta) 252 The zigzag anterior margin of the retina of the eye.

orifice (OR-i-fis) 480 An opening, entrance, or outlet of a body cavity or structure.

oropharynx (or'-ō-FAR-inx) 364 The portion of the throat located posterior to the mouth.

osmosis (os-MŌ-sis) 35 The diffusion of water through a semipermeable membrane from a region of greater concentration of solvent to one of lesser concentration.

osseous (OS-ē-us) 47 A supporting connective tissue; bone tissue.

osteocyte 87 A bone cell.

ovary (Ō-ver-ē) 479 A paired primary reproductive organ in the female that produces ova and female sex hormones.

oviduct (Ō-vi-dukt) 479 fallopian tube.

ovulation (ō'-vyoo-LĀ-shun or ov'-yoo-LĀ-shun) 499 The discharge of an ovum from an ovarian (Graafian) follicle; occurs at midpoint (day 14) of the average menstrual cycle.

ovum (Ō-vum) 488 An egg cell; a female gamete.

oxygen saturation The percentage of available oxygen-carrying sites on hemoglobin molecules that are bound to oxygen.

oxyphilic (ok'-si-FIL-ik) **cells** 284 Cells found in parathyroid tissue that are capable of producing parathyroid hormone.

oxytocin (ok'-si-TOK-sin) 280 481 A hormone stored by the neurohypophysis that stimulates uterine contractions and milk ejection.

P wave 348 A wave form of the ECG that represents electrical depolarization of the atria.

palate (PAL-ut) 389 The roof of the mouth; hard palate comprised of maxilla, soft palate primarily composed of skeletal muscle.

palatine (PAL-uh-tīn) 91 A paired facial bone; forms posterior one-third of roof of mouth.

palatine tonsil 307 One of two paired masses of lymphatic tissue embedded in mucous membrane laterally at the back of the throat, functions to combat infections of the pharyngeal region.

pancreas (PAN-krē-us) 278 A tubuloacinar gland lying behind and below the stomach in the curve of the duodenum, having the exocrine function of secreting digestive enzymes and the endocrine function of secreting the hormones insulin and glucagon.

papillae (pa-PIL-ē) 267 Indentations in the renal medulla at the base of medullary rays; elevations on the surface of the tongue that contain taste buds.

papillary muscles 313 Cylindrical muscular bundles attached to the muscular wall of the ventricle of the heart.

paranasal (par'-a-NĀ-zul) **sinuses** 364 Air- or mucus-filled cavities surrounding the nose.

parathyroid (par-a-THĪ-royd) **gland** 278 Several (usually four) endocrine glands located on the posterior surface of the lateral lobes of the thyroid gland.

parathyroid hormone or **parathormone** (PTH) 281 A hormone secreted by the parathyroid glands that increases serum calcium levels.

parietal (puh-RĪ-e-tul) **cells** 401 Cells of the stomach mucosa that secrete hydrochloric acid and intrinsic factor.

parotid (puh-ROT-id) **gland** 389 A paired salivary gland located anterior and inferior to the ear.

pars distalis (dis-TAL-is) 278 Anterior lobe of the pituitary gland.

pars intermedia (in'-ter-MĒ-dē-uh) 282 The intermediate lobe of the pituitary gland; difficult to distinguish in humans.

pars nervosa (ner-VŌ-suh) 278 The neurohypophysis.

penis (PE-nis) 483 The male copulatory organ and urinary outlet.

pepsin 401 A protease in the stomach that hydrolyzes proteins into proteoses and peptones.

pepsinogen (pep-SIN-ō-jen) 401 The precursor to pepsin.

peptidase (PEP-ti-dās) 416 An enzyme that hydrolyzes peptides into amino acids.

peptone 416 A partially digested protein.

pericardium (per'-i-KAR-dē-um) 317 A protective sac of fibrous and serous tissue that encloses the heart.

perimysium (per'-i-MIS-ē-um) 150 A connective tissue surrounding a muscle fascicle.

periodontal (per'-ē-ō-DON-tul) **membrane** 394 A membrane separating the alveolar bone of the jaw from the root of a tooth.